Countable Systems of Differential Equations

Countable Systems of Differential Equations

A.M. Samoilenko and Yu.V. Teplinskii

///VSP///

Utrecht • Boston, 2003

VSP
an imprint of Brill Academic Publishers
P.O. Box 346
3700 AH Zeist
The Netherlands

Tel: +31 30 692 5790
Fax: +31 30 693 2081
vsppub@compuserve.com
www.vsppub.com
www.brill.nl

First published in 2003

ISBN 90-6764-393-9

Printed in The Netherlands by Ridderprint bv, Ridderkerk.

Contents

PREFACE

The present book is devoted to the solution of various problems in the theory of differential equations in a space $\mathfrak{M}$ of bounded numerical sequences (called countable systems). In particular, we deal with the general theory of countable systems, the theory of oscillating solutions, and the theory of countable systems with pulse action.

Our main attention is given to the generalization of the results of numerous authors obtained in recent years for finite-dimensional systems of differential equations to the case of systems from the analyzed class.

It is clear that countable systems form a class of systems of differential equations in Banach spaces and, thus, many issues of the theory of equations of this sort are well known. At the same time, differential equations in the space $\mathfrak{M}$ have many specific features, which caused the necessity of development of the theory of these equations originated by A. N. Tikhonov and K. P. Persidskii [Tik, Per1, Per2].

As a result of the systematic investigation of countable systems, K. P. Persidskii [Per4] proved theorems on the existence and uniqueness of a solution of the Cauchy problem, developed methods for their approximate solution, and created the theory of stability of these systems. Extensive subsequent investigations of these systems were carried out by O. A. Zhautykov and K. G. Valeev [Zha1–Zha3, VaZ]. A great number of problems connected with the development of asymptotic methods and the investigation of oscillatory processes was solved by the followers of the scientific school of N. N. Bogolyubov and Yu. A. Mitropolsky.

The proposed monograph is based on the original results obtained by the authors in [Sat1–Sat8, STA, STL, STT1, STT2, Tep1–Tep10, TeA1–TeA3, TeT, TeL] and consists of four chapters.

The first chapter is devoted to the presentation of the general concepts of the theory of infinite systems of differential equations and is partially based on the results obtained by K. P. Persidskii.

The second chapter deals with the theory of covariant toroidal manifolds of linear and nonlinear countable systems similar to the theory of extensions of dynamical systems on a torus developed recently for finite-dimensional systems by Yu. A. Mitropolsky, A. M. Samoilenko, and V. L. Kulik [MSK2, Sam10, Sam8, SaK2, SaK3].

The third chapter contains results on the reducibility of linear systems with periodic and quasiperiodic coefficients.

In the fourth chapter, the theory of pulse systems [SaP1–SaP3, Per] is generalized to the case of countable systems of differential equations.

Throughout the book, we use double enumeration of formulas and statements (definitions, lemmas, theorems, etc.). The first number is the number of a given section and the second number is the ordinal number of a formula or statement (definition, lemma, theorem, etc.) in this section.

The Authors

1. GENERAL CONCEPTS OF THE THEORY OF INFINITE SYSTEMS OF DIFFERENTIAL EQUATIONS

In Chapter 1, we present various facts from the theory of countable systems of differential equations used in the subsequent chapters. The major part of these facts can be found in the works [Per1–Per4, Tik], which are well known to the experts in various fields of mathematics, physics, and engineering dealing with systems with infinitely many degrees of freedom.

At present, there are several directions of investigation of these systems, including their general theory, systems of partial differential equations, characteristic numbers and the stability of solutions, averaging, differential equations in normed spaces, "multiperiodic" solutions, etc. [Bro, VaZ, Gor, DaK, DaKr, Erm, Zha1–Zha3, Res, Kha1, Kha2, Kha].

In our book, we present only some results from the theory of systems of ordinary differential equations: theorems on the existence and uniqueness of solutions, properties of solutions of linear and nonlinear systems (including conditions for the continuous dependence of solutions on the initial data), and matrizants of linear systems and their properties. Moreover, we consider analogs of some properties of normal autonomous systems [Pon] and present the fundamentals of the numerical-analytic method aimed at finding periodic solutions of nonlinear countable systems [SaR].

1.1. Theorems on Existence and Uniqueness of Solutions

Consider a system of differential equations of the form

$$\frac{dy_i}{dx} = f_i(x, y_1, y_2, \ldots, y_n, \ldots), \quad i = 1, 2, \ldots . \tag{1.1}$$

Definition 1.1. *A function $f(y_1, y_2, \ldots)$ is called strongly continuous if, for any $\varepsilon > 0$, there exist N_0 and $\delta > 0$ such that the inequality $|y_i' - y_i''| < \delta$, $i = 1, 2, \ldots, N_0$, implies the estimate*

$$|f(y_1', y_2', \ldots) - f(y_1'', y_2'', \ldots)| < \varepsilon.$$

Theorem 1.1. *Assume that the right-hand sides of the system of equations (1.1)*

(a) are defined for any $y_i \in R^1$, $i = 1, 2, \ldots$, and all $x \in T_0 = [x_0, x_0 + a] \subset R^1$;

(b) are strongly continuous in $y_1, y_2, \ldots$ for fixed x and measurable in x for fixed y_i, $i = 1, 2, \ldots$;

(c) satisfy the inequalities

$$|f_i(x, y_1, y_2, \ldots)| < M_i(x)$$

for all $i = 1, 2, \ldots$, where $M_i(x)$ are functions summable on the segment T_0.

Then, for any vector $(y_1^0, y_2^0, \ldots)$ with real coordinates, there exists at least one solution $(y_1(x), y_2(x), \ldots)$ of the system of equations (1.1) such that $y_i(x_0) = y_i^0$, $i = 1, 2, \ldots$.

Proof. We replace the system of equations (1.1) by the following system of integral equations:

$$y_i(x) = y_i^0 + \int_{x_0}^{x} f_i(x, y_1(x), y_2(x), \ldots)\, dx\,, \quad i = 1, 2, \ldots\,, \tag{1.2}$$

and consider a mapping

$$z_i(x) = y_i^0 + \int_{x_0}^{x} f_i(x, y_1(x), y_2(x), \ldots)\, dx\,, \quad i = 1, 2, \ldots\,, \tag{A}$$

which establishes a correspondence between an arbitrary countable system of continuous functions $(y_1(x), y_2(x), \ldots)$ and another system of this sort $(z_1(x), z_2(x), \ldots)$. Note that if $f(x, y_1, \ldots, y_n)$ is a continuous function of finitely many variables $y_1, \ldots, y_n$ measurable with respect to x for fixed y_i, $i = \overline{1, n}$, then the function $\Phi(x) = f(x, \varphi_1(x), \ldots, \varphi_n(x))$ is measurable if $\varphi_i(x)$, $i = \overline{1, n}$, are measurable. Thus, the function $\Psi_n(x) = f(x, \varphi_1(x), \ldots, \varphi_n(x), 0, 0, \ldots)$ is measurable and, therefore, the function $f(x, \varphi_1(x), \varphi_2(x), \ldots) = \Psi(x)$ is also measurable because

$$\Psi(x) = \lim_{n \to \infty} \Psi_n(x),$$

which readily follows from the condition of strong continuity. The requirement of summability follows from condition (c) of Theorem 1.1.

We consider a system of functions $(y_1(x), y_2(x), \ldots)$ as a point P of an abstract space R. If there exists a point P invariant under mapping (A), then it specifies a solution of the system of equations (1.2) and, hence, of system (1.1).

Consider a set M_0 formed by three points P for which y_k satisfy the conditions

$1°$. $|y_k(x) - y_k^0| \le \int\limits_{x_0}^{x} M_k(x)\, dx$,

$2°$. $|y_k(x') - y_k(x'')| \le \int\limits_{x'}^{x''} M_k(x)\, dx$, $\quad k = 1, 2, \ldots$.

It is easy to see that mapping (A) maps the set M_0 into itself. We now introduce mapping (B) by putting every point P in correspondence with a set of numbers

$$\begin{array}{llll} \dfrac{a_1'}{N_1}, & \ldots, & \dfrac{q_1^n}{N_1}, & \ldots, \\ \ldots & \ldots & \ldots & \ldots, \\ \dfrac{a_n^1}{nN_n}, & \ldots, & \dfrac{a_n^n}{nN_n}, & \ldots, \\ \ldots & \ldots & \ldots & \ldots, \end{array} \tag{1.3}$$

where $N_i = y_i^0 + \int\limits_{x_0}^{x_0+a} M_i(x) dx$ and the numbers $a_n^1, \ldots, a_n^n, \ldots$ are the coefficients of the Fourier expansion of a function $y_n(x)$ in a certain complete orthogonal system of functions on the segment T_0. By ordering the set of numbers (1.3), we obtain a numerical sequence $b_1, b_2, \ldots, b_n, \ldots$. Moreover, we have

$$\begin{aligned} \sum_{k=1}^{\infty} (a_n^k)^2 &= \int\limits_{x_0}^{x_0+a} (y_n(x))^2\, dx \\ &\le \int\limits_{x_0}^{x_0+a} \Big(y_n^0 + \int\limits_{x_0}^{x} M_k(x) dx\Big)^2\, dx \end{aligned}$$

$$\leq \int_{x_0}^{x_0+a} N_n^2\, dx = aN_n^2,$$

whence it follows that

$$\sum_{i=1}^{\infty} b_i^2 = \sum_{n=1}^{\infty}\sum_{k=1}^{\infty}\Big(\frac{a_n^k}{nN_n}\Big)^2 \leq a\sum_{n=1}^{\infty}\frac{1}{n^2} = \frac{a\pi^2}{6}.$$

Thus, mapping (B) maps the set M_0 into a subset M_0^* of the Hilbert space l_2. Therefore, mapping (A) induces a mapping (A^*) of the set M_0^* into itself. Further, if mapping (A^*) has a fixed point $P^* \in M_0^*$, then the corresponding point $P \in M_0$ determines the solution of equation (1.2) [and, hence, (1.1)].

To use the Schauder theorem, it suffices to show that the set M_0^* is compact and convex. If $P'^* = (b_1', \ldots, b_n', \ldots)$ and $P''^* = (b_1'', \ldots, b_n'', \ldots)$ are points from M_0^*, then the point

$$\alpha P'^* + \beta P''^* = (\alpha b_1' + \beta b_1'',\, \alpha b_2' + \beta b_2'', \ldots), \quad \alpha+\beta = 1,\ \alpha > 0,\ \beta > 0,$$

belongs to M_0^* because it corresponds to the system of functions

$$\alpha y_1'(x) + \beta y_1''(x),\ \alpha y_2'(x) + \beta y_2''(x), \ldots$$

specifying a point from the set M_0. Indeed,

$$\begin{aligned}|\alpha y_k'(x) + \beta y_k''(x) - y_k^0| &= |\alpha(y_k'(x) - y_k^0) + \beta(y_k''(x) - y_k^0)| \\ &\leq (\alpha+\beta)\int_{x_0}^{x} M_k(x)\, dx = \int_{x_0}^{x} M_k(x)\, dx,\end{aligned}$$

i.e., condition 1° is satisfied. Similarly, the inequality

$$|\alpha y_k'(x') + \beta y_k''(x') - \alpha y_k'(x'') - \beta y_k''(x'')| \leq (\alpha+\beta)\int_{x_0}^{x} M_k(x)\, dx$$

implies condition 2°. Hence, the set M_0^* is convex. In this set, we choose an arbitrary sequence of points P_i^*. This sequence corresponds to the sequence of points $P_i(y_1^{(i)}(x), y_2^{(i)}(x), \ldots)$ in the set M_0. According to conditions 1° and 2°, the sequence $y_1^{(i)}(x)$, $i = 1, 2, \ldots$, is uniformly bounded and equicontinuous and, consequently, it contains a subsequence $y_1^{(\alpha_1)}(x), y_1^{(\alpha_2)}(x), \ldots, y_1^{(\alpha_s)}(x), \ldots$ that converges uniformly in $x \in T_0$. However, the sequence $y_2^{(\alpha_s)}(x)$, $s \to \infty$, is also uniformly bounded and equicontinuous and, hence, it also contains a convergent subsequence

$$y_2^{(\beta_1)}(x), y_2^{(\beta_2)}(x), \dots, y_2^{(\beta_s)}(x), \dots .$$

This process can be continued infinitely. We compose the table

$$\begin{array}{ccccc} y_1^{(\alpha_1)}(x) & y_1^{(\alpha_2)}(x) & \cdots & y_1^{(\alpha_s)}(x) & \cdots \\ y_2^{(\beta_1)}(x) & y_2^{(\beta_2)}(x) & \cdots & y_2^{(\beta_s)}(x) & \cdots \\ y_3^{(\gamma_1)}(x) & y_3^{(\gamma_2)}(x) & \cdots & y_3^{(\gamma_3)}(x) & \cdots \\ \cdots & \cdots & \cdots & \cdots & \cdots \end{array}$$

and rewrite the set of sequences row by row:

$$\begin{array}{cccc} y_1^{(\alpha_1)}(x) & y_1^{(\beta_2)}(x) & y_1^{(\gamma_3)}(x) & \cdots \\ y_2^{(\alpha_1)}(x) & y_2^{(\beta_2)}(x) & y_2^{(\gamma_3)}(x) & \cdots \\ y_3^{(\alpha_1)}(x) & y_3^{(\beta_2)}(x) & y_3^{(\gamma_3)}(x) & \cdots \\ \cdots & \cdots & \cdots & \cdots . \end{array}$$

Each of these sequences converges as a subsequence of a convergent sequence supplemented by finitely many elements. Thus, the sequence of points

$$P_{\alpha_1}, P_{\beta_2}, P_{\gamma_3}, \dots \subset M_0 \tag{1.4}$$

converges weakly (coordinatewise) to a point $P_0 \in M_0$ (uniformly in $x \in T_0$). For the sake of convenience, we rewrite sequence (1.4) as

$$P_1, P_2, P_3, \dots, P_n, \dots .$$

Let us show that the sequence of the corresponding points P_1^*, P_2^*, P_3^*, ... from the set M_0^* converges to the point $P_0^* \in M_0^*$ in the norm of the Hilbert space l_2. Indeed, the distance between the points P'^* and P''^* from M_0^* is given by the formula

$$\rho(P'^*, P''^*) = \sqrt{\sum_{i=1}^{\infty} (b_i' - b_i'')^2} = \sqrt{\sum_{n=1}^{\infty} \frac{1}{n^2 N_n^2} \int_{x_0}^{x_0+a} (y_n' - y_n'')^2 \, dx} \, ,$$

whence it follows that

$$\rho(P_0^*, P_k^*) \leq \sqrt{\sum_{n=1}^{n_0} \frac{1}{n^2 N_n^2} \int_{x_0}^{x_0+a} (y_n^0 - y_n^k)^2 dx + a \sum_{n=n_0}^{\infty} \frac{1}{n^2}}$$

is arbitrarily small for sufficiently large n_0 and k. This means that the set M_0^* is compact.

Note that one can easily prove that mapping (B) is a homeomorphism, i.e., the sets M_0 and M_0^* are topologically equivalent.

Theorem (1.1) is proved.

Example. Consider a system of equations

$$\frac{dy_i}{dx} = y_{i+1}, \quad i = 1, 2, 3, \ldots. \tag{1.5}$$

In the domain $H : -\infty < x < \infty$, $\sup\{|y_1|, |y_2|, \ldots\} < \infty$, all conditions of Theorem 1.1 are satisfied. Therefore, at least one solution of the system of equations (1.5) passes through every point $(x_0, y_1^0, y_2^0) \in H$.

It is easy to see that the series

$$y_s(x) = y_s^0 + (x + x_0)y_{s+1}^0 + \frac{(x - x_0)^2}{2} y_{s+2}^0 + \ldots, \qquad s = 1, 2, \ldots,$$

define the required solution. Thus, in particular, the solution of this sort that passes through the point $(0; 0, 0, 0, \ldots) \in H$ has the form $y_s(x) \equiv 0$, $s = 1, 2, \ldots$.

Consider a function

$$f(x) = \begin{cases} \exp\{-\frac{1}{x^2}\}, & x \neq 0, \\ 0, & x = 0. \end{cases}$$

The vector $y_{1s}(x) = \dfrac{d^s f(x)}{dx^s}$, $s = 1, 2, \ldots$, is also a solution of the system of equations (1.5) that passes through the point $(0; 0, 0, 0, \ldots) \in H$. The presented solutions have the same initial conditions but do not coincide, i.e., the theorem does not guarantee the uniqueness of a solution for certain initial conditions.

The following theorem is presented here without proof:

Theorem 1.2. *Let the system of equations (1.1) be such that*

1. $|f_n(x, y_1', \ldots, y_n', \ldots) - f_n(x, y_1'', \ldots, y_n'', \ldots)| \leq \sum_{i=1}^{\infty} K_{ni}|y_i' - y_i''|$, $n = 1, 2, \ldots$;

2. $\sum\limits_{n=1}^{\infty} K_{ni} = A_i < A < \infty,\ K_{ni} = \text{const} > 0,\ A = \text{const} > 0,\ A_i = \text{const} > 0,\ i = 1, 2, \ldots.$

Then there exists at most one solution $(y_1(x),\ y_2(x),\ \ldots)$ *of the system of equations (1.1) satisfying given initial conditions and such that*

$$\sum_{n=1}^{\infty} |y_n(x)| < B = \text{const} < \infty\,.$$

We now consider a domain h defined by the inequalities $0 \le x \le r$, $|y_s| \le R^0$, $s = 1, 2, \ldots$, and the domain H defined in the example considered above.

Assume that the right-hand sides of the system of equations (1.1) in the domain H satisfy the following conditions:

1*. The functions f_i, $i = 1, 2, \ldots$, are continuous in x at any point, i.e.,

$$f_i(x + \Delta x, y_1, y_2, \ldots) - f_i(x, y_1, y_2, \ldots) \to 0 \quad \text{as} \quad \Delta x \to 0.$$

2*. The functions f_i, $i = 1, 2, \ldots$, satisfy the Cauchy–Lipschitz condition in variables $y_1, y_2, \ldots$, i.e.,

$$|f_i(x, y_1', y_2', \ldots) - f_i(x, y_1'', y_2'', \ldots)| \le \alpha(x)\Delta y,$$

where $\Delta y = \sup\{|y_1' - y_1''|,\ |y_2' - y_2''|,\ \ldots\}$ and $\alpha(x)$ is a continuous function of x.

3*. The following inequalities hold for $y_1 = y_2 = \ldots = 0:$

$$|f_i(x, 0, 0, \ldots)| \le \beta(x)\,, \quad i = 1, 2, \ldots\,,$$

where $\beta(x)$ is a continuous function of x.

A vector function $(y_1(x),\ y_2(x),\ \ldots)$ is called bounded in H if, for any segment σ specified in advance, there exists a finite number $T(\sigma)$ such that

$$\sup\{|y_1(x)|,\ |y_2(x)|,\ \ldots\} \le T(\sigma)$$

for all $x \in \sigma$.

Theorem 1.3. *Assume that the system of equations (1.1) satisfies conditions 1*–3* in the domain* H. *Then, for any point* $(x_0,\ y_1^0,\ y_2^0,\ \ldots) \in H$, *there exists a* *unique bounded solution* $(y_1(x),\, y_2(x),\, \ldots)$ *of this system of equations such that* $y_i(x_0) = y_i^0$ *for all* $i = 1, 2, 3, \ldots$. *Furthermore, this solution is equicontinuous and exists for all finite values of* x.

Proof. We seek a solution of the system of equations (1.1) passing through a given point $(x_0,\, y_1^0,\, y_2^0,\, \ldots)$ by the method of successive approximations. For this purpose, we set

$$y_s^{(1)}(x) = y_s^0 + \int_{x_0}^{x} f_s(\tau, y_1^0, y_2^0, \ldots)\, d\tau,$$

$$y_s^{(m)}(x) = y_s^0 + \int_{x_0}^{x} f_s(\tau, y_1^{(m-1)}(\tau), y_2^{(m-1)}(\tau), \ldots)\, d\tau,$$

$$s = 1, 2, \ldots, \qquad m = 2, 3, 4, \ldots. \tag{1.6}$$

It is easy to see that conditions 1*–3* in the domain H guarantee the validity of the following statements:

4*. $|f_s(x, y_1, y_2, \ldots)| \le \alpha(x) \sup\{|y_1|, |y_2|, \ldots\} + \beta(x), \quad s = 1, 2, \ldots$.

5*. The functions f_s, $s = 1, 2, \ldots$, are continuous in the collection of their arguments, i.e.,

$$f_s(x + \Delta x, y_1 + \Delta y_1, y_2 + \Delta y_2, \ldots) - f_s(x, y_1, y_2, \ldots) \to 0$$

as $|\Delta x| + \Delta y \to 0$, where $\Delta y = \sup\{|\Delta y_1|, |\Delta y_2|, \ldots\}$;

6*. If the functions $y_1(x)$, $y_2(x)$, ... are equicontinuous in x, then the functions $f_s(x,\, y_1(x),\, y_2(x),\, \ldots)$, $s = 1, 2, \ldots$, are continuous in x.

By using conditions 1*–6*, we conclude that, for any $m = 1, 2, \ldots$, the functions $y_1^{(m)}(x)$, $y_2^{(m)}(x)$, ... exist for all finite values of x, are equicontinuous, and satisfy the inequalities

$$|y_s^{(1)}(x) - y_s^0| \le \int_{x_0}^{x} [\beta(\tau) + y_0 \alpha(\tau)]\, d\tau,$$

$$|y_s^{(m+1)}(x) - y_s^{(m)}(x)| \le \int_{x_0}^{x} \alpha(\tau) \Delta y^{(m)}(\tau)\, d\tau, \quad m = 1, 2, \ldots, \tag{1.7}$$

if, for definiteness, we assume that $x \geq x_0$ and set

$$\Delta y^{(m)}(x) = \sup\{|y_1^{(m)}(x) - y_1^{(m-1)}(x)|, \ldots\}, \quad y_0 = \sup\{|y_1^0|, |y_2^0|, \ldots\}.$$

Let $[a, b]$ be an arbitrary segment containing the indicated points x_0 and x. We denote its length by δ and set

$$L = \max\{\beta(x) + y_0\alpha(x)\}, \quad l = \max \alpha(x)$$

on the segment $[a, b]$. Then, by virtue of (1.7), we can write

$$|y_s^{(1)}(x) - y_s^0| \leq L|x - x_0| \leq L\delta,$$

$$|y_s^{(2)}(x) - y_s^{(1)}(x)| \leq Ll\,\frac{|x - x_0|^2}{2} \leq Ll\,\frac{\delta^2}{2},$$

$$\ldots\ldots\ldots\ldots\ldots\ldots\ldots\ldots$$

$$|y_s^{(m+1)}(x) - y_s^{(m)}(x)| \leq Ll^m\frac{|x - x_0|^{m+1}}{(m+1)!} \leq Ll^m\frac{\delta^{m+1}}{(m+1)!},$$

$$\ldots\ldots\ldots\ldots\ldots\ldots\ldots\ldots$$

This implies that the series

$$y_s(x) = y_s^0 + (y_s^{(1)} - y_s^0) + (y_s^{(2)}(x) - y_s^{(1)}(x)) + \ldots, \quad s = 1, 2, \ldots, \tag{1.8}$$

are absolutely convergent for all finite values of x. In this case, for any segment $[a, b]$ specified in advance, the convergence of the series is uniform in x and equicontinuous with respect to s, i.e.,

$$|y_s^{(m)}(x) - y_s(x)| < \varepsilon, \quad s = 1, 2, \ldots,$$

provided that $m \geq N(\varepsilon)$. Consequently, the functions

$$y_1(x),\ y_2(x),\ \ldots,\ y_n(x),\ \ldots \tag{1.9}$$

specified by series (1.8) are equicontinuous for all finite x and, moreover,

$$y_s(x) = \lim_{m\to\infty} y_s^{(m)}(x), \quad s = 1, 2, \ldots.$$

This means that the functions $f_s(x,\, y_1(x),\, y_2(x),\, \ldots)$ are continuous for all finite x and, thus, one can pass to the limit under the integral in inequalities (1.6). Therefore,

$$y_s(x) = y_s^0 + \int_{x_0}^{x} f_s(\tau, y_1(\tau), y_2(\tau), \ldots)\, d\tau,$$

$$\frac{dy_s(x)}{dx} = f_s(x, y_1(x), y_2(x), \ldots), \quad s = 1, 2, \ldots,$$

i.e., the vector function (1.9) is an equicontinuous solution of the system of equations (1.1) that passes through a given point $(x_0, y_1^0, y_2^0, \ldots)$ of the domain H. This solution is bounded because the function $y(x) = \sup\{|y_1(x)|, |y_2(x)|, \ldots\}$ is continuous and bounded on any finite segment.

It is easy to see that every bounded solution of the system of equations (1.1) is equicontinuous. Indeed, if $\sup\{|y_1(x)|, |y_2(x)|, \ldots\} \leq T(\sigma)$, then, in view of estimate 4*, we obtain

$$\left|\frac{dy_s(x)}{dx}\right| = |f_s(x, y_1(x), y_2(x), \ldots)| \leq \alpha(x)T(\sigma) + \mu(x),$$

i.e., the derivatives $y_1'(x), y_2'(x), \ldots$ are uniformly bounded. Hence, the analyzed solution is equicontinuous. Condition 2* implies that an equicontinuous solution of the system of equations (1.1) with given initial conditions is unique. This completes the proof of Theorem 1.3.

Corollary 1.1. *Under conditions 1*–3*, the plot of the unique bounded solution of the system of equations (1.1) passes through every point of the domain h. Moreover, this solution exists for $x \in [0, r]$ such that*

$$\sup\{\,|y_1(x)|, |y_2(x)|, \ldots\} \leq R^0.$$

For a linear system of differential equations

$$\frac{dy_s}{dx} = \sum_{i=1}^{\infty} p_{si}(x)y_i + q_s(x), \quad s = 1, 2, \ldots, \tag{1.10}$$

the following statement is true:

Theorem 1.4. *Assume that the system of equations (1.10) is such that the functions*

$$p_{si}(x), \quad p_s(x) = \sum_{j=1}^{\infty} |p_{sj}(x)|, \quad q_s(x)$$

are continuous in the domain H for all $s, i = 1, 2, \ldots$ and satisfy the inequalities $p_s(x) \leq \alpha(x)$ and $|q_s(x)| \leq \beta(x)$, where $\alpha(x)$ and $\beta(x)$ are continuous functions of x.

Then, for every point $(x_0,\ y_1^0,\ y_2^0,\ \ldots) \in H$, *there exists a unique bounded (equicontinuous) solution* $(y_1(x),\ y_2(x),\ \ldots)$ *of the system of equations (1.10) such that* $y_s(x_0) = y_s^0$, $s = 1, 2, \ldots$.

The proof of Theorem 1.4 reduces to the verification of conditions 1*–3* for the system of equations (1.10). In what follows, we consider only bounded solutions of the system of equations (1.1).

1.2. Truncation Method

Assume that the right-hand sides of the system of equations (1.1) satisfy the strengthened Cauchy–Lipschitz condition in the domain H:

$$|f_s(x, y_1, .., y_m, y'_{m+1}, y'_{m+2}, \ldots) - f_s(x, y_1, \ldots, y_m, y''_{m+1}, y''_{m+2})|$$

$$\leq \alpha(x) \Delta y \varepsilon_s(m)\,, \quad s = 1, 2, \ldots, \tag{2.1}$$

where $(x, y_1, \ldots, y_m, y'_m, y'_{m+1}, y'_{m+2}, \ldots)$ and $(x, y_1, \ldots, y_m, y''_{m+1}, y''_{m+2}, \ldots)$ are any two points of the domain H, $\alpha(x)$ is a continuous function of x,

$$\Delta y = \sup\{\, |y'_{m+1} - y''_{m+1}|, |y'_{m+2} - y''_{m+2}|, \ldots\},$$

and $\varepsilon_s(m) \to 0$ as $m \to \infty$.

Let

$$\frac{du_{sm}}{dx} = f_s(x, u_{1m}, u_{2m}, \ldots, u_{mm}, 0, 0, \ldots)\,, \quad s = 1, 2, \ldots, \tag{2.2}$$

be the truncated system of differential equations corresponding to system (1.1).

Suppose that

$$y_s = y_s(x, x_0, y_1^0, y_2^0, \ldots)\,, \quad s = 1, 2, \ldots, \tag{2.3}$$

is a solution of the system of equations (1.1) with initial conditions $(x_0,\ y_1^0,\ y_2^0, \ldots) \in H$, and $u_{sm} = u_{sm}(x)$, $s = \overline{1, m}$, is a solution of the system of equations (2.2), where $u_{sm}(x_0) = y_s^0$, $s = 1, 2, \ldots m$, and $u_{sm} = 0$ for $s = m+1, m+2, \ldots$.

We set $y_0 = \sup\{|y_1^0|,\ |y_2^0|, \ldots\}$ and $y(x) = \sup\{|y_1(x)|,\ |y_2(x)|,\ \ldots\}$. Let δ be the length of an arbitrarily chosen segment σ containing the point x_0, let $\alpha = \max\limits_{\sigma} \alpha(x)$, and let $\beta = \max\limits_{\sigma} \beta(x)$.

Lemma 2.1. *For any $x \in \sigma$, the following inequality is true:*

$$y(x) \le y_0 \exp\{\delta\alpha\} + \frac{\beta}{\alpha}(\exp\{\delta\alpha\} - 1)\,. \tag{2.4}$$

Indeed, in view of condition 4* in Sec. 1.1, we have

$$y(x) \le y_0 + \int_{x_0}^{x} [\beta(\nu) + \alpha(\nu)y(\nu)]\, d\nu \le y_0 + \int_{x_0}^{x} [\beta + \alpha y(\nu)]\, d\nu$$

if, for the sake of definiteness, we set $x \ge x_0$. Hence, $y(x) \le u(x)$, where $u(x)$ is a solution of the equation

$$u(x) = y_0 + \int_{x_0}^{x} (\beta + \alpha u(\nu))\, d\nu$$

of the form

$$u(x) = y_0 \exp\{(x - x_0)\alpha\} + \frac{\beta}{\alpha}(\exp\{(x - x_0)\alpha\} - 1).$$

This immediately implies the assertion of Lemma 2.1.

Theorem 2.1. *Assume that the system of equations (1.1) satisfies the conditions of Theorem 1.3 and estimate (2.1). Then a solution of the truncated system of equations (2.2) weakly converges to solution (2.3) of the system of equations (1.1), i.e.,*

$$\lim_{m \to \infty} u_{sm}(x) = y_s(x)\,, \quad s = 1, 2, \dots\,. \tag{2.5}$$

In this case, for any natural number g specified in advance and any arbitrarily small $\varepsilon > 0$, there exists a number $N(g, \varepsilon)$ such that, for $m \ge N$,

$$\omega_g(m) = \max_{x \in \sigma}\{|y_1(x) - u_{1m}(x)|, \dots, |y_g(x) - u_{gm}(x)|\} < \varepsilon, \tag{2.6}$$

where σ is an arbitrarily chosen segment containing the point x_0.

Proof. It is easy to see that a solution of the truncated system of equations (2.2) satisfies inequality (2.4) for any natural m and s and any $x \in \sigma$:

$$|u_{sm}(x)| \le y_0 \exp\{\delta\alpha\} + \frac{\beta}{\alpha}(\exp\{\delta\alpha\} - 1) = \gamma, \tag{2.7}$$

where δ, α, β, σ, and y_0 are the same as in inequality (2.4); moreover,

$$|u_{sm}(x+\Delta x) - u_{sm}(x)| \leq (\alpha\gamma + \beta)|\Delta x|. \tag{2.8}$$

The last estimate follows from the inequalities

$$\begin{aligned}
&|u_{sm}(x+\Delta x) - u_{sm}(x)| \\
&\quad \leq \int_x^{x+\Delta x} |f_s(\nu, u_{1m}(\nu), \dots, u_{mm}(\nu), 0, 0, \dots)|\, d\nu \\
&\quad \leq \int_x^{x+\Delta x} \alpha(\nu)\sup\{|u_{1m}(\nu)|, \dots, |u_{mm}(\nu)|\}\, d\nu + \int_x^{x+\Delta x} \beta(\nu)\, d\nu
\end{aligned}$$

deduced by using condition 4* from Sec. 1.1.

Now assume that condition (2.6) is not satisfied. Then there exist a natural number $g = g_0$, a real number $\varepsilon = \varepsilon_0$, and a sequence of natural numbers

$$m_1, m_2, \dots, m_n, \dots \to \infty \tag{2.9}$$

such that $\omega_{g_0}(m_n) \geq \varepsilon_0$, $n = 1, 2, \dots$.

At the same time, the sequence of functions

$$u_{sm_1}(x), u_{sm_2}(x), \dots, u_{sm_n}(x), \dots$$

satisfies inequalities (2.7) and (2.8) for all $x \in \sigma$ and $s = 1, 2, \dots$. By using the Arzelà theorem and the procedure of diagonalization described in the proof of Theorem 1.1, we extract a subsequence of this sequence

$$v_{s,1}(x), v_{s,2}(x), v_{s,3}(x), \dots, \qquad s = 1, 2, \dots,$$

that converges uniformly to a function $u_s(x)$ on the segment σ for any natural s. Furthermore, it is clear that the function $u_s(x)$ satisfies inequalities (2.7) and (2.8).

We fix the number s. Then we get

$$J_{sn} = |f_s(x, v_{1,n}(x), v_{2,n}(x), \dots) - f_s(x, u_1(x), u_2(x), \dots)| \leq J'_{sn} + J''_{sn},$$

where

$$\begin{aligned}
J'_{sn} = {} & |f_s(x, v_{1,n}(x), v_{2,n}(x), \dots) \\
& - f_s(x, u_1(x), \dots, u_g(x), v_{g+1,n}(x), v_{g+2,n}(x), \dots)|,
\end{aligned}$$

$$J''_{sn} = |f_s(x, u_1(x), \ldots, u_g(x), v_{g+1,n}(x), v_{g+2,n}(x), \ldots)$$
$$- f_s(x, u_1(x), u_2(x), \ldots)|.$$

Inequalities (2.1) and (2.7) yield

$$J''_{sn} \leq \alpha(x)\sup\{|v_{g+1,n}(x) - u_{g+1,n}(x)|, \ldots\}\varepsilon_s(g) \leq 2\alpha\gamma\varepsilon_s(g)$$

for any $x \in \sigma$. Since $\varepsilon_s(g) \to 0$ as $g \to \infty$, for any fixed $\nu > 0$ there exists a number $g' = g(\nu)$ such that $\varepsilon_s(g') < \nu$. We set $g = g'$. One can also find a number $N(s, \nu)$ such that

$$J'_{sn} \leq \alpha \sup\{|u_1(x) - v_{1,n}(x)|, \ldots, |u_g(x) - v_{g,n}(x)|\} < \alpha\nu$$

for $n \geq N(s, \nu)$. Then, for $n \geq N(s, \nu)$, we have

$$J_{sn} < (2\alpha\gamma + \alpha)\nu$$

for any $x \in \sigma$. This means that, for any $x \in \sigma$,

$$\lim_{n\to\infty} \int_{x_0}^{x} f_s(\tau, v_{1,n}(\tau), v_{2,n}(\tau), \ldots)\, d\tau = \int_{x_0}^{x} f_s(\tau, u_1(\tau), u_2(\tau), \ldots)\, d\tau.$$

Consequently,

$$u_s(x) = y_s^0 + \int_{x_0}^{x} f_s(\tau, u_1(\tau), u_2(\tau), \ldots)\, d\tau\,, \quad s = 1, 2, \ldots .$$

This yields $u_s(x) = y_s(x)$, $s = 1, 2, \ldots$, i.e., for any fixed natural g,

$$\max_{x\in\sigma}(\sup\{|y_1(x) - v_{1,n}(x)|, \ldots, |y_g(x) - v_{g,n}(x)|\}) \to 0$$

as $n \to \infty$, which contradicts inequality (2.9).

Theorem 2.1 is proved.

Corollary 2.1. *Assume that $\sigma \subset [0, r]$ and the quantities x_0 and σ are chosen so that the quantity γ satisfies the condition $\gamma \leq R$. Then Theorem 2.1 is true in the domain h.*

1.3. Solutions of the Linear System

Consider a linear system of equations of the form

$$\frac{dx_s}{dt} = \sum_{i=1}^{\infty} p_{si}(t)x_i\,, \quad s = 1, 2, \ldots, \tag{3.1}$$

satisfying the following conditions [called conditions $(\widehat{P})$]:

In the domain H, the functions $p_{si}(t)$, $p_s(t) = \sum_{j=1}^{\infty} |p_{sj}(t)|$ are continuous for all $s, i = 1, 2, \ldots$, and satisfy the inequalities $p_s(t) \leq \alpha(t)$, where $\alpha(t)$ is a continuous function of t.

Assume that a sequence of vector functions

$$\bar{x}_1(t) = \begin{pmatrix} x_{11}(t) \\ x_{21}(t) \\ x_{31}(t) \\ \vdots \end{pmatrix}, \quad \bar{x}_2(t) = \begin{pmatrix} x_{12}(t) \\ x_{22}(t) \\ x_{32}(t) \\ \vdots \end{pmatrix}, \quad \ldots, \quad \bar{x}_g(t) = \begin{pmatrix} x_{1g}(t) \\ x_{2g}(t) \\ x_{3g}(t) \\ \vdots \end{pmatrix}, \quad \ldots$$

is a countable set of solutions of the system of equations (3.1).

Theorem 3.1. *Assume that the system of equations (3.1) satisfies conditions* $(\widehat{P})$ *and*

$$|x_{s1}(t_0)| + |x_{s2}(t_0)| + |x_{s3}(t_0)| + \ldots \leq \beta = \text{const} < \infty, \quad s = 1, 2, \ldots, \tag{3.2}$$

for some $t = t_0$. *Then the series*

$$x_s(t) = c_1 x_{s1}(t) + \ldots + c_n x_{sn}(t) + \ldots, \quad s = 1, 2, \ldots, \tag{3.3}$$

specify a solution $x(t) = (x_1(t), \ldots)$ *of the system of equations (3.1). If* $\sup\{|c_1|, |c_2|, \ldots\} = c < \infty$, *where* c_i *are arbitrary real numbers, then these series can be integrated and differentiated term by term.*

Proof. Consider a system of functions

$$x_s^{(m)} = x_s^{(m)}(t) = \sum_{g=1}^{m} c_g x_{sg}(t), \quad s = 1, 2, \ldots .$$

It is obvious that this system specifies the vector function $x^{(m)} = (x_1^{(m)}(t), x_2^{(m)}(t), \ldots)$ which is a solution of the system of equations (3.1) and, moreover,

$$\sup\{|x_1^{(m)}(t_0)|, \ldots\} \leq c\beta.$$

Let σ be an arbitrary segment containing the point t_0. The matrix $[p_{ij}(t)]_{i,j=1}^{\infty}$ is denoted by P. The relation

$$x^{(m)}(t) = x^{(m)}(t_0) + \int_{t_0}^{t} P(\tau) x^{(m)}(\tau)\, d\tau$$

yields

$$\|x^{(m)}(t)\| \leq \|x^{(m)}(t_0)\| + \int_{t_0}^{t} \|P(\tau)\|\, \|x^{(m)}(\tau)\|\, d\tau,$$

where we have introduced the following compatible norms of vectors and matrices:

$$\|x^{(m)}(t)\| = \sup\{|x_1^{(m)}(t)|, \ldots\} \quad \text{and} \quad \|P(t)\| = \sup_i \sum_{j=1}^{\infty} |p_{ij}(t)|.$$

By using the Gronwall–Bellman inequality, we arrive at the formula

$$\|x^{(m)}(t)\| \leq c\beta \exp\Big\{\Big|\int_{t_0}^{t} \alpha(\tau)\, d\tau\Big|\Big\}, \quad t \in \sigma, \quad m = 1, 2, \ldots .$$

Consequently,

$$|x_s^{(m)}(t)| \leq c\beta \exp\Big\{\Big|\int_{t_0}^{t} \alpha(\tau)\, d\tau\Big|\Big\} \tag{3.4}$$

for all $t \in \sigma$. In expression (3.4), we fix the values of s and t. Without changing the absolute values of $c_1, \ldots, c_m$, we choose them to guarantee the validity of the inequalities $c_g x_{sg}(t) \geq 0$, $g = 1, 2, \ldots, m$. Then

$$\sum_{i=1}^{m} |c_i|\, |x_{si}(t)| \leq c\beta \exp\Big\{\Big|\int_{t_0}^{t} \alpha(\tau)\, d\tau\Big|\Big\},$$

whence

$$\sum_{i=1}^{m} |x_{si}(t)| \leq \beta \exp\Big\{\Big|\int_{t_0}^{t} \alpha(\tau)\, d\tau\Big|\Big\}. \tag{3.5}$$

Since the quantities s and t were chosen arbitrarily, inequality (3.5) is true for all $s = 1, 2, \ldots$ and $t \in \sigma$.

By virtue of (3.5), we obtain

$$|x_s(t)| \leq \sum_{i=1}^{\infty} |c_i|\,|x_{si}(t)| \leq c\beta \exp\Big\{\Big|\int_{x_0}^{t} \alpha(\tau)\,d\tau\,\Big|\Big\}, \qquad s = 1, 2, \dots,$$

and, consequently,

$$\|x(t)\| \leq c\beta \exp\Big\{\Big|\int_{t_0}^{t} \alpha(\tau)\,d\tau\,\Big|\Big\}, \qquad t \in \sigma. \tag{3.6}$$

This means that series (3.3) are absolutely convergent and, in view of the well-known Lebesgue theorem, can be integrated term by term. This yields

$$\begin{aligned} \sum_{g=1}^{m} c_g\Big(\sum_{k=1}^{\infty} p_{sk}(t)x_{kg}(t)\Big) &= \sum_{g=1}^{m} c_g x'_{sg}(t) \\ &= \frac{dx_g^{(m)}(t)}{dt} = \sum_{k=1}^{\infty} p_{sk}(t) \sum_{g=1}^{m} c_g x_{kg}(t), \end{aligned} \tag{3.7}$$

$$\sum_{k=1}^{\infty} |p_{sk}(t)| \sum_{g=1}^{\infty} |c_g|\,|x_{kg}(t)| \leq \alpha(t)c\beta \exp\Big\{\Big|\int_{t_0}^{t} \alpha(\tau)\,d\tau\Big|\Big\}. \tag{3.8}$$

By analogy with inequality (3.6), we get

$$\sum_{g=1}^{\infty} |c_g|\,|x'_{sg}(t)| \leq \alpha(t)c\beta \exp\Big\{\Big|\int_{t_0}^{t} \alpha(\tau)\,d\tau\Big|\Big\} \tag{3.9}$$

for all $t \in \sigma$. Hence, the series

$$\sum_{g=1}^{\infty} c_g x'_{sg}(t)\,, \qquad s = 1, 2, \dots,$$

are absolutely convergent, and, by virtue of (3.9), they can be integrated term by term.

As a result, we obtain

$$\sum_{g=1}^{\infty} c_g x_{sg}(t_0) + \int_{t_0}^{t} \Big(\sum_{g=1}^{\infty} c_g x'_{sg}(\tau)\Big)\, d\tau$$

$$= \sum_{g=1}^{\infty} c_g x_{sg}(t_0) + \sum_{g=1}^{\infty} c_g \int_{t_0}^{t} x'_{sg}(\tau)\, d\tau$$

$$= \sum_{g=1}^{\infty} c_g x_{sg}(t) = x_s(t). \tag{3.10}$$

In view of (3.9), we conclude that the functions $x_s(t)$, $s = 1, 2, \dots$, are equicontinuous for $t \in \sigma$. Moreover, it is clear that

$$\sum_{g=1}^{\infty} c_g x'_{sg}(t) = \sum_{g=1}^{\infty} c_g \sum_{k=1}^{\infty} p_{sk}(t) x_{kg}(t). \tag{3.11}$$

Therefore,

$$J_s = \sum_{k=1}^{\infty} p_{sk}(t) \sum_{g=1}^{\infty} c_g x_{kg}(t) - \sum_{k=1}^{\infty} p_{sk}(t) \sum_{g=1}^{m} c_g x_{kg}(t)$$

$$= \sum_{k=1}^{\infty} p_{sk}(t) \sum_{g=m+1}^{\infty} c_g x_{kg}(t)$$

$$= \sum_{k=1}^{n} p_{sk}(t) \sum_{g=m+1}^{\infty} c_g x_{kg}(t) + \sum_{k=n+1}^{\infty} p_{sk}(t) \sum_{g=m+1}^{\infty} c_g x_{kg}(t). \tag{3.12}$$

We fix the values of s and t. Condition $(\widehat{P})$ and inequality (3.6) imply that

$$\left| \sum_{k=n+1}^{\infty} p_{sk}(t) \Big(\sum_{g=m+1}^{\infty} c_g x_{kg}(t) \Big) \right| < \varepsilon c \beta \exp \left\{ \left| \int_{t_0}^{t} \alpha(\tau)\, d\tau \right| \right\}$$

for any fixed $\varepsilon > 0$, provided that $n \geq n(\varepsilon)$. In relation (3.12), we set $n = n(\varepsilon)$. Then, in view of the absolute convergence of series (3.8), one can find a number $m(\varepsilon)$ such that

$$\left| \sum_{g=m+1}^{\infty} c_g x_{kg}(t) \right| < \varepsilon\,, \quad k = 1, 2, \dots, n(\varepsilon),$$

for any $m \geq m(\varepsilon)$, and, therefore,

$$\left|\sum_{k=1}^{n(\varepsilon)} p_{sk}(t)\Big(\sum_{g=m+1}^{\infty} c_g x_{kg}(t)\Big)\right| \leq (\varepsilon)\alpha(t),$$

provided that $m \geq m(\varepsilon)$. Consequently,

$$|J_s| \leq \varepsilon\left(c\beta \exp\left\{\left|\int_{t_0}^{t} \alpha(\tau)\,d\tau\right|\right\} + \alpha(t)\right)$$

for $m \geq m(\varepsilon)$, i.e.,

$$\begin{aligned}
\lim_{m\to\infty} \sum_{k=1}^{\infty} p_{sk}(t) \sum_{g=1}^{m} c_g x_{kg}(t) &= \sum_{k=1}^{\infty} p_{sk}(t) \sum_{g=1}^{\infty} c_g x_{kg}(t) \\
&= \sum_{g=1}^{\infty} c_g x'_{sg}(t) \\
&= \sum_{g=1}^{\infty} c_g \sum_{k=1}^{\infty} p_{sk}(t) x_{kg}(t),
\end{aligned} \tag{3.13}$$

where the last equalities follow from (3.7).

Since the values of s and t were chosen arbitrarily, equalities (3.13) take place for any $s = 1, 2, \ldots$ and $t \in \sigma$. Since the series $p_s(t) = |p_{s1}(t)| + |p_{s2}(t)| + \ldots$ is uniformly convergent on any fixed segment $\Delta \subset \sigma$ and series (3.3) are continuous functions, the function

$$\Psi_s(t) = \sum_{k=1}^{\infty} p_{sk}(t) \sum_{g=1}^{\infty} c_g x_{kg}(t), \quad s = 1, 2, \ldots,$$

is continuous on the segment σ and, therefore, series (3.13) are continuous functions of t for $t \in \sigma$.

Hence, by virtue of (3.10), we find

$$\Big(\sum_{g=1}^{\infty} c_g x_{sg}(t)\Big)' = \sum_{g=1}^{\infty} c_g x'_{sg}(t), \quad s = 1, 2, \ldots, \tag{3.14}$$

i.e., series (3.3) admit term-by-term differentiation.

Equalities (3.13) and (3.14) imply that these series solve the system of differential equations (3.1).

1.4. Matrizant of a Linear System

In the present section, we also consider the linear system of equations (3.1) satisfying condition $(\widehat{P})$ introduced in Sec. 1.3. Clearly, this system satisfies the conditions of Theorem 1.4. Then, for any point $(t_0,\ x_1^0,\ x_2^0,\ \ldots) \in H$, there exists a unique solution $(x_1(t), x_2(t), \ldots)$ of system (3.1) such that $x_s(t_0) = x_s^0$, $s = 1, 2, 3, \ldots$.

We now construct a matrix $\Omega(t)$ whose ith column is occupied by the solution $x^{(i)}(t) = (x_1^{(i)}(t), x_2^{(i)}(t), \ldots)$ with the initial conditions t_0, $(\underbrace{0,\ 0,\ \ldots\ ,}_{i-1}\ 0,\ 1,\ 0,\ 0,\ \ldots)$. This means that $\Omega(t_0) = E$, where E is the identity matrix. The matrix thus constructed is denoted by $\Omega_{t_0}^t$. It is called the matrizant of the system of equations (3.1). We now consider some properties of the matrizant important for the subsequent presentation. First, we prove the following statement generalizing the well-known Gronwall–Bellman lemma to the entire real axis:

Lemma 4.1. *Assume that two nonnegative continuous scalar functions $u(t)$ and $f(t)$ defined for all $t \in R^1$ satisfy the inequality*

$$u(t) \le c + \left| \int_{t_0}^{t} f(\tau)u(\tau)\, d\tau \right|,$$

where c is a positive constant. Then

$$u(t) \le c \exp\Big\{ \Big| \int_{t_0}^{t} f(\tau)\, d\tau \Big| \Big\}$$

for all $t \in R^1$.

Proof. For $t \ge t_0$, the proof can be found, e.g., in [MiL]. Assume that $t \le t_0$. Then

$$u(t) \le c - \int_{t_0}^{t} f(\tau)u(\tau)\, d\tau, \tag{4.1}$$

whence

$$\int_{t}^{t_0} \frac{f(t)u(t)}{c - \int_{t_0}^{t} f(\tau)u(\tau)\, d\tau} dt \le \int_{t}^{t_0} f(\tau)\, d\tau.$$

In view of the fact that

$$\frac{d}{dt}\Big(c - \int_{t_0}^{t} f(\tau)u(\tau)d\tau\Big) = -f(t)u(t),$$

we arrive at the inequality

$$\ln\Big(c - \int_{t_0}^{t} f(\tau)u(\tau)d\tau\Big)\Big|_t^{t_0} \geq -\int_t^{t_0} f(\tau)\,d\tau.$$

Therefore,

$$c - \int_{t_0}^{t} f(\tau)u(\tau)d\tau \leq c\exp\Big\{\int_t^{t_0} f(\tau)\,d\tau\Big\}.$$

The last inequality and inequality (4.1) imply the estimate

$$u(t) \leq c\exp\Big\{-\int_{t_0}^{t} f(\tau)\,d\tau\Big\}, \quad t \leq t_0,$$

which completes the proof of Lemma 4.1.

Theorem 4.1. *The matrizant $\Omega_{t_0}^t$ is a matrix solution of the system of equations (3.1), i.e.,*

$$\frac{d\Omega_{t_0}^t}{dt} = P(t)\Omega_{t_0}^t. \tag{4.2}$$

The proof follows directly from the definition of matrizant.

Theorem 4.2. *For any constant matrix $C = [c_{ij}]_{i,j=1}^{\infty}$ such that $\|C\| < \infty$, the product $\Omega_{t_0}^t C$ exists and is a solution of the matrix equation (3.1).*

Proof. Relation (4.2) yields

$$\|\Omega_{t_0}^t\| \leq 1 + \int_{t_0}^{t} \|P(\tau)\|\,\|\Omega_{t_0}^\tau\|\,d\tau.$$

Therefore,

$$\|\Omega_{t_0}^t\| \leq \exp\Big\{\int_{t_0}^{t} \|P(\tau)\|\,d\tau\Big\} \leq \exp\{\alpha(t)(t - t_0)\},$$

where, for definiteness, we set $t \geq t_0$. This means that the product $\Omega_{t_0}^t C$ indeed exists for all t and is bounded in norm uniformly in t on any finite segment (see Lemma 4.1).

Since the columns of the matrizant are solutions of the system of equations (3.1) for which inequality (3.2) holds with $\beta = 1$, by using Theorem 3.1 we can easily show that the columns of the matrix $\Omega_{t_0}^t C$ are also solutions of the system of equations (3.1). This immediately implies the statement of Theorem 4.2.

Corollary 4.1. *For any solution $x(t) = (x_1(t), x_2(t), \dots)$ of the system of equations (3.1), there exists a constant vector $c = (c_1, c_2, \dots)$ such that*

$$x(t) = \Omega_{t_0}^t c$$

identically with respect to t.

To prove this assertion, it suffices to choose c in the form $c = x(t_0) = (x_1(t_0), x_2(t_0), \dots)$.

Corollary 4.2.

$$\Omega_{t_0}^t = \Omega_{t_1}^t \Omega_{t_0}^{t_1}, \quad t_0, t_1, t \in R^1. \tag{4.3}$$

Since $\Omega_{t_0}^t$ and $\Omega_{t_1}^t$ are matrix solutions of the system of equations (3.1), one can easily see that there exists a constant matrix C such that $\Omega_{t_0}^t = \Omega_{t_1}^t C$. If, in this equality, we set $t = t_1$, then we get $C = \Omega_{t_0}^{t_1}$, which yields equality (4.3).

Theorem 4.3. *Let the elements of the matrix P be continuous on a segment $T \subset R^1$ and let*

$$\sum_{j=1}^{\infty} \max_T |p_{ij}(t)| \leq p_0 = \text{const} < \infty\,, \quad i = 1, 2, \dots . \tag{4.4}$$

Then the matrizant of the system of equations (3.1) admits a representation in the form of the Peano series

$$\Omega_{t_0}^t = E + \int_{t_0}^{t} P(t)\,dt + \int_{t_0}^{t} \Big\{P(t) \int_{t_0}^{t} P(t)dt\Big\}\,dt + \dots \tag{4.5}$$

which is absolutely and uniformly convergent on the segment T in the strong sense.

Proof. Successive approximations for the matrizant can be found from the recurrence relations

$$\frac{dX_k}{dt} = P(t)X_{k-1}\,, \quad k = 1, 2, \ldots,$$

by choosing X_0 in the form of the identity matrix E. As a result, we obtain

$$X_0 = E, \quad X_1 = E + \int_{t_0}^{t} P(t)\,dt,$$

$$X_2 = E + \int_{t_0}^{t} P(t)\,dt + \int_{t_0}^{t} \Big\{P(t) \int_{t_0}^{t} P(t)\,dt\Big\}\,dt, \;\ldots\,. \tag{4.6}$$

The formal expressions (4.6) are, in fact, meaningful because $\int_{t_0}^{t} P(t)dt$ is a matrix bounded in norm and continuous in the segment T and $P(t) \int_{t_0}^{t} P(t)dt$ is a matrix with the same properties, as follows from condition (4.4).

Thus, it is necessary to show that series (4.5) is a solution of equation (3.1). We denote the length of a segment T containing the point $t = t_0$ by δ. In this case, it is easy to see that the matrix series (4.5) can be majorized by the following convergent number series:

$$1 + p_0\delta + \frac{p_0^2\delta^2}{2!} + \frac{p_0^3\delta^3}{3!} + \ldots + \frac{p_0^k\delta^k}{k!} + \ldots$$

and, therefore, it is absolutely and uniformly convergent.

We now differentiate each term of series (4.5). As a result, we obtain the matrix series

$$P(t) + P(t)\int_{t_0}^{t} P(t)\,dt + P(t)\int_{t_0}^{t} \Big\{P(t)\int_{t_0}^{t} P(t)dt\Big\}\,dt + \ldots, \tag{4.7}$$

which can be majorized by a convergent number series

$$p_0 + p_0^2\delta + \frac{p_0^3\delta^2}{2!} + \ldots + \frac{p_0^{k+1}\delta^k}{k!} + \ldots$$

and, therefore, it also absolutely and uniformly convergent in the strong sense on the segment T. Thus, we can perform the term-by-term differentiation of series (4.5).

Let us now show that the matrix $P(t)$ in series (4.7) can be separated as a common factor. For the sake of simplicity, we rewrite this series in the form

$$P(t) + P(t)P^{(1)}(t) + P(t)P^{(2)}(t) + \dots,$$

where $P^{(i)}(t) = [p_{sk}^{(i)}(t)]_{s,k=1}^{\infty}$, $i = 1, 2, \dots$. Obviously, to prove the required statement, it suffices to show that

$$\sum_{s=1}^{\infty}\sum_{i=1}^{\infty} p_{ks}(t)p_{sj}^{(i)}(t) = \sum_{i=1}^{\infty}\sum_{s=1}^{\infty} p_{ks}(t)p_{sj}^{(i)}(t)$$

for any natural k and j. At the same time, we have already proved that the repeated series on the left-hand side of this relation is absolutely convergent for all $t \in T$. This means that the repeated series on the right-hand side of this relation is absolutely convergent and the corresponding sums are equal.

We denote the sum of series (4.5) by $P^*(t) = [p_{ij}^*(t)]_{i,j=1}^{\infty}$. Then

$$\frac{dP^*(t)}{dt} = P(t)P^*(t)$$

for all $t \in T$. It is clear that $P^*(t_0) = E$ and, therefore, $P^*(t) \equiv \Omega_{t_0}^t$ on the segment T just as required

We rewrite the system of equations (1.10) in the matrix form

$$\frac{dx}{dt} = P(t)x + f(t), \tag{4.8}$$

where $f(t) = (f_1(t), f_2(t), \dots)$ is a countably dimensional vector function.

Let us now show that the solution of the system of equations (4.8) can be expressed by using the matrizant of the homogeneous system of equations.

Theorem 4.4. *Let the system of equations (4.8) be such that*

(i) the functions $p_{ij}(t)$ *and* $f_i(t)$ *are continuous in* t *on the segment* T*;*

(ii) $\sum_{j=1}^{\infty} \max_T |p_{ij}(t)| \le p_0 = \text{const} < \infty$ *and* $\|f(t)\| = \sup_i\{|f_i(t)|\} \le f^0 = \text{const} < \infty$, $i = 1, 2, 3, \dots$.

Then the solution of the system of equations (4.8) with initial conditions $t_0 \in T$, x^0 *can be represented in the form*

$$x(t) = \Omega_{t_0}^t x^0 + \int_{t_0}^{t} K(t, \tau) f(\tau)\, d\tau,$$

where $K(t, \tau) = \Omega_{t_0}^t (\Omega_{t_0}^{\tau})^{-1}$ *is the so-called Cauchy kernel.*

Proof. By using Corollary 4.2, one can easily see that $(\Omega_{t_0}^t)^{-1}$ exists and has the form $\Omega_t^{t_0}$. Moreover, this matrix is formed by elements continuous in t for $t \in T$, which follows from the continuity of the elements of the matrix Ω_τ^t in τ for $\tau \in T$. The last property is a consequence of the fact that the dependence of the solutions on initial data is continuous (see Theorem 5.3). The required solution is sought in the form

$$x(t) = \Omega_{t_0}^t z(t), \tag{4.9}$$

where $z(t) = (z_1(t), \dots)$ is an unknown vector function. The equation for this function takes the form

$$\frac{d}{dt}(\Omega_{t_0}^t z(t)) = P(t)(\Omega_{t_0}^t z(t)) + f(t). \tag{4.10}$$

Moreover, the function $z(t)$ must satisfy the following conditions on the segment T:

(a) $\|z(t)\| \le z_0 = \text{const} < \infty$,

(b) $\left|\dfrac{dz(t)}{dt}\right| \le z^0 = \text{const} < \infty$.

Denote

$$\|P\|_0 = \sup_i \sum_{j=1}^{\infty} \max_T |p_{ij}(t)|, \quad \Omega_{t_0}^t = [\omega_{ij}(t)]_{i,j=1}^{\infty}.$$

Thus, we get $\|P\|_0 \le p_0$, $\|\Omega_{t_0}^t\|_0 \le e^{p_0\delta}$, where δ is the length of the segment T, and

$$\left\|\frac{d\Omega_{t_0}^t}{dt}\right\|_0 \le p_0 \exp\{p_0\delta\}, \quad \|(\Omega_{t_0}^t)^{-1}\|_0 \le \exp\{p_0\delta\} = \varphi_0.$$

It is easy to see that the series

$$\sum_{i=1}^{\infty} \omega_{si}(t) z_i'(t) \quad \text{and} \quad \sum_{i=1}^{\infty} \omega_{si}'(t) z_i(t) \tag{4.11}$$

can be majorized by the convergent number series

$$\sum_{i=1}^{\infty} \max_T |\omega_{si}(t)| z^0 \quad \text{and} \quad \sum_{i=1}^{\infty} \max_T |\omega_{si}'(t)| z_0\,,$$

respectively, for any $s = 1, 2, 3, \ldots$. In this case, series (4.11) are absolutely convergent on the segment T and, hence, the series

$$\sum_{i=1}^{\infty}(\omega'_{si}(t)z_i(t) + \omega_{si}(t)z'_i(t)), \quad s = 1, 2, \ldots,$$

are uniformly convergent on this segment. Therefore, the equality

$$\frac{d}{dt}(\Omega^t_{t_0}z(t)) = \frac{d}{dt}(\Omega^t_{t_0})z(t) + \Omega^t_{t_0}\frac{dz(t)}{dt}$$

is true for a function $z(t)$ satisfying conditions (a) and (b).

We now pass from equation (4.10) to an equation of the form

$$(P(t)\Omega^t_{t_0})z(t) + \Omega^t_{t_0}\frac{dz(t)}{dt} = P(t)(\Omega^t_{t_0}z(t)) + f(t).$$

Since $(P(t)\Omega^t_{t_0})z(t) = P(t)(\Omega^t_{t_0}z(t))$, we arrive at the equation

$$\frac{dz}{dt} = \Omega^{-1}(t_0, t)f(t),$$

or

$$z(t) = \int_{t_0}^{t} \Omega^{-1}(t_0, \tau)f(\tau)\, d\tau + c,$$

where c is an arbitrary constant vector such that $\|c\| < \infty$. In this case, we have

$$\|z(t)\| \le \int_{t_0}^{t} \varphi_0 f^0\, d\tau + \|c\| \le \varphi_0 f^0 \delta + \|c\| < \infty$$

and

$$\left\|\frac{dz(t)}{dt}\right\| \le \varphi_0 f^0 = \text{const} < \infty,$$

i.e., the function $z(t)$ satisfies conditions (a) and (b).

As a result, we easily arrive at the expression for the required solution

$$x(t) = \Omega^t_{t_0}x_0 + \Omega^t_{t_0}\int_{t_0}^{t} \Omega^{t_0}_{\tau} f(\tau)\, d\tau,$$

and, therefore, prove the statement of the theorem (in view of the fact that uniformly convergent series admit term-by-term integration).

1.5. Normal Autonomous Systems

In this section, we consider a system of equations of the form

$$\frac{dx_s}{dt} = f_s(x_1, x_2, \ldots, x_n, \ldots), \quad s = 1, 2, \ldots .$$

In the vector form, it can be represented as

$$\frac{dx}{dt} = f(x), \tag{5.1}$$

where $x = (x_1, x_2, \ldots)$ and $f(x) = (f_1(x),\ f_2(x), \ldots)$.

We consider equation (5.1) in a set $\mathfrak{M}$ of bounded numerical sequences with ordinary norm $\|x\| = \sup\{|x_1|, |x_2|, \ldots\}$, which turns the set $\mathfrak{M}$ into a complete linear normed space.

Theorem 5.1. *Let equation (5.1) be such that the following conditions are satisfied in the domain H:*

(i) $\|f(x)\| \leq \beta = \text{const} < \infty$;

(ii) $\|f(x') - f(x'')\| \leq \alpha \|x' - x''\|$, *where* $\alpha = \text{const} < \infty$.

Then, for any $t_0 \in (-\infty, \infty)$ and $x^{(0)} \in \mathfrak{M}$, there exists a unique solution $x(t)$ of equation (5.1) such that $x(t_0) = x^{(0)}$. Moreover, this solution is defined for all t and belongs to the space $\mathfrak{M}$.

The formulated theorem is an obvious consequence of Theorem 1.3.

The following statements are true for equation (5.1):

1. If equation (5.1) has two solutions $x = x(t)$ and $x = x_1(t)$, then these solutions either have no common points or coincide.

2. If a solution $x = x(t)$ of equation (5.1) is such that there exist $t_1 \neq t_2$ for which $x(t_1) = x(t_2)$, then this solution either corresponds to the equilibrium state or is periodic.

The proof of these statements does not differ from the proof of similar statements for finite-dimensional systems of differential equations (see, e.g., [Pon]). We now prove the following assertion:

Theorem 5.2. *Under the conditions of Theorem 5.1, equation (5.1) determines a dynamical system in the space $\mathfrak{M}$.*

We check the validity of all three conditions specifying a dynamical system in a metric space $\mathfrak{M}$ for the family of mappings $\varphi(x,t)$, where $\varphi(x,t)$ is a solution of equation (5.1) such that $\varphi(x,0) = x$, $x \in \mathfrak{M}$.

We recall these conditions:

1. Initial condition: $\varphi(x,0) = x$.

2. Continuity in the collection of variables x and t.

3. Group property: $\varphi(\varphi(x,t_1),t_2) = \varphi(x,t_1+t_2)$ for any $x \in \mathfrak{M}$ and any real t_1 and t_2.

The first property is obvious.

The proof of the second condition reduces to the proof of the following statement: Given an arbitrary point $x^0 \in \mathfrak{M}$ and an arbitrary number $t_0 \in (-\infty,\infty)$, for any $\varepsilon > 0$ there exists $\delta > 0$ such that

$$\rho\left(\varphi(x,t),\varphi(x^0,t_0)\right) < \varepsilon$$

whenever $\rho(x,x^0) < \delta$ and $|t-t_0| < \delta$. The following inequality is satisfied:

$$\begin{aligned}&\rho\left(\varphi(x^0+\Delta x^0,t),\varphi(x^0,t_0)\right)\\ &\qquad \le \rho\left(\varphi(x^0+\Delta x^0,t),\varphi(x^0,t)\right) + \rho\left(\varphi(x^0,t),\varphi(x^0,t_0)\right).\end{aligned} \tag{5.2}$$

In view of the fact that, for $x, y \in \mathfrak{M}$, the distance $\rho(x,y)$ is defined as follows:

$$\rho\,(x,y) = \|x-y\|,$$

we obtain

$$\begin{aligned}\rho(\varphi(x^0,t),\varphi(x^0,t_0)) &= \left\|x^0 + \int_0^t f(\varphi(x^0,\tau))\,d\tau - x^0 - \int_0^{t_0} f(\varphi(x^0,\tau))\,d\tau\right\|\\ &\le \left|\int_{t_0}^t \|\, f(\varphi(x^0,\tau))\,\|\,d\tau\right| \le \beta\,|t-t_0|.\end{aligned} \tag{5.3}$$

Similarly,

$$\begin{aligned}
&\rho(\varphi(x^0+\Delta x^0,t),\varphi(x^0,t))\\
&\quad = \Big\| \Delta x^0 + \int_0^t \{ f(\varphi(x^0+\Delta x^0,\tau)) - f(\varphi(x^0,\tau)) \}\, d\tau \Big\| \\
&\quad \le \|\Delta x^0\| + \int_0^t \alpha \| \varphi(x^0+\Delta x^0,\tau) - \varphi(x^0,\tau) \|\, d\tau.
\end{aligned}$$

The last inequality implies the estimate

$$\begin{aligned}
\rho(\varphi(x^0+\Delta x^0,t),\varphi(x^0,t)) \;&\le\; \|\Delta x^0\| \exp\Big\{ \Big| \int_0^t \alpha\, d\tau \Big| \Big\} \\
&=\; \|\Delta x^0\| \exp\{\alpha|t|\}. \qquad (5.4)
\end{aligned}$$

In view of relations (5.2)–(5.4), we conclude that the positive number δ must be chosen to guarantee that the inequality

$$\beta|t-t_0| + \|\Delta x^0\| \exp\{\alpha|t|\} < \varepsilon \qquad (5.5)$$

is satisfied for all $\|\Delta x^0\| < \delta$ and $|t-t_0| < \delta$.

Clearly, we can take

$$\delta \le \min\Big\{1, \frac{\varepsilon}{\beta + \exp\{\alpha(1+|t_0|)\}}\Big\}. \qquad (5.6)$$

Indeed,

$$\begin{aligned}
\beta|t-t_0| + \|\Delta x^0\| \exp\{\alpha|t|\} \;&<\; \beta\delta + \delta \exp\{\alpha|t_0| + \delta\alpha\} \\
&\le \varepsilon \frac{\beta + \exp\{\alpha(|t_0|+1)\}}{\beta + \exp\{\alpha(|t_0|+1)\}} = \varepsilon.
\end{aligned}$$

Thus, it remains to prove the third condition. We denote $\varphi(x,t_1) = y$ and consider the solution $\psi(y,t)$. Since equation (5.1) is autonomous, the function $\varphi(x,t+t_1)$ is its solution.

For $t = 0$, we have $\psi(y,0) = y = \varphi(x,t_1)$ and $\varphi(x,0+t_1) = \varphi(x,t_1)$, i.e., $\psi(y,0) = \varphi(x,0+t_1)$. Hence, the initial values of these solutions coincide. Thus, in view of the uniqueness of a solution for all $t \in R^1$, we get

$$\psi(y,t) = \varphi(x,t+t_1),$$

whence, for $t = t_2$, we obtain the required equality.

Remark. Theorem 5.1 remains valid if its first condition is replaced by

$$\|f(0)\| \leq \beta = \text{const} < \infty.$$

Theorem 5.2 also remains valid in this case. Indeed, in estimate (5.3), we find

$$\Big|\int_{t_0}^{t} \| f(\varphi(x^0,\tau)) \| \, d\tau\Big| \leq \Big|\int_{t_0}^{t} \{\alpha \, \|\varphi(x^0,\tau)\| + \|f(0)\|\} \, d\tau\Big|$$

$$\leq \beta|t - t_0| + \alpha \int_{t_0}^{t} \| \varphi(x^0,\tau) \| \, d\tau.$$

By σ we denote a segment containing the points t_0 and t. According to the definition of the boundedness of the solution $\varphi(x^0, t)$, we conclude that there exists a positive constant $T(x^0, \sigma)$ such that $\|\varphi(x^0,t)\| \leq T(x^0,\sigma)$. Then

$$\rho(\varphi(x^0,t), \varphi(x^0,t_0)) \leq \beta|t - t_0| + T(x^0,\sigma)|t - t_0|, \quad t \in \sigma.$$

Inequality (5.5) now takes the form

$$(\beta + T(x^0,\sigma))|t - t_0| + \|\Delta x^0\| \exp\{\alpha|t|\} < \varepsilon. \tag{5.7}$$

We denote the length of the segment σ by l. Note that l can be an arbitrarily large finite number.

In this case, to satisfy estimate (5.7), it suffices to choose δ from the interval

$$\delta \leq \min\Big\{1; \frac{\varepsilon}{\beta + T(x^0,\sigma) + \exp\{\alpha(1 + |t_0|)\}}; l\Big\}.$$

The last statement is checked in exactly the same way as in the case of choosing the corresponding value of δ according to condition (5.6) to satisfy inequality (5.5) in the proof of Theorem 5.2.

Note that the reasoning used to prove Theorem 5.2 can also be used for the analysis of the dependence of solutions on the initial data in the general case of a system of equations of the form (1.1).

Theorem 5.3. *Let the system of equations (1.1) satisfy the conditions of Theorem 1.3. Then the solution* $y_s = y_s(x, x_0, y_1^0, y_2^0, \ldots)$, $s = 1, 2, \ldots,$ *of this system is a continuous function of the initial data* $x_0, y_1^0, y_2^0, \ldots$.

Proof. Assume that the solution $y_s = \bar{y}_s(x),\ s = 1, 2, \ldots,$ of the system of equations (1.1) passes through a point $(x_0 + \Delta x_0,\ y_1^0 + \Delta y_1^0,\ y_2^0 + \Delta y_2^0, \ldots)$ of the domain H. Then

$$\begin{aligned}\Delta y_s(x) &= \bar{y}_s(x) - y_s(x)\\ &= \Delta y_s^0 + \int_{x_0}^{x} [f_s(\tau, \bar{y}_1(\tau), \bar{y}_2(\tau), \ldots) - f_s(\tau, y_1(\tau), y_2(\tau), \ldots)]\, d\tau\\ &\quad + \int_{x_0+\Delta x_0}^{x_0} f_s(\tau, \bar{y}_1(\tau), \bar{y}_2(\tau), \ldots)\, d\tau, \quad s = 1, 2, \ldots. \end{aligned} \tag{5.8}$$

By σ we denote the minimum segment containing the points x_0 and $x_0 + \Delta x_0$. Its length is denoted by δ. We also recall that $\|y_0\| = \sup\{|y_1^0|, |y_2^0|, \ldots\}$ and $\|\Delta y_0\| = \sup\{|\Delta y_1^0|, |\Delta y_2^0|, \ldots\}$.

Then, by virtue of inequality (2.4) proved in Lemma 2.1, we find

$$\|\bar{y}(x)\| \le \|y_0 + \Delta y_0\| \exp\{\delta\alpha\} + \frac{\beta}{\alpha}(\exp\{\delta\alpha\} - 1) = \eta,$$

where α and β are the same parameters as in Lemma 2.1.

In view of condition 4* in Sec. 1.1, we obtain

$$\Big| \int_{x_0+\Delta x_0}^{x_0} f_s(\tau, \bar{y}_1(\tau), \bar{y}_2(\tau), \ldots)\, d\tau \Big| \le (\alpha\eta + \beta)\delta = g.$$

By using equality (5.8) and condition 2 in Sec. 1.1, we get

$$\|\Delta y(x)\| \le \|\Delta y_0\| + g + \int_{x_0}^{x} \alpha(\nu) \|\Delta y(\nu)\|\, d\nu$$

or, equivalently,

$$\|\, \bar{y}(x) - y(x)\, \| \le \|\Delta y_0\| + g + \int_{x_0}^{x} \alpha(\nu) \|\, \bar{y}(\nu) - y(\nu)\, \|\, d\nu,$$

where, for the sake of definiteness, we set $x \ge x_0$.

Hence, for any $x \in R^1$, we have

$$\| \bar{y}(x) - y(x) \| \leq (\|\Delta y_0\| + g) \exp\Big\{\Big| \int_{x_0}^{x} \alpha(\nu)\, d\nu \Big|\Big\}.$$

This completes the proof of Theorem 5.3.

Corollary 5.1. *The matrizant Ω_τ^t of the system of equations (3.1) is a matrix whose elements are continuous in τ.*

1.6. Periodic Solutions

For countable systems, there exists a relatively well developed theory of periodic solutions. Here, we present some results obtained by using the numerical-analytic method of successive periodic approximations.

First, we introduce some concepts and notation. Let $\mathbf{M} = \{M_1, M_2, \ldots\}$ be vectors with nonnegative coordinates (points of the space $\mathfrak{M}$), let $\mathbf{K}, \mathbf{P}, \ldots$ be infinite matrices $\mathbf{K} = [K_{ij}]_{i,j=1}^{\infty}, \ldots$ with nonnegative elements (K_{ij} is the element of the ith row and jth column of the corresponding matrix), let $M, K, \ldots$ be the norms of the vectors $\mathbf{M}, \mathbf{K}, \ldots$, and let E be the identity matrix.

By $\mathcal{K}, \ldots$ we denote the linear operators from the space $\mathfrak{M}$ into space $\mathfrak{M}$ generated by the matrices $\mathbf{K}, \ldots$. Let $\|\mathcal{K}\|, \ldots$ be the norms of these operators.

If

$$\sup_i \sum_{j=1}^{\infty} K_{ij} < \infty, \tag{6.1}$$

then the matrix $\mathbf{K}$ generates an operator $\mathcal{K}$ and, moreover,

$$\|\mathcal{K}\| = \|\mathbf{K}\| = \sup_i \sum_{j=1}^{\infty} K_{ij}. \tag{6.2}$$

A linear operator $\mathcal{K}$ is called completely regular if

$$\| \mathcal{K} \| \leq g < 1. \tag{6.3}$$

Let D be a bounded domain of the space $\mathfrak{M}$ considered together with its boundary. For positive $\mathbf{d} = (d_1, d_2, \ldots, d_n, \ldots)$, $D - \mathbf{d}$ is defined as a set of points contained in D together with their $\mathbf{d}$-neighborhood. In this case, the $\mathbf{d}$-neighborhood of a point $\bar{x} \in \mathfrak{M}$ is the set of points x for which $|\bar{x}_i - x_i| < d_i$, $i = 1, 2, 3, \ldots$. By $D + \mathbf{d}$ we denote the $\mathbf{d}$-neighborhood of the domain D.

We write $\mathbf{K} \leq \mathbf{K}'$ if either $K_{ij} \leq K'_{ij}$ or $K_i \leq K'_i$. Points $|x|$ belong to the space $\mathfrak{M}$ and have the form $|x| = (|x_1|, |x_2|, \ldots, |x_n|, \ldots)$. A vector function $f(t,x) = (f_1(t,x), f_2(t,x), \ldots)$ of time t and $x \in D$ takes values in the space $\mathfrak{M}$.

If a function $f(t,x)$ is periodic in time with period T, then its integral time average is denoted by $\overline{f(t,x)}$:

$$\overline{f(t,x)} = \frac{1}{T}\int\limits_0^T f(t,x)\,dt.$$

The meaning of the remaining notation is either obvious or will be explained separately.

We consider a countable system of differential equations

$$\frac{dx_i}{dt} = f_i(t, x_1, x_2, \ldots), \quad i = 1, 2, \ldots, n, \ldots. \tag{6.4}$$

Assume that the right-hand side of system (6.4) is defined for any real t and all $x = (x_1, x_2, \ldots)$ from a domain D. Moreover, for any fixed t and x from their domains of variation, the function $f(t,x) = (f_1(t,x), f_2(t,x), \ldots)$ is a point of the space $\mathfrak{M}$. We represent the system of equations (6.4) in the form of a differential equation in the space $\mathfrak{M}$:

$$\frac{dx}{dt} = f(t,x). \tag{6.5}$$

The solution of (6.5) is understood as a function $x(t) = (x_1(t), x_2(t), \ldots)$ defined for t from an interval (a,b), continuously differentiable with respect to t in this interval, satisfying equation (6.5), and belonging to the space $\mathfrak{M}$ for all $t \in (a,b)$.

Assume that the function $f(t,x)$ is a continuous periodic function of t with period T that satisfies the inequalities

$$|f(t,x)| \leq \mathbf{M}, \; |\, f(t,x') - f(t,x'')\,| \leq \mathbf{K}\,|x' - x''| \tag{6.6}$$

for all t, x, x', x'' from the domain

$$-\infty < t < \infty, \quad x, x', x'' \in D. \tag{6.7}$$

Assume that the domain D, vector $\mathbf{M}$, and matrix $\mathbf{K}$ satisfy the following conditions:

A. The set $D - \dfrac{\mathbf{M}T}{2}$ is not empty.

B. The operator $Q = \dfrac{\mathcal{K}T}{3.1}$ is completely regular.

We now introduce the following concepts:

1. A T-system in the domain D of the space $\mathfrak{M}$ is defined as a system of equations (6.5) with continuous right-hand side periodic in t with period T that satisfies inequalities (6.6) and conditions A and B in domain (6.7).

2. A Δ-constant at the point τ, x_0 relative to system (6.8) is defined as a value of the parameter Δ for which the solution of the equation

$$\frac{dx}{dt} = f(t,x) - \Delta\,, \quad \Delta \in \mathfrak{M}\,, \tag{6.8}$$

that passes through the point $x = x_0$ for $t = \tau$ is T-periodic, provided that the indicated value of Δ is unique.

Lemma 6.1. *Assume that a sequence of functions $\alpha_0 = 1$, $\alpha_1(t)$, $\alpha_2(t)$, ... is defined by the recurrence relations*

$$\alpha_{m+1}(t) = \left(1 - \frac{t-\tau}{T}\right) \int_{\tau}^{t} \alpha_m(s)\, ds + \frac{t-\tau}{T} \int_{t}^{\tau+T} \alpha_m(s)\, ds\,, \tag{6.9}$$

$$m = 0, 1, 2, \ldots\,.$$

Then there exist positive constants

$$q_0 = 1,\ q_1 = \frac{1}{3},\ q_2 = \frac{1}{10},\ \ldots,\ \frac{1}{5^m} < q_m \le \frac{1}{10^{[\frac{m}{2}]} \cdot 3^{2(\frac{m}{2})}},\ \ldots, \tag{6.10}$$

where $\left[\dfrac{m}{2}\right]$ and $\left(\dfrac{m}{2}\right)$ are, respectively, the integral and fractional parts of the number $\dfrac{m}{2}$, such that

$$\alpha_{m+1}(t) \le T^m q_m 2(t-\tau)\left(1 - \frac{t-\tau}{T}\right) = T^m q_m \alpha_1(t) \tag{6.11}$$

for all t from the segment $[\tau, \tau + T]$.

Proof. The results of direct calculations show that

$$\alpha_1(t) = 2(t-\tau)\left(1 - \frac{t-\tau}{T}\right),$$

$$\alpha_2(t) = \alpha_1(t)\left(\frac{T}{6} + \frac{\alpha_1(t)}{3}\right) \le \frac{T}{3}\alpha_1(t),$$

$$\alpha_3(t) = \alpha_1(t)\left(\frac{T^2}{20} + \frac{T}{15}\alpha_1(t) + \frac{1}{15}\alpha_1^2(t)\right) = \frac{T^2}{10}\alpha_2(t) \tag{6.12}$$

for all $t \in [\tau, \tau + T]$. Thus, for $m = 0, 1, 2$, inequality (6.11) is satisfied. It follows from relations (6.9) and (6.12) that, for $t \in [\tau, \tau + T]$,

$$\alpha_4(t) \le \frac{T^2}{10}\alpha_2(t) = \frac{T^3}{30}\alpha_1(t),$$

$$\alpha_m(t) \le \frac{T^2}{10}\alpha_{m-2}(t), \quad m \ge 3. \tag{6.13}$$

Let us now show that inequalities (6.13) imply the estimate

$$\alpha_m(t) \le \frac{T^{m-1}}{10^{[\frac{m-1}{2}]} \cdot 3^{2(\frac{m-1}{2})}}\alpha_1(t), \quad m = 2, 3, \dots . \tag{6.14}$$

Indeed, for $m = 2$, inequality (6.14) is true. Assume that it is true for $m \le l$. Setting $m = l + 1$ and using (6.13) and (6.14), we find that

$$\alpha_{l+1}(t) \le \frac{T^2}{10}\alpha_{l-1}(t) \le \frac{T^2}{10}\frac{T^{l-2}}{10^{[\frac{l}{2}-1]} \cdot 3^{2(\frac{l}{2}-1)}}\alpha_1(t)$$

$$= \frac{T^l}{10^{[\frac{l}{2}]} \cdot 3^{2(\frac{l}{2})}}\alpha_1(t), \quad t \in [\tau, \tau + T]. \tag{6.15}$$

Finally, by induction, we conclude that inequality (6.14) holds for all $m \ge 2$. Therefore,

$$\alpha_{m+1}(t) \le T^m q_m \alpha_1(t) \tag{6.16}$$

for all $m \ge 0$ and $\tau \le t \le \tau + T$ whenever

$$q_m \le \frac{1}{10^{[\frac{m}{2}]} \cdot 3^{2(\frac{m}{2})}}, \quad m = 2, 3, \dots . \tag{6.17}$$

To find a lower bound for q_m, we use relations (6.12) and write

$$\alpha_3(t) \ge \frac{T}{5}\alpha_1(t)\left(\frac{T}{4} + \frac{1}{3}\alpha_1(t)\right) = \frac{T}{5}\alpha_1(t)\left(\frac{T}{6} + \frac{1}{3}\alpha_1(t) + \frac{T}{12}\right)$$

$$= \frac{T}{5}\alpha_2(t) + \frac{T^2}{5 \cdot 12}\alpha_1(t).$$

Hence,

$$\alpha_4(t) \geq \frac{T}{5}\alpha_3(t) + \frac{T^2}{60}\alpha_2(t) \geq \frac{T^2}{25}\alpha_2(t) + \frac{T^2}{60}\alpha_2(t)$$
$$= \frac{17}{300}T^2\alpha_2(t) \geq \frac{T^2}{18}\alpha_2(t).$$

By induction, we conclude that

$$\alpha_m(t) \geq \frac{T^2}{18}\alpha_{m-2}(t), \quad m = 4, 5, \ldots . \tag{6.18}$$

By analogy with the case of deducing estimate (6.14) from (6.13), inequality (6.18) implies that

$$\alpha_m(t) \geq \frac{T^{m-2}}{18^{[\frac{m}{2}-1]} \cdot 3^{2(\frac{m}{2}-1)}}\alpha_2(t)$$
$$= \frac{T^{m-2}}{18^{[\frac{m}{2}-1]} \cdot 3^{2(\frac{m}{2}-1)}}\alpha_1(t)\left(\frac{T}{6} + \frac{\alpha_1(t)}{3}\right)$$
$$\geq \frac{T^{m-1}}{6 \cdot 18^{[\frac{m}{2}-1]} \cdot 3^{2(\frac{m}{2}-1)}}\alpha_1(t), \quad m \geq 4.$$

Further, for q_m, we obtain

$$T^m q_m \alpha_1(t) \geq \frac{T^m}{6 \cdot 18^{[\frac{m+1}{2}-1]} \cdot 3^{2(\frac{m+1}{2}-1)}}\alpha_1(t). \tag{6.19}$$

This inequality implies the estimate

$$q_m \geq \frac{1}{5^m}, \tag{6.20}$$

which completes the proof of Lemma 6.1.

Lemma 6.2. *Suppose that a continuous function $f(t, x)$ defined in domain (6.7) is periodic in t with period T, takes values in $\mathfrak{M}$, and satisfies inequalities (6.6) and conditions A and B. Then the sequence of T-periodic functions of t*

$$x_m(t, \tau, x_0) = x_0 + \int_\tau^t \left(f(t, x_{m-1}(t, \tau, x_0)) - \overline{f(t, x_{m-1}(t, \tau, x_0))}\right) dt,$$
$$x_0 = x_0(t, \tau, x_0), \tag{6.21}$$

converges to a function $x_\infty(t,\tau,x_0)$ defined in the domain

$$(-\infty,\infty)\times(-\infty,\infty)\times\left(D-\frac{\mathrm{M}T}{2}\right) \tag{6.22}$$

uniformly in

$$(t,\tau,x_0)\in(-\infty,\infty)\times(-\infty,\infty)\times\left(D-\frac{\mathrm{M}T}{2}\right).$$

Moreover, the function $x_\infty(t,\tau,x_0)$ is periodic in t and τ with period T and satisfies the equation

$$x(t,\tau,x_0) = x_0 + \int_\tau^t \Big(f(t,x(t,\tau,x_0)) - \overline{f(t,x(t,\tau,x_0))}\Big)\,dt. \tag{6.23}$$

Proof. In what follows, we denote the set of points from D contained in D together with their $\left(\frac{\mathrm{M}T}{2}\right)$-neighborhoods by D_f. Any function $f(t)$ continuous on the segment $[\tau,\tau+T]$ satisfies the inequality

$$\left|\int_\tau^t \Big(f(t) - \frac{1}{T}\int_\tau^{\tau+T} f(s)\,ds\Big)\,dt\right|$$

$$\le \left|\left(1-\frac{t-\tau}{T}\right)\int_\tau^t f(s)ds\right| + \left|\frac{t-\tau}{T}\int_t^{\tau+T} f(s)\,ds\right|, \quad t\in[\tau,\tau+T].$$

This yields

$$\left|\int_\tau^t (f(t) - \overline{f(t)})\,dt\right| \le 2\,|f(t)|_0(t-\tau)\left(1-\frac{t-\tau}{T}\right),$$

where

$$|f(t)|_0 = \max_{t\in[\tau,\tau+T]} |f(t)|.$$

In this case, for $x_0\in D$, we have

$$|x_1(t,\tau,x_0) - x_0| \le 2(t-\tau)\left(1-\frac{t-\tau}{T}\right)\mathrm{M} \le \frac{T}{2}\mathrm{M},$$

whence it follows that $x_1(t,\tau,x_0) \in D$ for $x_0 \in D_f$. Further, if $x_{m-1}(t,\tau,x_0) \in D$, then equation (6.21) implies that

$$|x_m(t,\tau,x_0) - x_0| \le \frac{T}{2}\mathbf{M},$$

i.e., $x_m(t,\tau,x_0) \in D$ for $x_0 \in D_f$. By induction, we conclude that, for all $m \ge 0$, $(t,\tau) \in (-\infty,\infty) \times (-\infty,\infty)$, and $x_0 \in D_f$, the functions $x_m(t,\tau,x_0)$ belong to the set D.

We now estimate the expression

$$r_{m+1} = |x_{m+1}(t,\tau,x_0) - x_m(t,\tau,x_0)|. \tag{6.24}$$

In view of inequalities (6.6) and relations (6.21), we obtain

$$r_{m+1}(t) \le \left(1 - \frac{t-\tau}{T}\right) \int_\tau^t \mathbf{K} r_m \, dt + \frac{t-\tau}{T} \int_t^{\tau+T} \mathbf{K} r_m \, dt.$$

We set

$$r_m(t) = \mathbf{K}^{m-1}\mathbf{M}\alpha_m(t). \tag{6.25}$$

Then, for $\alpha_m(t)$, we can write

$$\alpha_{m+1}(t) \le \left(1 - \frac{t-\tau}{T}\right) \int_\tau^t \alpha_m(t)\, dt + \frac{t-\tau}{T} \int_t^{\tau+T} \alpha_m(t)\, dt, \tag{6.26}$$

where $m \ge 1$ and $t \in [\tau, \tau+T]$. Since, by virtue of (6.21),

$$\alpha_1(t) \le 2(t-\tau)\left(1 - \frac{t-\tau}{T}\right),$$

we conclude that α_1 satisfies inequality (6.26) for $\alpha_0 = 1$. By applying Lemma 6.1 to the functions $\alpha_0, \alpha_1, \ldots$, we find that

$$\alpha_{m+1}(t) \le 2(t-\tau)\left(1 - \frac{t-\tau}{T}\right) T^m q_m \tag{6.27}$$

for all $m \ge 0$ and $t \in [\tau, \tau+T]$. Taking into account relations (6.25) and (6.27), we finally get

$$r_{m+1}(t) \le \mathbf{K}^m \mathbf{M} T^m q_m 2(t-\tau)\left(1 - \frac{t-\tau}{T}\right) \tag{6.28}$$

for $m \ge 0$ and $\tau \le t \le \tau + T$.

By using the estimate for the quantity q_m deduced in Lemma 6.1 and inequality (6.28), we obtain

$$|x_{m+k} - x_m| < 2\,(t-\tau)\left(1 - \frac{t-\tau}{T}\right)\left(\frac{\mathbf{K}T}{3.1}\right)^m \sum_{i=0}^{k-1}\left(\frac{\mathbf{K}T}{3.1}\right)^i \mathbf{M}$$

for all $m \geq 2$ and $k = 0, 1, \ldots$.

Since the operator $Q = \dfrac{KT}{3.1}$ is completely regular, the last inequality implies the estimate

$$\|x_{m+k}(t,\tau,x_0) - x_m(t,\tau,x_0)\| \leq \frac{T}{2}\,q^m \sum_{i=0}^{k-1} q^i M, \tag{6.29}$$

where q is a constant, $q < 1$, and $m \geq 2$.

Inequality (6.29) implies that the sequence $x_m(t,\tau,x_0)$ converges to a function $x_\infty(t,\tau,x_0)$ uniformly for all t, τ, x_0 from domain (6.22). The periodicity of the function $x_\infty(t,\tau,x_0)$ in t follows from the periodicity of the functions $x_m(t,\tau,x_0)$.

Passing to the limit in equality (6.21) as $m \to \infty$, we see that $x_\infty(t,\tau,x_0)$ satisfies equation (6.23).

Lemma 6.2 is proved.

We now estimate the deviation of the function $x_\infty(t,\tau,x_0)$ from $x_n(t,\tau,x_0)$. It follows from equation (6.23) that

$$|x_\infty(t,\tau,x_0) - x_0| < \frac{MT}{2}.$$

For $n \geq 2$, inequalities (6.28) and (6.29) imply that

$$\|x_\infty(t,\tau,x_0) - x_1(t,\tau,x_0)\| < q\left(\frac{1}{1-q} + \frac{1}{30}\right)\frac{MT}{2},$$

$$\|\, x_\infty(t.\tau,x_0) \;-\; x_n(t,\tau,x_0)\,\| \leq \frac{q^n}{1-q}\,\frac{MT}{2},$$

whence it follows that

$$\|\, x_\infty(t,\tau,x_0) \;-\; x_n(t,\tau,x_0)\,\| < \frac{q^n MT}{2}\left(\frac{1}{1-q} + \frac{\delta_{1n}}{30}\right) \tag{6.30}$$

for all $n \geq 0,\ t, \tau \in (-\infty, \infty)$, and $x_0 \in D - \dfrac{\mathbf{M}T}{2}$. Here, δ_{1n} is the Kronecker symbol:

$$\delta_{1n} = \begin{cases} 0 & \text{for} \quad n \neq 1, \\ 1 & \text{for} \quad n = 1. \end{cases}$$

Theorem 6.1. *Any point* $(\tau, x_0) \in (\infty, \infty) \times \left(D - \dfrac{\mathbf{MT}}{2} \right)$ *has a* Δ*-constant relative to a* T*-system specified in the domain* D *of the space* $\mathfrak{M}$.

Proof. Since the right-hand side of a T-system, i.e., the function $f(t, x)$, satisfies the conditions of Lemma 6.2, the sequence of periodic functions (6.21) converges uniformly to the function $x_\infty(t, \tau, x_0)$.

We set

$$\Delta(\tau, x_0) = \overline{f(t, x_\infty(t, \tau, x_0))}. \tag{6.31}$$

Let us show that this function specifies the Δ-constant at a point τ, x_0 relative to a T-system defined in the domain D. To prove the last statement, it suffices to show that the equation

$$\frac{dx}{dt} = f(t, x) - u\,, \quad u \in \mathfrak{M}\,, \tag{6.32}$$

has a T-periodic solution $x = x(t, \tau, x_0)$ such that $x(\tau, \tau, x_0) = x_0$ for a single value of the parameter u.

Assume the contrary. Then there exist constants u_1 and u_2, $u_1 \neq u_2$, such that the solution $x(t, \tau, x_0, u)$ is periodic for $u = u_1$ and $u = u_2$.

For the difference $x(t, \tau, x_0, u_2) - x(t, \tau, x_0, u_1)$, we obtain the identity

$$x(t, u_2) - x(t, u_1) = \int\limits_\tau^t (f(t, x(t, u_2)) - f(t, x(t, u_1)) + \overline{f(t, x(t, u_1))} - \overline{f(t, x(t, u_2))})\, dt. \tag{6.33}$$

This yields

$$|x(t, u_2) - x(t, u_1)| \leq \mathbf{K} \left[\left(1 - \frac{t - \tau}{T}\right) \int\limits_\tau^t |x(t, u_2) - x(t, u_1)|\, dt + \frac{t - \tau}{T} \int\limits_t^{\tau + T} |x(t, u_2) - x(t, u_1)|\, dt \right] \tag{6.34}$$

and, finally,

$$|x(t,u_2)-x(t,u_1)| < \mathbf{K}^n \alpha_n(t)\, |x(t,u_2)-x(t,u_1)|_0,$$

where $n = 0, 1, \ldots,$ $|x(t,u_2)-x(t,u_1)|_0 = \sup\limits_t |x(t,u_2)-x(t,u_1)|$, and $\alpha_n(t)$ is a positive function defined by relations (6.26).

By using the estimate for $\alpha_n(t)$, we obtain

$$|x(t,u_2)-x(t,u_1)| < \frac{\mathbf{K}^n T^n}{3.1^n}\frac{T}{2}\,|x(t,u_2)-x(t,u_1)|_0 \,.$$

At the same time, in view of the fact that the operator Q is completely regular, we can write

$$|x(t,u_2)-x(t,u_1)|_0 < q^n \frac{T}{2} |x(t,u_2)-x(t,u_1)|_0 \,, \quad 0 < q < 1,$$

which is possible only for $x(t,u_2) = x(t,u_1)$. This contradiction proves Theorem 6.1.

By using Lemma 6.2 and estimate (6.30), one can easily prove the following corollary of Theorem 6.1:

Corollary 6.1. *The solution $x = x(t,\tau,x_0)$ of a T-system passing through a point $x_0 \in D - \dfrac{MT}{2} = Df$ at $t = \tau$ is T-periodic if and only if the Δ-constant at the point τ, x_0 is equal to zero. This solution $x(t,\tau,x_0)$ satisfies the identity*

$$x(t,\tau,x_0) \;=\; x_\infty(t,\tau,x_0),$$

where $x_\infty(t,\tau,x_0)$ is the limit of the sequence of periodic functions (6.21), and the following estimate is true:

$$\|\, x(t,\tau,x_0) - x_n(t,\tau,x_0)\,\| < q^n \left(\frac{1}{1-q} + \frac{\delta_{1n}}{30}\right)\frac{MT}{2},$$

where $n \geq 0$ and $t \in (-\infty, \infty)$.

This corollary relates the problem of the existence of periodic solutions of a T-system to the problem of zeros of the function $\Delta = \Delta(\tau, x_0)$ and the problem of finding these solutions to the evaluation of the functions $x_n(t,\tau,x_0)$.

The periodicity of the indicated functions reduces their evaluation to the determination of definite integrals and, hence, allows one to use computers.

The functions $x_n(t,\tau,x_0)$ can be especially easily found in the case where $f(t,x)$ is a polynomial function with respect to x whose coefficients are trigonometric polynomials.

In this case, the expressions $x_n(t,\tau,x_0)$ are also trigonometric polynomials and their evaluation reduces to elementary operations, such as the operation of multiplication of polynomials.

If the functions $x_n(t,\tau,x_0)$ are already determined, then we can find the Δ-constant at the point τ, x_0. Indeed, it is clear from the proof of Theorem 6.1 that the Δ-constant at a point $(\tau,x_0)\in(-\infty,\infty)\times Df$ is determined by using the formula

$$\Delta(\tau,x_0) = \overline{f(t,x_\infty(t,\tau,x_0))}\,. \tag{6.35}$$

If we substitute the function $x_n(t,\tau,x_0)$ for $x_\infty(t,\tau,x_0)$ in relation (6.35), then we get

$$\Delta_n(\tau,x_0) = \overline{f(t,x_n(t,\tau,x_0))}. \tag{6.36}$$

This function is regarded as the nth approximation to the Δ-constant.

It is easy to see that $\Delta_n\to\Delta$ as $n\to\infty$ and, moreover,

$$\|\Delta(\tau,x_0)-\Delta_n(\tau,x_0)\|\le\frac{3.1}{3}q^{n+1}\left(\frac{1}{1-q}+\frac{\delta_{1n}}{30}\right)M. \tag{6.37}$$

By using the computed values of the Δ-constant, one can solve, in many cases, the problem of the existence and location of points at which the Δ-constant is equal to zero and, hence, to solve the problem of the existence and determination of periodic solutions of the T-system.

Thus, in finding the location of points at which the Δ-constant is equal to zero, one can use the following statement:

Lemma 6.3. *Let $\Delta'_n(\tau,x_0)$ be the already calculated nth approximation to the Δ-constant $\Delta(\tau,x_0)$ at the point $(\tau,x_0)\in(-\infty,\infty)\times\left(D-\frac{\mathrm{M}T}{2}\right)=Df$ and let*

$$\|\,\Delta_n(\tau,x_0)\;-\;\Delta'_n(\tau,x_0)\,\|\le\eta.$$

If, in addition, a closed domain $D_1\subseteq D-\frac{\mathrm{M}T}{2}$ contains a point at which the Δ-constant is equal to zero, then, for any $n\ge 0$,

$$\|\Delta'_n(\tau,x_0)\|<q\left(1+\frac{3.1}{3}\frac{q}{1-q}+\frac{3.1}{30}q^2\right)d+\frac{3.1}{3}q^{n+1}\left(\frac{1}{1-q}+\frac{\delta_{1n}}{30}\right)M+\eta,$$

$$d=\sup_{x\in D_1}\|x-x_0\|.$$

By using this lemma and the values of Δ-constants computed at finitely many points, one can select a subset of D_1 containing all points at which the Δ-constant can be equal to zero and, hence, specify the set containing all possible points through which periodic solutions of the T-system can pass at $t = \tau$.

We now proceed to the proof of Lemma 6.1. Assume that the Δ-constant is equal to zero at a point $\tau, x', x' \in D_1$. Then

$$\begin{aligned} \|\Delta_n'(\tau, x_0)\| \ &\leq \ \|\Delta_n'(\tau, x_0) - \Delta_n(\tau, x_0)\| + \|\Delta_n(\tau, x_0) - \Delta(\tau, x_0)\| \\ &\quad + \|\Delta(\tau, x_0) - \Delta(\tau, x')\| \,, \end{aligned} \tag{6.38}$$

or, in view of inequality (6.37),

$$\begin{aligned} \|\Delta_n'(\tau x_0)\| \ &\leq \ \|\Delta(\tau, x_0) - \Delta(\tau, x')\| \\ &\quad + \eta + \frac{3.1}{3} q^{n+1} \left(\frac{1}{1-q} + \frac{\delta_{1n}}{30} \right) M. \end{aligned} \tag{6.39}$$

We estimate the norm $\|\Delta(\tau, x_0) - \Delta(\tau, x')\|$. Relation (6.35) implies that

$$|\Delta(\tau, x_0) - \Delta(\tau, x')| \leq \frac{\mathbf{K}}{T} \int_0^T |x_\infty(t, \tau, x_0) - x_\infty(t, \tau, x')| \, dt.$$

For $|x_\infty(t, \tau, x_0) - x_\infty(t, \tau, x')| = r(t)$, we have the following estimate:

$$\begin{aligned} r(t) \ &\leq \ (E + \mathbf{K}\alpha_1(t) + \mathbf{K}^2\alpha_2(t) + \ldots + \mathbf{K}^n\alpha_n(t)) \, |x_0 - x'| \\ &\quad + \ \mathbf{K}^{n+1}\alpha_{n+1}|r(t)|_0 \,, \end{aligned} \tag{6.40}$$

where $\alpha_1(t), \alpha_2(t), \ldots$ are positive functions satisfying relation (6.9). Indeed, in view of the fact that $x_\infty(t, \tau, x_0)$ satisfies equation (6.23), we find

$$\begin{aligned} &|x_\infty(t, \tau, x_0) - |x_\infty(t, \tau, x')| \\ &\qquad \leq |x_0 - x'| + \mathbf{K} \left(\left(1 - \frac{t-\tau}{T}\right) \int_\tau^t |x_\infty(s, \tau, x_0) - x_\infty(s, \tau, x')| \, ds \right. \\ &\qquad\qquad \left. + \frac{t-\tau}{T} \int_t^{\tau+T} |x_\infty(s, \tau, x_0) - x_\infty(s, \tau, x')| \, ds \right). \end{aligned}$$

As a result of the solution of this inequality, we arrive at estimate (6.40). Further, by virtue of Lemma 6.1, we obtain

$$r(t) \leq \left[E + \sum_{n=0}^{\infty} \frac{(\mathbf{K}T)^n \mathbf{K}}{10^{[\frac{n}{2}]} \cdot 3^{2(\frac{n}{2})}} \alpha_1(t)\right] |x_0 - x'|, \tag{6.41}$$

where $\left[\frac{n}{2}\right]$ and $\left(\frac{n}{2}\right)$ are the integral and fractional parts of the number $\frac{n}{2}$, respectively.

By using inequality (6.41), we obtain

$$\|\Delta(\tau, x_0) - \Delta(\tau, x')\| \leq q \left(1 + \frac{3.1}{3} \frac{q}{1-q} + \frac{3.1}{90} q^2\right) \|x_0 - x'\|. \tag{6.42}$$

In view of (6.39), this completes the proof of Lemma 6.3.

In what follows, we denote the image of the set D under the mapping A by AD.

Let us now study the problem of the existence of periodic solutions of a T-system. We first prove the following auxiliary statement:

Lemma 6.4. *Let D be a closed bounded domain in the space $\mathfrak{M}$ and let A_0 and A be continuous mappings from D into $\mathfrak{M}$ such that $\|Ax - A_0 x\| < \varepsilon$.*

If, in this case, A_0 is a topological mapping, then AD contains $A_0 D - \varepsilon$: $A_0 D - \varepsilon \subset AD$.

Proof. Since A_0 is a topological mapping, the continuous mapping AA_0^{-1} defined on $A_0 D$ maps the set $A_0 D$ onto AD. By setting $A_0 D = D_1$ and $AA_0^{-1} = A_1$, we reduce the proof of Lemma 6.4 to the case where $A_0 = E$ (E is the identity mapping). Thus, let $A_0 = E$. It is necessary to prove that $D - \varepsilon \subset AD$. We take $x_0 \in D - \varepsilon$. Then the ball $T_\varepsilon(x_0)$: $\|x - x_0\| \leq \varepsilon$ belongs to D: $T_\varepsilon(x_0) \subset D$. We define a continuous operator

$$A^* x = x_0 + (x - Ax) \tag{6.43}$$

on $T_\varepsilon(x_0)$ and consider this operator as a mapping in the space C^∞ defined as the space of number sequences $x = (x_1, x_2, \ldots)$ uniformly bounded by a number ε. The norm in this space is introduced as follows:

$$|\mathbf{x}| = \sum_{k=1}^{\infty} \frac{1}{2^k} |x_k| \,. \tag{6.44}$$

For $x_0 = 0$, the ball $T_\varepsilon(x_0)$ coincides with C^∞, and the operator A^* transforms C^∞ into itself: $|A^*x - x_0| = |x - Ax| < \varepsilon$.

Moreover, the family of kth coordinates $v_k = \{A^*x\}_k$ of the mapping A^* is compact for any $k = 1, 2, \ldots$, which, in view of the theorem on the compactness of a set in the space C^∞, implies that the set of points $\{v_k\}$ is compact as a set in the space C^∞. By virtue of the Schauder theorem, we conclude that the operator A^* has a fixed point in C^*:

$$A^*x^* = x^*. \tag{6.45}$$

In view of relation (6.43), this yields

$$Ax^* = x_0. \tag{6.46}$$

Equality (6.46) means that x_0 is the image of a point x^* under the mapping A. Moreover, relation (6.46) holds for any point $x_0 \in D - \varepsilon$, and the point x^* belongs to the ball $T_\varepsilon(x_0)$ and, hence, to the set D. Therefore, relation (6.46) means that the entire set $D - \varepsilon$ is the image of the set D under the mapping A. Thus, the required inclusion $D - \varepsilon \subset AD$ is true.

The case $x_0 \neq 0$ is reduced to the analyzed case by a shift of coordinates.

We now prove the following theorem on periodic solutions of T-systems:

Theorem 6.2. *Assume that a T-system*

$$\frac{dx}{dt} = f(t, x) \tag{6.47}$$

defined in a domain D of the space $\mathfrak{M}$ satisfies the following conditions:

(i) for some fixed $\tau \in (-\infty, \infty)$ and integer $n = 0, 1, \ldots$, the mapping $\Delta_n : D - \dfrac{\mathrm{M}T}{2} \to \mathfrak{M}$, $\Delta_n(x_0) = \overline{f(t, x_n(t, \tau, x_0))}$, has a singular point x_0 : $\Delta_n(x^0) = 0$;

(ii) there exists a closed bounded domain $D_1 \subset D - \dfrac{\mathrm{M}T}{2}$ containing the point x^0 and such that Δ_n topologically maps D_1 onto $\Delta_n D_1$;

(iii) the inequality

$$\inf_{x \in \Gamma_{D_1}} \|\Delta_n(x)\| \geq \frac{3.1}{3} q^{n+1} \left(\frac{1}{1-q} + \frac{\delta_{1n}}{30} \right) M \tag{6.48}$$

holds on the boundary Γ_{D_1} of the domain D_1.

Then T-system (6.47) possesses a T-periodic solution $x = x(t)$ such that $x(\tau) \in D_1$.

For $n = 0$, one has $\Delta_0(x) = \overline{f(t,x)}$, and Theorem 6.2 establishes the existence of periodic solutions of system (6.47) relative to the averaged system

$$\frac{dx}{dt} = \overline{f(t,x)}.$$

Proof. By virtue of estimate (6.37) and Lemma 6.4, the set

$$\Delta_n D_1 - \frac{3.1}{3} q^{n+1} \left(\frac{1}{1-q} + \frac{\delta_{1n}}{30} \right) M$$

is contained in ΔD_1. If this set contains the origin of the coordinate system $\Delta = (\Delta^1, \ldots, \Delta^n, \ldots)$, then the set ΔD_1 also contains the origin of coordinates. This property is sufficient for the T-system to have T-periodic solutions.

Therefore, to prove the theorem, it suffices to show that

$$0 \in \Delta_n D_1 - \frac{3.1}{3} q^{n+1} \left(\frac{1}{1-q} + \frac{\delta_{1n}}{30} \right) M.$$

Since the mapping $\Delta_n(x)$ is topological, the set $\Delta_n D_1$ is a domain and, moreover, the boundary $\Gamma_{\Delta_n D_1}$ of the domain $\Delta_n D_1$ is the image of the boundary of the domain D_1:

$$\Gamma_{\Delta_n D_1} = \Delta_n \Gamma_{D_1}. \tag{6.49}$$

By virtue of condition (i) in Theorem 6.2, the set $\Delta_n D_1$ contains the singular point of the system $\Delta = (\Delta^1, \ldots, \Delta^n, \ldots)$. This point belongs to the set $\Delta_n D_1 - r$ if the distance between it and all points z of the boundary $\Gamma_{\Delta_n D_1}$ of the set $\Delta_n D$ is not smaller than r. This means that the origin belongs to the set

$$\Delta_n D_1 - \frac{3.1}{3} q^{n+1} \left(\frac{1}{1-q} + \frac{\delta_{1n}}{30} \right) M$$

whenever

$$\inf_{x \in \Gamma_{D_1}} \|z\| \geq \frac{3.1}{3} q^{n+1} \left(\frac{1}{1-q} + \frac{\delta_{1n}}{30} \right) M. \tag{6.50}$$

By virtue of relation (6.49), we get the identity

$$z = \Delta_n(x) \Big|_{x \in \Gamma_{D_1}}.$$

Therefore, inequality (6.50) takes the form

$$\inf_{x \in \Gamma_{D_1}} \|\Delta_n(x)\| \geq \frac{3.1}{3} q^{n+1} \left(\frac{1}{1-q} + \frac{\delta_{1n}}{30} \right) M. \tag{6.51}$$

This enables us to conclude that the T-system possesses T-periodic solutions, provided that inequality (6.51) is satisfied.

Theorem 6.2 is proved.

2. INVARIANT TORI

In the present chapter, our main attention is focused on the investigation of the invariant toroidal manifold of a countable system of differential equations either in the space $\mathfrak{M}$ of bounded number sequences or in the products $\mathfrak{M} \times \mathcal{F}_m$ or $\mathfrak{M} \times \mathcal{F}_\infty$, where $\mathcal{F}_m$ and $\mathcal{F}_\infty$ are m-dimensional and countably-dimensional tori, respectively. In our presentation, we mainly deal with conditions for the existence of invariant toroidal manifolds for linear systems in $\mathfrak{M} \times \mathcal{F}_m$ and $\mathfrak{M} \times \mathcal{F}_\infty$, the perturbation theory of these manifolds for nonlinear systems, the properties of smoothness and stability of these manifolds, and the behavior of trajectories in small vicinities of invariant toroidal manifolds. Whenever possible, we try to reduce the analyzed problems to the case of finite-dimensional systems of differential equations with increasing dimensionality.

The problems discussed in this chapter were studied in numerous works, among which we especially mention [Sam10, MSK2, BMS, Kul1, Kul2, MiL, MiK2, Sam1, Sam2, Sam5–Sam9, Sam13, SaK1–SaK3, SaP1].

2.7. Green Function

Let $\mathfrak{M}$ be the space of bounded number sequences $x = (x_1, x_2, \ldots)$ equipped with the norm $\|x\| = \sup n\{\,|\,x_n\,|\,\}$. The metric induced by this norm turns $\mathfrak{M}$ into a Banach space. Consider an infinite matrix $A = [\,a_{ij}\,]_{i,j=1}^{\infty}$ such that $\sup\limits_{i} \sum\limits_{j=1}^{\infty} |\,a_{ij}\,| = \alpha < \infty$. It is easy to see that the operation of multiplication of the matrix A by $x \in \mathfrak{M}$ specifies a linear bounded operator in the set $\mathfrak{M}$ whose norm is equal to α. Indeed,

$$\|Ax\| = \sup_i \left| \sum_{k=1}^{\infty} a_{ik} x_k \right| \le \sup_i \sum_{k=1}^{\infty} |\,a_{ik}\,|\,|\,x_k\,| \le \alpha \sup_k |\,x_k\,|\ ,$$

whence $\dfrac{\|Ax\|}{\|x\|} \le \alpha$ for $\|x\| \ne 0$.

Let us show that $\alpha = \sup\limits_{\|x\|\neq 0} \dfrac{\|Ax\|}{\|x\|}$. We choose an arbitrarily small real number $\varepsilon > 0$ and find an element $x^0 \in \mathfrak{M}$ such that $\dfrac{\|Ax^0\|}{\|x^0\|} > \alpha - \varepsilon$.

Obviously, there exists a number p such that

$$\sum_{k=1}^{\infty} |\, a_{pk} \,| > \alpha - \varepsilon.$$

We choose $x^0 = (x_1^0, x_2^0, \ldots)$ such that $|\, x_1^0 \,| = |\, x_2^0 \,| = \ldots = |\, x_k^0 \,| = \ldots$ and, moreover, $|\, a_{pk}\, x_k^0 \,| = a_{pk}\, x_k^0$ for $k = 1, 2, 3, \ldots$. Then

$$\frac{\|Ax^0\|}{\|x^0\|} = \frac{\sup\limits_i \left| \sum\limits_{k=1}^{\infty} a_{ik} x^0 \right|}{|\, x_1^0 \,|} \geq \frac{\left| \sum\limits_{k=1}^{\infty} a_{pk} x^0 \right|}{|\, x_1^0 \,|}$$

$$= \frac{|\, x_1^0 \,| \sum\limits_{k=1}^{\infty} |\, a_{pk} \,|}{|\, x_1^0 \,|} = \sum_{k=1}^{\infty} |\, a_{pk} \,| > \alpha - \varepsilon\,.$$

This operator norm is, at the same time, the matrix norm associated with the vector norm in the space $\mathfrak{M}$.

Consider a system of differential equations of the form

$$\frac{d\varphi}{dt} = a(\varphi), \quad \frac{dx}{dt} = P(\varphi)x, \tag{7.1}$$

where $\varphi = (\varphi_1, \ldots, \varphi_m)$ is an m-dimensional vector, $x = (x_1, x_2, \ldots) \in \mathfrak{M}$, $P(\varphi) = [p_{ij}(\varphi)]_{i,j=1}^{\infty}$ is an infinite matrix, and $a(\varphi) = \{\, a_1(\varphi), \ldots, a_m(\varphi) \,\}$. The first equation in system (7.1) is finite-dimensional. Hence, if $a(\varphi)$ satisfies the Lipschitz condition, then it possesses a unique solution $\varphi = \varphi(t, \varphi) = \varphi_t(\varphi)$ such that $\varphi_0(\varphi) = \varphi$, where φ is an arbitrary constant vector from the set R^m. Substituting this solution in the second equation in system (7.1), we obtain

$$\frac{dx}{dt} = P(\varphi_t(\varphi))x. \tag{7.2}$$

If the elements of the matrix $P(\varphi)$ are continuous in φ, the functions $p_s(\varphi) = \sum\limits_{j=1}^{\infty} |p_{sj}(\varphi)|$ are continuous in φ for all $s = 1, 2, \ldots$, and $\|P(\varphi)\| \leq p^0 = \text{const} < \infty$, then equation (7.2) satisfies conditions $(\widehat{P})$ in Sec. 1.3 for all initial values φ of the solution $\varphi_t(\varphi)$.

Then the matrizant of equation (7.2) dependent on the initial values φ exists. We denote this matrizant by $\Omega_\tau^t(\varphi)$ and set $\Omega_\tau^\tau(\varphi) = E$, where E is the identity matrix.

Lemma 7.1. *For any $t, \tau, \theta \in R^1$, the following identity is true:*

$$\Omega_\tau^t(\varphi_\theta(\varphi)) = \Omega_{\tau+\theta}^{t+\theta}(\varphi). \tag{7.3}$$

Indeed, in the identity

$$\frac{d\Omega_\tau^t(\varphi)}{dt} = P(\varphi_t(\varphi))\Omega_\tau^t(\varphi)\,,$$

we replace φ by $\varphi_\theta(\varphi)$. This gives

$$\frac{d\Omega_\tau^t(\varphi_\theta(\varphi))}{dt} = P(\varphi_t(\varphi_\theta(\varphi)))\Omega_\tau^t(\varphi_\theta(\varphi))\,.$$

Further, in view of the group property of solutions of an autonomous system of equations, we arrive at the identity

$$\frac{d\Omega_\tau^t(\varphi_\theta(\varphi))}{dt} = P(\varphi_{t+\theta}(\varphi))\Omega_\tau^t(\varphi_\theta(\varphi))$$

valid for all $t, \tau, \theta \in R^1$ and $\varphi \in R^m$. Hence, $\Omega_\tau^t(\varphi_\theta(\varphi))$ is the matrizant of the equation

$$\frac{dx}{dt} = P(\varphi_{t+\theta}(\varphi))x \tag{7.4}$$

equal to E for $t = \tau$. At the same time, the matrizant $\Omega_{\tau+\theta}^{t+\theta}(\varphi)$ of equation (7.4) possesses the same property and, hence, in view of the uniqueness of a solution, identity (7.3) is proved.

It is easy to see that if $a(\varphi)$ and $P(\varphi)$ are periodic in φ_i, $i = \overline{1, m}$, then

$$\varphi_t(\varphi + 2\pi) - \varphi_t(\varphi) = 2\pi \tag{7.5}$$

and the matrizant $\Omega_\tau^t(\varphi)$ is a matrix periodic in φ_i, $i = \overline{1, m}$, with period 2π.

We assume that $C(\varphi)$ is an infinite matrix whose elements $c_{ij}(\varphi)$, $i, j = 1, 2, \ldots$, are continuous in φ and satisfy the inequality

$$\sup_i \sum_{j=1}^{\infty} \sup_\varphi |\, c_{ij}(\varphi) \,| < \infty.$$

The set of matrices satisfying the last estimate is denoted by C_φ.

A function

$$G_0(\tau,\varphi)=\begin{cases}\Omega_\tau^0(\varphi)C(\varphi_\tau(\varphi)), & \tau\le 0,\\ \Omega_\tau^0(\varphi)[C(\varphi_\tau(\varphi))-E], & \tau>0,\end{cases}\tag{7.6}$$

such that

$$\int_{-\infty}^{\infty}\|G_0(\tau,\varphi)\|\,d\tau\le K=\text{const}<\infty$$

uniformly in φ is called the Green function of the system of equations (7.1).

As follows from its definition, the Green function $G_0(\tau,\varphi)$ is 2π-periodic with respect to φ_i, $i=\overline{1,m}$, provided that $a(\varphi)$, $P(\varphi)$, and $C(\varphi)$ are also periodic and, moreover,

$$G_0(0,\varphi)-G_0(+0,\varphi)=E.$$

Lemma 7.2. *Consider the matrix $G_t(\tau,\varphi)$ defined by the formula*

$$G_t(\tau,\varphi)=\begin{cases}\Omega_\tau^t(\varphi)C(\varphi_\tau(\varphi)), & \tau\le 0,\\ \Omega_\tau^t(\varphi)[C(\varphi_\tau(\varphi))-E], & \tau>0.\end{cases}\tag{7.7}$$

Then

$$G_0(\tau,\varphi_t(\varphi))=G_t(\tau+t,\varphi).\tag{7.8}$$

Proof. First, we note that the products of infinite matrices appearing in relation (7.7) exist because the matrix $C(\varphi)$ is bounded in norm according to the condition and the matrizant $\Omega_\tau^t(\varphi)$ satisfies the inequality

$$\|\Omega_\tau^t(\varphi)\|\le\exp\{p^0\,|\,t-\tau\,|\},\quad t,\ \tau\in R^1$$

(see Theorem 4.1). By using relation (7.3), we obtain

$$\begin{aligned}G_0(\tau,\varphi_t(\varphi)) &= \begin{cases}\Omega_\tau^0(\varphi_t(\varphi))C(\varphi_\tau(\varphi_t(\varphi))), & \tau\le 0,\\ \Omega_\tau^0(\varphi_t(\varphi))[C(\varphi_\tau(\varphi)))-E], & \tau>0,\end{cases}\\ &= \begin{cases}\Omega_{\tau+t}^t(\varphi)C(\varphi_{\tau+t}(\varphi)), & \tau+t\le t,\\ \Omega_{\tau+t}^t(\varphi)[C(\varphi_{\tau+t}(\varphi))-E], & \tau+t>t,\end{cases}\\ &= G_t(\tau+t,\varphi).\end{aligned}$$

Lemma 7.3. *If* $\|G_0(\tau,\varphi)\| \leq K\exp\{-\gamma\,|\,t\,|\,\}$, *where the constants* $K =$ const > 0 *and* $\gamma =$ const > 0 *are independent of* φ, *then*

$$\|G_t(0,\varphi)\| \leq K\exp\{-\gamma\,|\,t\,|\,\}. \tag{7.9}$$

Indeed, by using relation (7.8), we get

$$\|G_z(\tau+z,\varphi)\| \leq K\exp\{-\gamma\,|\,\tau+z-z\,|\,\},$$

since, according to the condition of Lemma 7.3,

$$\|G_0(\tau,\varphi_z(\varphi))\| \leq K\exp\{-\gamma\,|\,\tau\,|\,\}.$$

If we now denote $z = t$ and $\tau + z = t_1$, then we find

$$\|G_t(t_1,\varphi)\| \leq K\exp\{-\gamma\,|\,t-t_1\,|\,\}.$$

By setting $t_1 = 0$ in this inequality, we arrive at inequality (7.9).

Consider a system of equations

$$\frac{d\varphi}{dt} = a(\varphi), \quad \frac{dx}{dt} = P(\varphi)x + f(x). \tag{7.10}$$

Note that the system of equations (7.1) is a partial case of this system. Here, the vector function $f(\varphi) = \{f_1(\varphi), f_2(\varphi), \ldots\}$ is such that the functions $f_i(\varphi)$ are continuous in φ for all $i = 1, 2, \ldots$ and, moreover, $\|f(\varphi)\| \leq f^0 = \text{const} < \infty$. This means that the equation

$$\frac{dx}{dt} = P(\varphi_t(\varphi))x + f(\varphi_t(\varphi))$$

satisfies the conditions of Theorem 4.1 in the domain H.

The theorem below presents conditions under which the system of equations (7.10) possesses a family of bounded solutions.

Theorem 7.1. *Assume that the system of equations (7.10) is such that the elements of the matrix* $P(\varphi) \in C_\varphi$ *are continuous in* φ *and the Green function* $G_0(\tau,\varphi)$ *of the corresponding homogeneous system exists and has the form (7.2). Then the expression*

$$h(t) = \int_{-\infty}^{\infty} G_t(\tau,\varphi) f(\varphi_\tau(\varphi))\, d\tau$$

defines a family of bounded solutions of the system of equations (7.10).

Proof. We denote

$$\Omega^t_\tau(\varphi) = [\omega_{ij}(t,\tau,\varphi)]^{\infty}_{i,j=1} \quad \text{and} \quad f(\varphi_\tau(\varphi)) = f(\tau).$$

Let us show that

$$\Big(\Omega^t_\tau(\varphi)C(\varphi_\tau(\varphi))\Big)f(\tau) = \Omega^t_\tau(\varphi)\Big(C(\varphi_\tau(\varphi))f(\tau)\Big). \tag{7.11}$$

Without loss of generality, we can prove this assertion for the first coordinates of the vector functions on the left-hand and right-hand sides of relation (7.11). Thus, it is necessary to prove that

$$\sum_{i=1}^{\infty}\sum_{j=1}^{\infty}\omega_{1j}(t,\tau,\varphi)c_{ji}(\varphi_\tau(\varphi)) = \sum_{j=1}^{\infty}\sum_{i=1}^{\infty}\omega_{1j}(t,\tau,\varphi)c_{ji}(\varphi_\tau(\varphi))f_i(\tau)$$

for any $\tau, t \in R^1$ and $\varphi \in R^m$. To do this, it suffices to show that one of these repeated series converges.

In view of the fact that

$$\|\Omega^t_\tau(\varphi)\| \le \exp\{p^0\,|\,t-\tau\,|\,\} \quad \text{and} \quad \|C(\varphi)\| \le C^0 = \text{const} < \infty,$$

we arrive at the inequality

$$\sum_{i=1}^{\infty}\sum_{j=1}^{\infty}|\,\omega_{1j}(t,\tau,\varphi)\,|\,|\,c_{ji}(\varphi_\tau)\,|\,|\,f_i(\tau)\,| \le f^0C^0\exp\{p^0\,|\,t-\tau\,|\,\},$$

which proves relation (7.11).

Similarly, one can show that

$$(\Omega^t_\tau(\varphi)[C(\varphi_\tau(\varphi)) - E])f(\tau) = \Omega^t_\tau(\varphi)([C(\varphi_\tau(\varphi)) - E]f(\tau)).$$

By using the equality

$$\frac{d\Omega^t_\tau(\varphi)}{dt} = P(\varphi_t(\varphi))\Omega^t_\tau(\varphi),$$

we easily see that

$$\frac{d}{dt}\Big\{\Omega^t_\tau(\varphi)(C(\varphi_\tau(\varphi))f(\tau))\Big\} = \frac{d\Omega^t_\tau(\varphi)}{dt}C(\varphi_\tau(\varphi))f(\tau),$$

$$\frac{d}{dt}\{\Omega^t_\tau(\varphi)(C(\varphi_\tau(\varphi)) - E)f(\tau)\} = \frac{d\Omega^t_\tau(\varphi)}{dt}(C(\varphi_\tau(\varphi)) - E)f(\tau).$$

Let us now show that the improper integral

$$J_1 = \int_{-\infty}^{t} \frac{d\Omega_\tau^t(\varphi)}{dt} C(\varphi_\tau(\varphi)) f(\tau)\, d\tau$$

converges uniformly in t and φ.

Indeed, by using relation (7.8), we obtain

$$\begin{aligned}
\|J_1\| \;\le\; & \int_{-\infty}^{t} \Big\|\frac{dG_t(\tau,\varphi)}{dt} f(\varphi_\tau(\varphi))\Big\|\, d\tau \le p^0 \int_{-\infty}^{t} \|G_t(\tau,\varphi) f(\varphi_\tau(\varphi))\| d\tau \\
\le\; & p^0 \int_{-\infty}^{\infty} \|G_t(\tau,\varphi) f(\varphi_\tau(\varphi))\| d\tau \\
=\; & p^0 \int_{-\infty}^{\infty} \|G_0(\tau,\varphi_t(\varphi))\|\, \|f(\varphi_{t+\tau}(\varphi))\| d\tau \le p^0 f^0 K.
\end{aligned}$$

A similar estimate is also true for the integral

$$\int_{t}^{\infty} \frac{d\Omega_\tau^t(\varphi)}{dt} (C(\varphi_\tau(\varphi)) - E) f(\tau)\, d\tau.$$

This enables us to write the following chain of equalities proving the formulated theorem:

$$\begin{aligned}
\frac{dh(t)}{dt} \;=\; & \int_{-\infty}^{t} \frac{d\Omega_\tau^t(\varphi)}{dt} C(\varphi_\tau(\varphi)) f(\tau)\, d\tau + \Omega_t^t(\varphi) C(\varphi_t(\varphi)) f(t) \\
& + \int_{t}^{\infty} \frac{d\Omega_\tau^t(\varphi)}{dt} [C(\varphi_\tau(\varphi)) - E] f(\tau)\, d\tau - \Omega_t^t(\varphi)[C(\varphi_t(\varphi)) - E] f(t) \\
=\; & P(\varphi_t(\varphi)) \Big\{ \int_{-\infty}^{t} G_t(\tau,\varphi) f(\tau)\, d\tau + \int_{t}^{\infty} G_t(\tau,\varphi) f(\tau)\, d\tau \Big\} + f(t) \\
=\; & P(\varphi_t(\varphi)) \int_{-\infty}^{\infty} G_t(\tau,\varphi) f(\tau)\, d\tau + f(t),
\end{aligned}$$

i.e.,

$$\frac{dh}{dt} = P(\varphi_t(\varphi))h(t) + f(\varphi_t(\varphi)).$$

Theorem 7.1 is proved.

We now consider a system of equations

$$\frac{d\varphi}{dt} = a(\varphi, p), \quad \frac{dx}{dt} = P(\varphi, p)x \tag{7.12}$$

dependent on a parameter $p \in [p_1, p_2] \subset R^1$. Here, $\varphi \in R^m$, $x \in \mathfrak{M}$, and $P(\varphi, p) = [p_{ij}(\varphi, p)]_{i,j=1}^{\infty}$ is an infinite matrix. If the function $a(\varphi, p)$ satisfies the Lipschitz condition with respect to φ, then the first equation in system (7.12) possesses a unique solution φ_t^p such that $\varphi_0^p(\varphi) = \varphi$, where φ is an arbitrary constant vector from R^m.

Consider an equation

$$\frac{dx}{dt} = P(\varphi_t^p(\varphi), p)x. \tag{7.13}$$

Let $\omega(z)$ be a continuous scalar function nondecreasing in the segment $[0, p_2 - p_1]$ and taking the value $\omega(0) = 0$ for $z = 0$.

Theorem 7.2. *Assume that the system of equations (7.12) satisfies the following conditions:*

(i) $\sum\limits_{j=1}^{\infty} \sup\limits_{\varphi, p} |\, p_{sj}(\varphi, p)\,| \le p^0 = \text{const} < \infty\,, \quad s = 1, 2, \ldots;$

(ii)
$$\|a(\varphi, p) - a(\bar{\varphi}, \bar{p})\| \le \alpha_1 \|\varphi - \bar{\varphi}\| + \alpha_2 \omega(\,|\,p - \bar{p}\,|\,)$$
and
$$\|P(\varphi, p) - P(\bar{\varphi}, \bar{p})\| \le \beta_1 \|\varphi - \bar{\varphi}\| + \beta_2 \omega(\,|\,p - \bar{p}\,|\,)$$
for any $\varphi, \bar{\varphi} \in R^m$ *and* $p, \bar{p} \in [p_1, p_2]$;

(iii) for any fixed $p \in [p_1, p_2]$ *and* $\varphi \in R^m$, *equation (7.13) possesses a unique solution bounded on the entire real axis;*

(iv) the Green function $G_0(\tau, \varphi, p)$ *of equation (7.12) exists and satisfies the estimate*

$$\|G_t(\tau, \varphi, p)\| \le K \exp\{-\gamma\,|\,t - \tau\,|\,\}, \tag{7.14}$$

where $\gamma > 0$ *and* $K > 0$ *are constants independent of* τ, φ, *and* p.

Then the Green function $G_0(\tau, \varphi, p)$ is continuous with respect to the collection of its variables φ and p.

Proof. We denote the difference $G_t(\tau, \varphi, p) - G_t(\tau, \bar{\varphi}, \bar{p}) = L(t)$ and differentiate it with respect to t for $t \neq \tau$. This gives

$$\begin{aligned}\frac{dL(t)}{dt} &= P(\varphi_t^p(\varphi), p)G_t(\tau, \varphi, p) - P(\varphi_t^{\bar{p}}(\bar{\varphi}), \bar{p})G_t(\tau, \bar{\varphi}, \bar{p}) \\ &= P(\varphi_t^p(\varphi), p) \times \{G_t(\tau, \varphi, p) - G_t(\tau, \bar{\varphi}, \bar{p})\} \\ &\quad + \{P(\varphi_t^p(\varphi), p) - P(\varphi_t^{\bar{p}}(\bar{\varphi}), \bar{p})\}G_t(\tau, \bar{\varphi}, \bar{p}).\end{aligned}$$

This means that $L(t)$ is a matrix solution of an inhomogeneous system of equations

$$\frac{dL(t)}{dt} = P(\varphi_t^p(\varphi), p)L(t) + F(t, \varphi), \tag{7.15}$$

where

$$F(t, \varphi) = \left\{P(\varphi_t^p, p) - P(\varphi_t^{\bar{p}}(\bar{\varphi}), \bar{p})\right\}G_t(\tau, \bar{\varphi}, \bar{p}).$$

It is clear that the last expression exists and satisfies the inequality

$$\|F(t, \varphi)\| \leq 2p^0 K \exp\{-\gamma\,|\,t - \tau\,|\} \leq 2Kp^0.$$

Following the proof of Theorem 7.1, one can easily show that the expression

$$\int_{-\infty}^{\infty} G_t(t, \varphi, p)F(t_1, \varphi)dt_1$$

is a bounded solution of equation (7.15). Thus, by virtue of condition (iii), this solution must coincide with $L(t)$. Hence, we have the identity

$$L(t) = \int_{-\infty}^{\infty} G_t(t_1, \varphi, p)\Big[P(\varphi_{t_1}^p(\varphi), p) - P(\varphi_{t_1}^{\bar{p}}(\bar{\varphi}), \bar{p})\Big]G_{t_1}(\tau, \bar{\varphi}, \bar{p})dt_1, \tag{7.16}$$

which implies the inequality

$$\|L(0)\| \leq \int_{-\infty}^{\infty} \|G_0(t_1, \varphi, p)\|\, \|P(\varphi_{t_1}^p(\varphi), p) - P(\varphi_{t_1}^{\bar{p}}(\bar{\varphi}), \bar{p})\|\, \|G_{t_1}(\tau, \bar{\varphi}, \bar{p})\|\, dt_1$$

$$\leq \int_{-\infty}^{\infty} K \exp\{-\gamma\,|\,t_1\,|\,\} K \exp\{-\gamma\,|\,t_1 - \tau\,|\,\}$$

$$\times \Big(\beta_1 \|\varphi^p_{t_1}(\varphi) - \varphi^{\bar p}_{t_1}(\bar\varphi)\| + \beta_2 \omega(\,|\,p - \bar p\,|\,)\Big) dt_1. \tag{7.17}$$

Let us now estimate the difference $\hat\varphi(t) = \varphi^p_t(\varphi) - \varphi^{\bar p}_t(\bar\varphi)$. Clearly, for $t \geq 0$, the equality

$$\hat\varphi(t) = \varphi - \bar\varphi + \int_0^t \Big(a(\varphi^p_{t_1}(\varphi), p) - a(\varphi^{\bar p}_{t_1}(\bar\varphi), \bar p)\Big) dt_1\,,$$

implies the inequality

$$\|\hat\varphi(t)\| \leq \|\varphi - \bar\varphi\| + \alpha_1 \int_0^t \|\hat\varphi(t_1)\|\, dt_1 + \alpha_2 \omega(\,|\,p - \bar p\,|\,) t.$$

Further, if we denote the right-hand side of this expression by $A(t)$, then we get

$$\frac{dA(t)}{dt} \leq \alpha_1 A(t) + \alpha_2 \omega(\,|\,p - \bar p\,|\,).$$

Integrating this inequality, we obtain

$$A(t) \leq A(0) + \int_0^t (\alpha_2 \omega(\,|\,p - \bar p\,|\,) + \alpha_1 A(t)) dt,$$

whence it follows that (see Lemma 2.1)

$$A(t) \leq \Big(\|\varphi - \bar\varphi\| + \frac{\alpha_2}{\alpha_1} \omega(\,|\,p - \bar p\,|\,)\Big) \exp\{\alpha_1 t\}.$$

Similarly, for $t < 0$, we arrive at the inequality

$$\|\varphi^p_t(\varphi) - \varphi^{\bar p}_t(\bar\varphi)\| \leq \left(\|\varphi - \bar\varphi\| + \frac{\alpha_2}{\alpha_1} \omega(\,|\,p - \bar p\,|\,)\right) \exp\{\alpha_1 \|t\|\}\,. \tag{7.18}$$

Thus, by using inequality (7.17), we can write

$$
\begin{aligned}
&\|L(0)\| \\
&\leq \int_{-\infty}^{\infty} K^2 \exp\{-\gamma\,|\,t_1\,| - \gamma\,|\,t_1 - \tau\,|\,\} \\
&\qquad \times\Big\{\beta_1\Big(\|\varphi-\bar{\varphi}\| + \frac{\alpha_2}{\alpha_1}\omega(\,|\,p-\bar{p}\,|\,)\Big)\exp\{\alpha_1\,|\,t_1\,|\,\} + \beta_2\omega(\,|\,p-\bar{p}\,|\,)\Big\}\,dt_1 \\
&\leq K^2\beta_1\Big(\|\varphi-\bar{\varphi}\| + \frac{\alpha_2}{\alpha_1}\omega(\,|\,p-\bar{p}\,|\,)\Big)J_1 + K^2\beta_2\omega(\,|\,p-\bar{p}\,|\,)J_2. \qquad (7.19)
\end{aligned}
$$

In the case where $\gamma > \alpha_1$, for the integrals J_1 and J_2 we obtain

$$
\begin{aligned}
J_1 &= \int_{-\infty}^{\infty} \exp\{-\gamma\,|\,t_1\,| - \gamma\,|\,t_1 - \tau\,| + \alpha_1\,|\,t_1\,|\,\}dt_1 \\
&\leq \int_{-\infty}^{\tau} \exp\{\gamma(t_1 - \tau) - (\gamma - \alpha_1)\,|\,t_1\,|\,\}dt_1 \\
&\qquad + \int_{\tau}^{\infty} \exp\{-\gamma(t_1 - \tau) - (\gamma - \alpha_1)\,|\,t_1\,|\,\}dt_1 \leq \frac{2}{\gamma},
\end{aligned}
$$

$$
\begin{aligned}
J_2 &= \int_{-\infty}^{\infty} \exp\{-\gamma\,|\,t_1\,| - \gamma\,|\,t_1 - \tau\,|\,\}dt_1 \\
&\leq \int_{-\infty}^{\tau} \exp\{\gamma(t_1 - \tau)\}dt_1 + \int_{\tau}^{\infty} \exp\{-\gamma(t_1 - \tau)\}dt_1 \leq \frac{2}{\gamma}.
\end{aligned}
$$

Thus, in view of inequalities (7.19), we get

$$
\begin{aligned}
&\|G_0(\tau,\varphi,p) - G_0(\tau,\bar{\varphi},\bar{p})\| \\
&\qquad \leq \frac{2K^2\beta_1}{\gamma}\Big(\|\varphi - \bar{\varphi}\| + \frac{\alpha_2}{\alpha_1}\omega(\,|\,p - \bar{p}\,|\,)\Big) + \frac{2K^2\beta_2}{\gamma}\omega(\,|\,p - \bar{p}\,|\,).
\end{aligned}
$$

Let $\nu > 0$ be a constant. By using condition (i) and inequality (7.18), we find

$$\|P(\varphi_t^p(\varphi),p) - P(\varphi_t^{\bar{p}}(\bar{\varphi}),\bar{p})\|$$
$$\leq (2p^0)^{\frac{\nu}{\nu+1}} \Big\{\beta_1\|\varphi-\bar{\varphi}\| + \frac{\beta_1\alpha_2+\beta_2\alpha_1}{\alpha_1}\omega(|p-\bar{p}|)\Big\}^{\frac{1}{\nu+1}}$$
$$\times \exp\Big\{\frac{\alpha_1}{\nu+1}|t|\Big\}. \tag{7.20}$$

Then

$$\|G_0(\tau,\varphi,p) - G_0(\tau,\bar{\varphi},\bar{p})\|$$
$$\leq \int_{-\infty}^{\infty} K\exp\{-\gamma|t_1|\}K\exp\{-\gamma|t_1-\tau|\}(2p^0)^{\frac{\nu}{\nu+1}}$$
$$\times \Big\{\beta_1\|\varphi-\bar{\varphi}\| + \frac{\beta_1\alpha_2+\beta_2\alpha_1}{\alpha_1}\omega(|p-\bar{p}|)\Big\}^{\frac{1}{\nu+1}}$$
$$\times \exp\Big\{\frac{\alpha_1}{\nu+1}|t_1|\Big\}dt_1$$
$$\leq (2p^0)^{\frac{\nu}{\nu+1}} \Big\{\beta_1\|\varphi-\bar{\varphi}\| + \frac{\beta_1\alpha_2+\beta_2\alpha_1}{\alpha_1}\omega(|p-\bar{p}|)\Big\}^{\frac{1}{\nu+1}} K^2$$
$$\times \int_{-\infty}^{\infty} \exp\Big\{-\gamma|t_1-\tau| + \Big(\frac{\alpha_1}{\nu+1}-\gamma\Big)|t_1|\Big\}dt_1.$$

By choosing ν from the solutions of the inequality

$$\gamma > \frac{\alpha_1}{\nu+1},$$

we finally obtain

$$\int_{-\infty}^{\infty} \exp\Big\{-\gamma|t_1-\tau| + \Big(\frac{\alpha_1}{\nu+1}-\gamma\Big)|t_1|\Big\}dt_1 \leq \frac{2}{\gamma - \dfrac{\alpha_1}{\nu+1}}.$$

This completes the proof of Theorem 7.2.

Let us now present conditions sufficient for the differentiability of the Green function with respect to the parameter.

Theorem 7.3. *Assume that the system of equations (7.12) is such that*

(i) conditions (i), (iii), and (iv) of Theorem 7.2 are satisfied;

(ii) the matrix $P(\varphi, p)$ and the function $a(\varphi, p)$ are continuously differentiable with respect to p and φ_i, $i = \overline{1, m}$, and, moreover,

$$\sum_{j=1}^{\infty}\Big\{\max_{\varphi,p}\Big|\frac{\partial p_{sj}(\varphi,p)}{\partial \varphi_i}\Big| + \max_{\varphi,p}\Big|\frac{\partial p_{sj}(\varphi,p)}{\partial p}\Big|\Big\} \le p_s,$$

$$\sum_{s=1}^{\infty} p_s < \infty, \quad i = \overline{1,m}.$$

Then the Green function $G_0(\tau, \varphi, p)$ of equation (7.12) is differentiable with respect to the parameter p whenever $\gamma > \alpha$ and α is chosen from the condition

$$\Big|\Big\langle \frac{\partial a(\varphi,p)}{\partial \varphi}\eta, \eta\Big\rangle\Big| \le \alpha\,|\,\eta\,|^2,$$

where $\langle \cdots \rangle$ is the Euclidean product and $|\,\eta\,| = |\,(\eta_1, \ldots, \eta_m)\,| = \Big(\sum\limits_{i=1}^{m} \eta_i^2\Big)^{\frac{1}{2}}$.

Proof. We divide the identity

$$G_0(\tau,\varphi,p) - G_0(\tau,\varphi,\bar{p})$$

$$= \int_{-\infty}^{\infty} G_0(t_1,\varphi,p)\Big(P(\varphi_{t_1}^{p}(\varphi),p) - P(\varphi_{t_1}^{\bar{p}}(\varphi),\bar{p})\Big)G_{t_1}(\tau,\varphi,\bar{p})dt_1$$

by $p - \bar{p}$ and pass to the limit as $\bar{p} \to p$.

Let us show that the limit

$$\lim_{\bar{p}\to p}\Big(G_0(t_1,\varphi,p)\frac{P(\varphi_{t_1}^{p}(\varphi),p) - P(\varphi_{t_1}^{\bar{p}}(\varphi),\bar{p})}{p-\bar{p}}G_{t_1}(\tau,\varphi,\bar{p})\Big)$$

$$= G_0(t_1,\varphi,p)\frac{\partial P(\varphi_{t_1}^{p}(\varphi),p)}{\partial p}G_{t_1}(\tau,\varphi,p) \tag{7.21}$$

exists in the sense of coordinatewise convergence. As earlier, $\varphi_t^p(\varphi)$ is the solution of the first equation in system (7.12) such that $\varphi_0^p(\varphi) = \varphi$. Denote

$$G_0(t_1,\varphi,p) = [g_{0ik}(t_1,\varphi,p)]_{i,k=1}^{\infty},$$

$$G_{t_1}(\tau,\varphi,p) = [g_{ik}(t_1,\tau,\varphi,p)]_{i,k=1}^{\infty}.$$

The element of the matrix under the limit sing in relation (7.21) located in the rth column and lth row has the form

$$W = \sum_{k=1}^{\infty} g_{0lk}(t_1, \varphi, p) \times \sum_{i=1}^{\infty} \frac{p_{ki}(\varphi_{t_1}^{p}(\varphi), p) - p_{ki}(\varphi_{t_1}^{\bar{p}}(\varphi), \bar{p})}{p - \bar{p}} g_{ir}(t_1, \tau, \varphi, \bar{p}). \quad (7.22)$$

We have

$$\begin{aligned} S &= \sum_{i=1}^{\infty} g_{0lk}(t_1, \varphi, p) \frac{p_{ki}(\varphi_{t_1}^{p}(\varphi), p) - p_{ki}(\varphi_{t_1}^{\bar{p}}(\varphi), \bar{p})}{p - \bar{p}} g_{ir}(t_1, \tau, \varphi, \bar{p}) \\ &\leq \sum_{i=1}^{\infty} K^2 \exp\{-\gamma\,|\,t_1\,| - \gamma\,|\,t_1 - \tau\,|\}\left|\frac{\partial p_{ki}(\varphi_{t_1}^{p}(\varphi), p)}{\partial p}\right|_{p=p_{ki}}, \end{aligned}$$

where $p_{ki} \in (p, \bar{p})$. At the same time,

$$\left|\frac{\partial p_{ki}(\varphi_{t_1}^{p}(\varphi), p)}{\partial p}\right| = \left|\sum_{j=1}^{m} \frac{\partial p_{ki}(\varphi_{t_1}^{p}(\varphi), p)}{\partial \varphi_{jt_1}^{p}(\varphi)} \cdot \frac{\partial \varphi_{jt_1}^{p}(\varphi)}{\partial p} + \frac{\partial p_{ki}(\varphi, p)}{\partial p}\right|.$$

By using the inequality

$$\left\|\frac{\partial \varphi_{jt_1}^{p}(\varphi)}{\partial p}\right\| \leq K_1 \exp\{\alpha\,|\,t_1\,|\}$$

established in [MSK2], we obtain

$$\begin{aligned} |\,S\,| &\leq K^2 \exp\{-\gamma\,|\,t_1\,| - \gamma\,|\,t_1 - \tau\,|\} \\ &\quad \times \sum_{i=1}^{\infty}\Big\{\sum_{j=1}^{m}\left|\frac{\partial p_{ki}(\varphi, p)}{\partial \varphi_j}\right|\left|\frac{\partial \varphi_{jt_1}^{p}(\varphi)}{\partial p}\right| + \left|\frac{\partial p_{ki}(\varphi, p)}{\partial p}\right|\Big\}_{p=p_{ki}} \\ &\leq mK^2 K_1 \exp\{-(\gamma - \alpha)\,|\,t_1 - \tau\,| - \gamma\,|\,t_1\,|\Big\} \sum_{i=1}^{\infty} \max_{\varphi,p}\left|\frac{\partial p_{ki}(\varphi, p)}{\partial \varphi_j}\right| \\ &\quad + K^2 \exp\{-\gamma\,|\,t_1\,| - \gamma\,|\,t_1 - \tau\,|\} \sum_{i=1}^{\infty} \max_{\varphi,p}\left|\frac{\partial p_{ki}(\varphi, p)}{\partial p}\right| \\ &\leq p_k K^2 (K_1 m + 1) \end{aligned}$$

for $\gamma > 0$ and, moreover, the series on the right-hand side of the expression for S converges uniformly with respect to the parameter p. Then

$$|W| \leq \sum_{k=1}^{\infty} |S| \leq \sum_{k=1}^{\infty} p_k K^2 (K_1 m + 1)$$

$$\leq K^2 (K_1 m + 1) \sum_{k=1}^{\infty} p_k < \infty,$$

and, therefore, one can pass to the limit as $\bar{p} \to p$ in expression (7.22) term by term. This proves equality (7.21).

In this case,

$$\lim_{\bar{p} \to p} \frac{G_0(\tau, \varphi, \bar{p}) - G_0(\tau, \varphi, p)}{\bar{p} - p}$$

$$= \int_{-\infty}^{\infty} G_0(s, \varphi, p) \frac{\partial P(\varphi_s^p(\varphi), p)}{\partial p} G_s(\tau, \varphi, p)\, ds \tag{7.23}$$

whenever the integral on the right-hand side is uniformly convergent. At the same time,

$$\left\| G_0(s, \varphi, p) \frac{\partial P(\varphi_s^p(\varphi), p)}{\partial p} G_s(\tau, \varphi, p) \right\|$$

$$\leq K^2 \exp\{-\gamma |s| - \gamma |s - \tau|\} \left\| \frac{\partial P(\varphi_s^p(\varphi), p)}{\partial p} \right\|$$

$$\leq \sup_s \{p_s\} \{m K_1 \exp\{\alpha |s|\} + 1\} K^2 \exp\{-\gamma |s| - \gamma |s - \tau|\}$$

$$\leq \widehat{K} \exp\{-\gamma |s - \tau| - (\gamma - \alpha) |s|\},$$

where $\widehat{K} = \sup_s \{p_s\} K^2 (m K_1 + 1)$, and it is clear that $\sup_s \{p_s\} < \infty$. This estimate guarantees the uniform convergence of the integral on the right-hand side of relation (7.23).

Thus, for $\tau \neq 0$, the expression

$$\lim_{\bar{p} \to p} \frac{G_0(\tau, \varphi, p) - G_0(\tau, \varphi, \bar{p})}{p - \bar{p}}$$

determines the matrix function $\frac{\partial}{\partial p}G_0(\tau,\varphi,p)$ satisfying the inequality

$$\left\|\frac{\partial G_0(\tau,\varphi,p)}{\partial p}\right\| \le \widehat{K}\int_{-\infty}^{\infty}\exp\{-\gamma\,|\,s-\tau\,|-(\gamma-\alpha)\,|\,s\,|\,\}ds$$

$$\le \widehat{\widehat{K}}\exp\{-(\gamma-\alpha)\,|\,\tau\,|\,\}, \tag{7.24}$$

where $\widehat{\widehat{K}} = \widehat{K}\dfrac{2\gamma+\alpha}{\alpha(2\gamma-\alpha)} = \text{const} > 0.$

Similarly, we can present conditions for the differentiability of the Green function with respect to φ_i, $i = \overline{1,m}$. This problem is studied in the next section.

Remark 7.1. In proving Theorems 7.2 and 7.3, we essentially used inequality (7.14). This inequality is equivalent to the inequality

$$\|G_0(\tau,\varphi)\| \le K\exp\{-\gamma\,|\,\tau\,|\,\},$$

where K and γ are positive constants independent of φ.

The last inequality follows from the equality

$$\begin{aligned}
G_t(\tau,\varphi) &= \begin{cases} \Omega_\tau^t(\varphi)C(\varphi_\tau(\varphi)), & \tau\le t,\\ \Omega_\tau^t(\varphi)[C(\varphi_\tau(\varphi))-E], & \tau>t,\end{cases}\\
&= \begin{cases} \Omega_{\tau-t}^0(\varphi)C(\varphi_\tau(\varphi)), & \tau\le t,\\ \Omega_{\tau-t}^0(\varphi)[C(\varphi_\tau(\varphi))-E], & \tau>t,\end{cases}\\
&= \begin{cases} \Omega_z^0(\varphi_t(\varphi))C(\varphi_z(\varphi_t(\varphi))), & z\le 0,\\ \Omega_z^0(\varphi_t(\varphi))[C(\varphi_z(\varphi_t(\varphi)))-E], & z>0,\end{cases}\\
&= G_0(z,\varphi_t(\varphi)),
\end{aligned}$$

where $z = \tau - t$, by virtue of the estimate

$$\|G_t(\tau,\varphi)\| = \|G_0(z,\varphi_t(\varphi))\| \le K\exp\{-\gamma\,|\,t-\tau\,|\,\}.$$

We now assume that, for any $\varphi \in R^m$, the space $\mathfrak{M}$ can be decomposed into a direct sum of subspaces $E_1(\varphi)$ and $E_2(\varphi)$, i.e., $\mathfrak{M} = E_1(\varphi)\oplus E_2(\varphi)$, so that the solution $x_t(\varphi,x_0) = \Omega_0^t(\varphi)x_0$ of equation (7.2) taking a value $x_0 \in E_1(\varphi)$ for $t = 0$ satisfies the inequality

$$\|x_t(\varphi,x_0)\| \le K\exp\{-\gamma(t-\tau)\}\|x_\tau(\varphi,x_0)\|, \quad t\ge\tau,\ \tau,t\in R^1,$$

and the solution taking a value $x_0 \in E_2(\varphi)$ satisfies the inequality

$$\|x_t(\varphi, x_0)\| \le K \exp\{\gamma(t-\tau)\}\|x_\tau(\varphi, x_0)\|, \quad t \le \tau,\ \tau, t \in R^1,$$

where K and γ are positive constants independent of φ.

In this case, equation (7.2) is called exponentially dichotomous on the entire real axis $R^1 = (-\infty, \infty)$.

In the case of exponential dichotomy on R^1, the only solution of equation (7.2) bounded on the entire real axis R^1 is trivial.

If there exists an infinite matrix $C_1(\varphi) \in C_\varphi$ projecting the space $\mathfrak{M}$ onto $E_1(\varphi)$, then equation (7.2) (exponentially dichotomous on R^1) is called an exponentially dichotomous equation with matrix projector.

In this case, the function

$$\tilde{G}_t(\tau, \varphi) = \begin{cases} \Omega_0^t(\varphi) C_1(\varphi) \Omega_\tau^0(\varphi), & t \ge \tau, \\ -\Omega_0^t(\varphi)[E - C_1(\varphi)]\Omega_\tau^0(\varphi), & t \le \tau, \end{cases}$$

satisfies the inequality [DaK, Sam10]

$$\|\tilde{G}_t(\tau, \varphi)\| \le K \exp\{-\gamma\,|\,t-\tau\,|\,\},$$

where K and γ are positive constants independent of φ and $t, \tau \in R^1$.

Theorem 7.4. *Assume that the system of equations (7.1) is such that*

(i) the matrix $P(\varphi) \in C_\varphi$,

(ii) $\|a(\varphi) - a(\bar{\varphi})\| \le \alpha\|\varphi - \bar{\varphi}\|$ *and* $\|P(\varphi) - P(\bar{\varphi})\| \le \beta\|\varphi - \bar{\varphi}\|$, $\alpha, \beta =$ const > 0,

(iii) the corresponding equation (7.2) is an exponentially dichotomous equation with matrix projector $C_1(\varphi)$.

Then the Green function $G_0(\tau, \varphi)$ *of the system of equations (7.1) exists and satisfies the inequality*

$$\|G_0(\tau, \varphi)\| \le K \exp\{-\gamma\,|\,\tau\,|\,\}, \tag{7.25}$$

where K *and* γ *are positive constants independent of* $\varphi \in R^m$ *and* $\tau \in R^1$.

Proof. We define the function

$$\begin{aligned} G_0(\tau, \varphi) &= \tilde{G}_t(0, \varphi_\tau(\varphi))\Big|_{t=-\tau} \\ &= \begin{cases} \Omega_\tau^0(\varphi) C_1(\varphi_\tau(\varphi)), & \tau \le 0, \\ \Omega_\tau^0(\varphi)[C_1(\varphi_\tau(\varphi)) - E], & \tau > 0. \end{cases} \end{aligned}$$

This function satisfies inequality (7.25). To prove the theorem, it suffices to show that the matrix $C_1(\varphi)$ is a continuous function of φ. For this purpose, we consider a matrix function

$$Z_t(\varphi,\bar{\varphi}) = \begin{cases} \Omega_0^t(\varphi)C_1(\varphi) - \Omega_0^t(\bar{\varphi})C_1(\bar{\varphi}), & t \geq 0, \\ -\Omega_0^t(\varphi)C_2(\varphi) + \Omega_0^t(\bar{\varphi})C_2(\bar{\varphi}), & t < 0, \end{cases}$$

where $C_2(\varphi) = E - C_1(\varphi)$. It is clear that $Z_0(\varphi,\bar{\varphi}) = C_1(\varphi) - C_1(\bar{\varphi})$. In this case,

$$\|Z_t(\varphi,\bar{\varphi})\| \leq 2K\exp\{-\gamma\,|\,t\,|\}, \quad t \in R^1.$$

By using the proof of Theorem 7.1, we show that the function $Z_t(\varphi,\bar{\varphi})$ is a solution of the equation

$$\frac{dZ}{dt} = P(\varphi_t(\varphi))Z + [P(\varphi_t(\varphi)) - P(\varphi_t(\bar{\varphi}))]\tilde{G}_t(0,\bar{\varphi}).$$

In view of condition (iii), we arrive at the following representation:

$$C_1(\varphi) - C_1(\bar{\varphi}) = \int\limits_{-\infty}^{\infty} \tilde{G}_0(\tau,\varphi)[P(\varphi_\tau(\varphi)) - P(\varphi_\tau(\bar{\varphi}))]\tilde{G}_\tau(0,\bar{\varphi})d\tau.$$

In the analyzed case, inequality (7.20) takes the form

$$\|P(\varphi_\tau(\varphi)) - P(\varphi_\tau(\bar{\varphi}))\| \leq (2p^0)^{\frac{\nu}{\nu+1}}(\|\varphi-\bar{\varphi}\|\beta)^{\frac{1}{\nu+1}}\exp\Big\{\frac{\alpha}{\nu+1}\,|\,\tau\,|\Big\}.$$

By using this inequality, we get

$$\begin{aligned} &\|C_1(\varphi) - C_1(\bar{\varphi})\| \\ &\quad\leq K^2(2p^0)^{\frac{\nu}{\nu+1}}(\|\varphi-\bar{\varphi}\|)^{\frac{1}{\nu+1}}\int\limits_{-\infty}^{\infty}\exp\Big\{-2\gamma\,|\,\tau\,| + \frac{\alpha}{\nu+1}\,|\,\tau\,|\Big\}\,d\tau \\ &\quad\leq \frac{2K^2}{2\gamma - \frac{\alpha}{\nu+1}}(2p^0)^{\frac{\nu}{\nu+1}}\beta^{\frac{1}{\nu+1}}\|\varphi-\bar{\varphi}\|^{\frac{1}{\nu+1}}, \end{aligned}$$

where ν is a positive constant chosen from the condition

$$2\gamma - \frac{\alpha}{\nu+1} > 0.$$

Theorem 7.4 is proved.

Taking into account Theorems 7.2–7.4, we can also prove the following statements:

Theorem 7.5. *Assume that the system of equations (7.12) satisfies conditions (i) and (ii) of Theorem 7.2. In this case, if the corresponding equation (7.13) is exponentially dichotomous for any* $\varphi \in R^m$ *and* $p \in [p_1, p_2]$ *with the matrix projector* $C_1(\varphi, p)$*, then the system of equations (7.12) possesses the Green function satisfying the inequality*

$$\|G_0(\tau, \varphi, p)\| \leq K \exp\{-\gamma \,|\, \tau \,|\, \},$$

where K *and* γ *are positive constants independent of* τ*,* φ*, and* $p \in [p_1, p_2]$*. The indicated Green function is continuous with respect to the collection of variables* φ *and* p*.*

Theorem 7.6. *Assume that the system of equations (7.12) satisfies condition (i) of Theorem 7.2 and condition (ii) of Theorem 7.3. If, in addition, equation (7.13) is exponentially dichotomous for any* $\varphi \in R^m$ *and* $p \in [p_1, p_2]$ *with the matrix projector* $C_1(\varphi, p)$*, then the Green function of the system of equations (7.12) is differentiable with respect to the parameter* p*, provided that* $\gamma > \alpha$*, where* α *is chosen from the same condition as in Theorem 7.3.*

Remark 7.2. Assume that the matrizant $\Omega_0^t(\varphi)$ of equation (7.2) satisfies the following condition for all $\varphi \in R^m$:

$$\|\Omega_0^t(\varphi)\| \leq K \exp\{-\gamma t\} \quad \text{for} \quad t \geq 0, \tag{7.26}$$

where K and γ are positive constants independent of φ.

As a result, equation (7.2) turns into an exponentially dichotomous equation with matrix projector. In this case, $E_1(\varphi) = \mathfrak{M}$, $E_2(\varphi) = \{0 \in \mathfrak{M}\}$, and the role of the projector C_1 is played by the identity matrix E. The Green function of the system of equations (7.1) exists and has the form

$$G_0(\tau, \varphi) = \begin{cases} \Omega_\tau^0, & \tau \leq 0, \\ 0, & \tau > 0, \end{cases}$$

because inequality (7.26) implies that

$$\|\Omega_\tau^0(\varphi)\| \leq K \exp\{\gamma \tau\}, \quad \tau \geq 0.$$

2.8. Existence of a Smooth Invariant Torus

We consider a system of equations

$$\frac{d\varphi}{dt} = a(\varphi), \qquad \frac{dx}{dt} = P(\varphi)x + c(\varphi), \tag{8.1}$$

where $c(\varphi) = (c_1(\varphi), c_2(\varphi), \ldots)$ is a countably dimensional vector function and the functions $a_i(\varphi)$, $p_{ij}(\varphi)$, and $c_i(\varphi)$ are 2π-periodic in φ_i, $i = \overline{1, m}$, for all $i, j = 1, 2, 3, \ldots$. Note that system (7.1) is the homogeneous part of this system.

Let $C^r(\mathcal{F}_m)$, $r = 0, 1, 2, \ldots$, be the sets of vector functions and matrices 2π-periodic in φ_i, $i = \overline{1, m}$, whose derivatives with respect to φ_i up to the rth order inclusive exist and are bounded in the norm $\|\cdot\|$ and continuous as functions of the coordinates φ.

Definition 8.1. *A surface $\mathcal{F}\colon x = u(\varphi) = \{u_1(\varphi), u_2(\varphi), \ldots\}$, $\varphi \in \mathcal{F}_m$, is called an invariant torus of the system of equations (8.1) if $u(\varphi) \in C^0(\mathcal{F}_m)$, the functions $u_i(\varphi_t(\varphi))$, $i = 1, 2, \ldots$, are continuously differentiable with respect to t, and*

$$\frac{du(\varphi_t(\varphi))}{dt} = P(\varphi_t(\varphi))u(\varphi_t(\varphi)) + c(\varphi_t(\varphi)) \tag{8.2}$$

for all $t \in R^1$ and $\varphi \in \mathcal{F}_m$.

Definition 8.2. *The Green function of the system of equations*

$$\frac{d\varphi}{dt} = a(\varphi), \qquad \frac{dx}{dt} = P(\varphi)x, \tag{8.3}$$

obtained from the system of equations (8.1) for $c(\varphi) \equiv 0$, is called the Green function of the problem of invariant torus for system (8.1) if the matrix $C(\varphi)$ appearing in an expression of the form (7.6) belongs to the space $C^0(\mathcal{F}_m) \cap C_\varphi$.

Here, by $\mathcal{F}_m$ we denote the m-dimensional torus.

In the space $C^0(\mathcal{F}_m)$, we select a subset $C^0_{\text{Lip}}(\mathcal{F}_m)$ whose elements satisfy the Lipschitz condition with respect to φ.

Theorem 8.1. *Let the system of equations (8.1) be such that*

(i) $a(\varphi) \in C^0_{\text{Lip}}(\mathcal{F}_m)$, $P(\varphi) \in C_\varphi \cap C^0(\mathcal{F}_m)$, and $c(\varphi) \in C^0(\mathcal{F}_m)$;

(ii) the Green function of the problem of invariant torus $G_0(\tau, \varphi)$ exists.

Then system (8.1) possesses an invariant torus $\mathcal{F}: x = u(\varphi)$, $\varphi \in \mathcal{F}_m$, *where*

$$u(\varphi) = \int_{-\infty}^{\infty} G_0(\tau, \varphi)c(\varphi_\tau(\varphi))d\tau. \tag{8.4}$$

Proof. It is easy to see that the integral on the right-hand side of equality (8.4) converges uniformly in φ and the function $u(\varphi)$ is 2π-periodic in φ_i, $i = \overline{1, m}$. By using the formulas presented in Remark 7.1, we obtain

$$\begin{aligned} u(\varphi_t(\varphi)) &= \int_{-\infty}^{\infty} G_0(\tau, \varphi_t(\varphi))c(\varphi_\tau(\varphi_t(\varphi)))d\tau \\ &= \int_{-\infty}^{\infty} G_t(\tau + t, \varphi)c(\varphi_{\tau+t}(\varphi))d\tau \\ &= \int_{-\infty}^{\infty} G_t(\tau, \varphi)c(\varphi_\tau(\varphi))d\tau. \end{aligned}$$

As a result, the proof of Theorem 8.1 is reduced to the proof of Theorem 7.1.

Theorem 8.2. *Assume that the system of equations (8.1) is such that*

(i) $a(\varphi)$, $c(\varphi) \in C^0_{\mathrm{Lip}}(\mathcal{F}_m)$, *and* $P(\varphi) \in C^0_{\mathrm{Lip}}(\mathcal{F}_m) \cap C_\varphi$;

(ii) an equation of the form (7.2) is an exponentially dichotomous equation with matrix projector.

Then the system of equations (8.1) possesses the invariant torus $\mathcal{F}$ *specified by function (8.4) and satisfying the Hölder condition with respect to* φ :

$$\|u(\varphi) - u(\bar{\varphi})\| \leq K^0 \|\varphi - \bar{\varphi}\|^{\frac{\nu}{2(\nu+1)}},$$

where $K^0 = \mathrm{const} > 0$, ν *is a constant taken from the inequality*

$$\frac{\gamma}{\alpha} > \frac{\nu}{\nu + 1},$$

α *is the Lipschitz constant of the function* $a(\varphi)$, *and* γ *is a constant from inequality (7.25).*

Proof. According to condition (ii) of the formulated theorem, the Green function of the problem of invariant tori for the system of equations (8.1) exists and satisfies the inequality

$$\|G_0(\tau, \varphi)\| \leq N \exp\{-\gamma \,|\, \tau \,|\,\}, \tag{8.5}$$

where γ and N are positive constants independent of τ and φ.

In this case, the following equality is true:

$$G_0(\tau, \varphi) - G_0(\tau, \bar{\varphi})$$
$$= \int_{-\infty}^{\infty} G_0(t_1, \varphi)[P(\varphi_{t_1}(\varphi)) - P(\varphi_{t_1}(\bar{\varphi}))]G_{t_1}(\tau, \bar{\varphi})dt_1. \tag{8.6}$$

According to the condition of the theorem, there exist positive constants α, β, and η independent of φ and such that

$$\|a(\varphi) - a(\bar{\varphi})\| \leq \alpha\|\varphi - \bar{\varphi}\|,$$
$$\|P(\varphi) - P(\bar{\varphi})\| \leq \beta\|\varphi - \bar{\varphi}\|,$$
$$\|c(\varphi) - c(\bar{\varphi})\| \leq \eta\|\varphi - \bar{\varphi}\|.$$

Then, for any $\nu > 0$,

$$\|P(\varphi) - P(\bar{\varphi})\|^{\nu} \leq \beta^{\nu}\|\varphi - \bar{\varphi}\|^{\nu}$$

and, since

$$\|P(\varphi) - P(\bar{\varphi})\| \leq 2p^0,$$

we have

$$\|P(\varphi) - P(\bar{\varphi})\|^{\nu+1} \leq 2p^0\beta^{\nu}\|\varphi - \bar{\varphi}\|^{\nu},$$

or, finally,

$$\|P(\varphi) - P(\bar{\varphi})\| \leq K\|\varphi - \bar{\varphi}\|^{\frac{1}{\nu+1}}, \quad K = (2p^0\beta^{\nu})^{\frac{1}{\nu+1}}.$$

Similarly, we obtain

$$\|c(\varphi) - c(\bar{\varphi})\| \leq K_1\|\varphi - \bar{\varphi}\|^{\frac{\nu}{\nu+1}}, \quad K_1 = (2c^0\eta^{\nu})^{\frac{1}{\nu+1}}.$$

By using the last two inequalities and the inequality

$$\|\varphi_t(\varphi) - \varphi_t(\bar{\varphi})\| \leq \exp\{\alpha \,|\, t \,|\,\}\|\varphi - \bar{\varphi}\|, \quad t \in R^1,$$

we now estimate the difference $u(\varphi) - u(\bar{\varphi})$. Thus, we have

$$\begin{aligned} u &= \|u(\varphi) - u(\bar{\varphi})\| \\ &\leq \int_{-\infty}^{\infty} \|G_0(\tau, \varphi) - G_0(\tau, \bar{\varphi})\| \, \|c(\varphi_\tau(\varphi))\| d\tau \\ &\quad + \int_{-\infty}^{\infty} \|G_0(\tau, \bar{\varphi})\| \, \|c(\varphi_\tau(\varphi)) - c(\varphi_\tau(\bar{\varphi}))\| d\tau \,. \end{aligned} \tag{8.7}$$

We denote the integrals on the right-hand side of this inequality by J_1 and J_2, respectively. Equality (8.6) implies that

$$\begin{aligned} &\|G_0(\tau, \varphi) - G_0(\tau, \bar{\varphi})\| \\ &\quad \leq N^2 K \|\varphi - \bar{\varphi}\|^{\frac{\nu}{\nu+1}} \int_{-\infty}^{\infty} \exp\Big\{-\gamma\,|\,t_1\,| - \gamma\,|\,t_1 - \tau\,| + \frac{\alpha\nu}{\nu+1}\,|\,t_1\,|\Big\} dt_1 \\ &\quad \leq \frac{2}{\gamma - \alpha\nu/(\nu+1)} N^2 K \|\varphi - \bar{\varphi}\|^{\frac{\nu}{\nu+1}}, \end{aligned}$$

whence we easily get

$$\begin{aligned} &\|G_0(\tau, \varphi) \; - \; G_0(\tau, \bar{\varphi})\| \\ &\qquad \leq \; \Big(\frac{2}{\gamma - \alpha\nu/(\nu+1)} N^3 K\Big)^{\frac{1}{2}} \|\varphi - \bar{\varphi}\|^{\frac{\nu}{2(\nu+1)}} \exp\Big\{-\frac{\gamma}{2}\,|\,\tau\,|\Big\}. \end{aligned}$$

Then

$$\begin{aligned} J_1 &\leq \int_{-\infty}^{\infty} c^0 \Big(\frac{2}{\gamma - \alpha\nu/(\nu+1)}\Big)^{\frac{1}{2}} \|\varphi - \bar{\varphi}\|^{\frac{\nu}{2(\nu+1)}} \exp\Big\{-\frac{\gamma}{2}\;|\,\tau\,|\Big\} \, d\tau \\ &\leq S_1 \|\varphi - \bar{\varphi}\|^{\frac{\nu}{2(\nu+1)}}, \end{aligned}$$

where

$$S_1 = \frac{4c^0}{\gamma} \Big(\frac{2}{\gamma - \alpha\nu/(\nu+1)} N^3 (2p^0 \beta^\nu)^{\frac{1}{\nu+1}}\Big)^{\frac{1}{2}} = \text{const} > 0.$$

By using the inequality

$$\|c(\varphi_\tau(\varphi)) - c(\varphi_\tau(\bar\varphi))\| \leq (2K_1c^0)^{\frac{1}{2}}\|\varphi - \bar\varphi\|^{\frac{\nu}{2(\nu+1)}} \exp\Big\{\frac{\alpha\nu}{2(\nu+1)}\,|\,\tau\,|\,\Big\},$$

we conclude that

$$\begin{aligned} J_2 \;&\leq \int\limits_{-\infty}^{\infty} N(2K_1c^0)^{\frac{1}{2}}\|\varphi - \bar\varphi\|^{\frac{\nu}{2(\nu+1)}} \exp\Big\{-\Big(\gamma - \frac{\alpha\nu}{2(\nu+1)}\Big)\,|\,\tau\,|\,\Big\}d\tau \\ &\leq\; S_2\|\varphi - \bar\varphi\|^{\frac{\nu}{2(\nu+1)}}, \end{aligned}$$

where

$$S_2 = \frac{2N}{\gamma - \dfrac{\alpha\nu}{2(\nu+1)}}\Big(2c^0(2c^0\eta^\nu)^{\frac{1}{\nu+1}}\Big)^{\frac{1}{2}} = \text{const} > 0.$$

The required estimate now follows from inequality (8.7):

$$u \leq (S_1 + S_2)\|\varphi - \bar\varphi\|^{\frac{\nu}{2(\nu+1)}}.$$

The following assertion establishes conditions for the differentiability of the invariant torus with respect to φ:

Theorem 8.3. *Assume that the system of equations (8.1) satisfies the conditions of Theorem 8.2 and* $a(\varphi)$, $P(\varphi)$, *and* $c(\varphi)$ *belong to* $C^1(\mathcal{F}_m)$. *Let the series*

$$\sum_{i=1}^{\infty} |\,c_i(\varphi)\,| \qquad \textit{and} \qquad \sum_{i=1}^{\infty}\Big|\frac{\partial c_i(\varphi)}{\partial\varphi_s}\Big|$$

be uniformly convergent for $s = \overline{1,m}$ *and let*

$$\sum_{i=1}^{\infty}\max_{\varphi}\Big|\frac{\partial p_{sj}(\varphi)}{\partial\varphi_i}\Big| \leq p_s = \text{const} < \infty, \qquad \sum_{s=1}^{\infty} p_s = p^* < \infty.$$

If, in addition,

$$\gamma > \alpha = \max_{\varphi}\max_{\|\eta\|=1}\Big\|\frac{\partial a(\varphi)}{\partial\varphi}\eta\Big\|, \quad \eta \in R^m, \tag{8.8}$$

then the invariant torus of the system of equations (8.1) is differentiable with respect to φ.

The proof of this assertion is similar to the proof of Theorem 7.3 on the differentiability of the Green function with respect to a parameter. We divide equality (8.6) by $\bar{\varphi}_r - \varphi_r$, assume that $\bar{\varphi}$ differs from φ only on the rth coordinates, and pass to the limit as $\bar{\varphi}_r \to \varphi_r$.

Let us now show that

$$\lim_{\bar{\varphi}_r \to \varphi_r} G_0(t, \bar{\varphi}) \frac{P(\varphi_{t_1}(\varphi)) - P(\varphi_{t_1}(\bar{\varphi}))}{\bar{\varphi}_r - \varphi_r} G_{t_1}(\tau, \varphi)$$
$$= G_0(t_1, \varphi) \frac{\partial P(\varphi_{t_1}(\varphi))}{\partial \varphi_r} G_{t_1}(\tau, \varphi) \tag{8.9}$$

in the weak sense.

The element standing in the pth column and lth row of the matrix under the limit sign on the left-hand side of this equality has the form

$$\sum_{k=1}^{\infty} \omega_{lpk} = \sum_{k=1}^{\infty} g_{0lk}(t_1, \bar{\varphi}) \sum_{i=1}^{\infty} \frac{p_{ki}(\varphi_{t_1}(\bar{\varphi})) - p_{ki}(\varphi_{t_1}(\bar{\varphi}))}{\bar{\varphi}_r - \varphi_r} g_{ip}(t_1, \tau, \varphi), \tag{8.10}$$

where

$$G_0(t_1, \varphi) = [g_{0ik}(t_1, \varphi)]_{i,k=1}^{\infty} \quad \text{and} \quad G_{t_1}(\tau, \varphi) = [g_{ik}(t_1, \tau, \varphi)]_{i,k=1}^{\infty}.$$

The following inequalities are true:

$$|\,\omega_{lpk}\,| \;\le\; \sum_{i=1}^{\infty} N \exp\{-\gamma\,|\,t_1\,|\} \sum_{j=1}^{m} \Big| \frac{\partial p_{ki}(\varphi_{t_1}(\varphi))}{\partial \varphi_{t_1 j}(\varphi)} \Big| \exp\{\alpha\,|\,t_1\,|\}$$
$$\times LN \exp\{-\gamma\,|\,t_1 - \tau\,|\}$$
$$\le\; LN^2 \sum_{j=1}^{m} \sum_{i=1}^{\infty} \Big| \frac{\partial p_{ki}(\varphi_{t_1}(\varphi))}{\partial \varphi_{t_{1j}}(\varphi)} \Big|,$$

$$\sum_{k=1}^{\infty} |\,\omega_{lpk}\,| \;\le\; LN^2 \sum_{j=1}^{m} \sum_{k=1}^{\infty} \sum_{i=1}^{\infty} \Big| \frac{\partial p_{ki}(\varphi_{t_1})}{\partial \varphi_{t_1 j}} \Big|$$
$$\le\; LN^2 \sum_{j=1}^{m} \sum_{k=1}^{\infty} p_k \;\le\; LN^2 \sum_{j=1}^{m} p^* = LN^2 m p^*.$$

They enable us to pass to the limit as $\bar{\varphi}_r - \varphi_r \to 0$ in the repeated series (8.10) term by term. In this case, $L = \text{const} < \infty$ is determined from the inequality [MSK2]

$$\Big\| \frac{\partial \varphi_t(\varphi)}{\partial \varphi_r} \Big\| \le L \exp\{\alpha\,|\,t\,|\}, \quad t \in R^1.$$

Hence, by passing to the limit, we arrive at relation (8.9).

Then

$$\lim_{\bar{\varphi}_r \to \varphi_r} \frac{G_0(\tau, \bar{\varphi}) - G_0(\tau, \varphi)}{\bar{\varphi}_r - \varphi_r} = \int_{-\infty}^{\infty} G_0(s, \varphi) \frac{\partial P(\varphi_s(\varphi))}{\partial \varphi_r} G_s(\tau, \varphi)\, ds$$

whenever the integral on the right-hand side converges uniformly. This condition is obvious.

The expression

$$\lim_{\bar{\varphi}_r \to \varphi_r} \frac{G_0(\tau, \bar{\varphi}) - G_0(\tau, \varphi)}{\bar{\varphi}_r - \varphi_r}$$

determines the function $\dfrac{\partial G_0(\tau, \varphi)}{\partial \varphi_r}$. For this function, we obtain

$$\begin{aligned} \left\| \frac{\partial G_0(\tau, \varphi)}{\partial \varphi_r} \right\| &\leq \int_{-\infty}^{\infty} N^2 \exp\{-(\gamma - \alpha)\,|\,\tau\,| - \gamma\,|\,s - \tau\,|\,\} ds \\ &\leq K \exp\{-(\gamma - \alpha)\,|\,\tau\,|\,\}, \end{aligned} \tag{8.11}$$

where $K = N^2 \rho^*$ and $\tau \in R^1$.

Let us show that

$$\begin{aligned} &\frac{\partial}{\partial \varphi_r} \int_{-\infty}^{\infty} G_0(\tau, \varphi) c(\varphi_\tau(\varphi))\, d\tau \\ &\quad = \int_{-\infty}^{\infty} \Big(\frac{\partial G_0(\tau, \varphi)}{\partial \varphi_r} c(\varphi_\tau(\varphi)) + G_0(\tau, \varphi) \frac{\partial c(\varphi_\tau(\varphi))}{\partial \varphi_r} \Big) d\tau. \end{aligned}$$

For this purpose, it suffices to prove that the integral

$$\int_{-\infty}^{\infty} \frac{\partial}{\partial \varphi_r} \Big(G_0(\tau, \varphi) c(\varphi_\tau(\varphi)) \Big) d\tau$$

is uniformly convergent and

$$\frac{\partial}{\partial \varphi_r} \Big(G_0(\tau, \varphi) c(\varphi_\tau(\varphi)) \Big) = \frac{\partial G_0(\tau, \varphi)}{\partial \varphi_r} c(\varphi_\tau(\varphi)) + G_0(\tau, \varphi) \frac{\partial c(\varphi_\tau(\varphi))}{\partial \varphi_r}.$$

We represent the lth coordinate of the vector $G_0(\tau,\varphi)c(\varphi_\tau(\varphi))$ in the form of a series

$$\sum_{i=1}^{\infty} g_{oli}(\tau,\varphi)c_i(\varphi_\tau(\varphi)).$$

This series can be differentiated term by term because

$$\sum_{i=1}^{\infty}\Big(\frac{\partial g_{oli}(\tau,\varphi)}{\partial\varphi_r}c_i(\varphi_\tau(\varphi)) + g_{oli}(\tau,\varphi)\frac{\partial c_i(\varphi_\tau(\varphi))}{\partial\varphi_r}\Big)$$

is uniformly convergent for all $l = 1,2,3,\dots$. This follows from the boundedness of the derivative $\dfrac{\partial P(\varphi)}{\partial\varphi_i}$ for $i = \overline{1,m}$ and the fact that the series

$$\sum_{i=1}^{\infty}|\,c_i(\varphi)\,| \quad\text{and}\quad \sum_{i=1}^{\infty}\Big|\frac{\partial c_i(\varphi)}{\partial\varphi_s}\Big|, \qquad s=\overline{1,m},$$

are uniformly convergent. Finally, by using inequalities (8.5) and (8.11) and the conditions of the theorem, we obtain

$$\Big\|\int_{-\infty}^{\infty}\frac{\partial}{\partial\varphi_i}\Big(G_0(\tau,\varphi)c(\varphi_\tau(\varphi))\Big)d\tau\Big\| \le \frac{K_1}{\gamma-\alpha},$$

where K_1 is a constant independent of τ and φ.

2.9. C^l-Differentiability of the Invariant Torus

In this section, we present conditions under which the invariant torus of the system of equations (8.1) has continuous derivatives with respect to φ up to the lth order inclusive.

Parallel with (8.1), we consider the following system of equations:

$$\frac{dx_{sn}}{dt} = \sum_{i=1}^{n} p_{si}(\varphi_t(\varphi))x_{in} + c_s(\varphi_t(\varphi)), \quad s=\overline{1,n}. \tag{9.1}$$

Let $\overset{(n)}{\Omega}{}^t_\tau(\varphi)$ be the matrizant of the corresponding homogeneous system

$$\frac{dx_{sn}}{dt} = \sum_{i=1}^{n} p_{si}(\varphi_t(\varphi))x_{in}, \quad s=\overline{1,n}\,. \tag{9.2}$$

Assume that system (9.1) possesses the invariant torus $\mathcal{F}^{(n)}$:

$$x^{(n)} = u^{(n)}(\varphi), \quad \varphi \in \mathcal{F}_m, \quad \text{where} \quad x^{(n)} = (x_{1n}, x_{2n}, \ldots, x_{nn}).$$

We now prove the following assertion:

Theorem 9.1. *Assume that the system of equations (8.1) satisfies the following conditions:*

(i) $a(\varphi) \in C^0_{\mathrm{Lip}}(\mathcal{F}_m)$ *and* $P(\varphi),\ c(\varphi) \in C^0(\mathcal{F}_m)$;

(ii) $\sum_{j=n+1}^{\infty} \max_{\varphi} |p_{sj}(\varphi)| \leq K\varepsilon(n)$, $s = 1, 2, \ldots$, *where* $K = \text{const} < \infty$ *and* $\varepsilon(n) \to 0$ *as* $n \to \infty$;

(iii) $\| \overset{(n)}{\Omega}{}^0_\tau(\varphi) \| \leq N \exp\{\gamma\tau\}$, $\tau \leq 0$, *where* N *and* γ *are positive constants independent of* n *and* φ.

Then the system of equations (8.1) possesses the invariant torus $\mathcal{F}\colon x = u(\varphi)$, *where*

$$\lim_{n\to\infty} u_j^{(n)}(\varphi) = u_j(\varphi), \quad j = 1, 2, 3, \ldots,$$

uniformly in $\varphi \in \mathcal{F}_m$.

Proof. We now show that, in the domain H, the right-hand side of the system of equations

$$\frac{dx}{dt} = P(\varphi_t(\varphi))x + c(\varphi_t(\varphi))$$

satisfies the strengthened Cauchy–Lipschitz condition with respect to x.

Indeed, let $x = (x_1, x_2, \ldots, x_n, x_{n+1}, \ldots)$ and $\bar{x} = (x_1, x_2, \ldots, x_n, \bar{x}_{n+1}, \bar{x}_{x+2}, \ldots)$ be two arbitrary points of the space $\mathfrak{M}$ whose coordinates coincide up to the number n inclusive. Denote $\Delta_n x = \sup\{ \,|x_{n+1} - \bar{x}_{n+2}\,|, \ldots\}$. For any $s = 1, 2, 3, \ldots$, the following inequalities are true:

$$\Big|\sum_{i=1}^{\infty} p_{si}(\varphi_t(\varphi))x_i - \sum_{i=1}^{n} p_{si}(\varphi_t(\varphi))x_i - \sum_{i=n+1}^{\infty} p_{si}(\varphi_t(\varphi))\bar{x}_i\Big|$$

$$\leq \sum_{i=n+1}^{\infty} |\,p_{si}(\varphi_t(\varphi))\,|\,|\,x_i - \bar{x}_i\,|$$

$$\leq \Delta_n x \sum_{i=n+1}^{\infty} \max |\,p_{si}(\varphi_t(\varphi))\,| \leq K\Delta_n x \varepsilon(n).$$

In this case, the following limit exists coordinatewise:

$$\lim_{n\to\infty} \overset{(n)}{\Omega}{}^t_\tau(\varphi) = \Omega^t_\tau(\varphi), \tag{9.3}$$

where $\Omega^t_\tau(\varphi)$ is the matrizant of the system of equations

$$\frac{dx}{dt} = P(\varphi_t(\varphi))x\,.$$

To show that

$$\|\Omega^0_\tau(\varphi)\| \leq N\exp\{\gamma\tau\}, \quad \tau \leq 0,\ N,\ \gamma = \text{const} > 0,$$

we assume the contrary. Then there exist $\tau \leq 0$ and $\varphi \in \mathcal{F}_m$ such that

$$\|\Omega^0_\tau(\varphi)\| = \sup_i \sum_{j=1}^{\infty} |\,\omega_{ij}(\tau, 0, \varphi)\,| > N\exp\{\gamma\tau\}.$$

Hence,

$$\sum_{j=1}^{\infty} |\,\omega_{i^*j}(\tau, 0, \varphi)\,| > N\exp\{\gamma\tau\}$$

for some i^*. But then there exists a number N^0 such that

$$\sum_{j=1}^{N^0} |\,\omega_{i^*j}(\tau, 0, \varphi)\,| > N\exp\{\gamma\tau\},$$

which is impossible because

$$\omega_{i^*j}(\tau, 0, \varphi) = \lim_{n\to\infty} \omega^{(n)}_{i^*j}(\tau, 0, \varphi)$$

and

$$\sum_{j=1}^{N^0} |\,\omega_{i^*j}(\tau, 0, \varphi)\,| = \lim_{n\to\infty} \sum_{j=1}^{N^0} \left|\omega^{(n)}_{i^*j}(\tau, 0, \varphi)\right| \leq N\exp\{\gamma\tau\},$$

where $\omega_{ij}(\tau, t, \varphi)$ and $\omega^{(n)}_{ij}(\tau, t, \varphi)$ are elements of the matrizants $\Omega^t_\tau(\varphi)$ and $\overset{(n)}{\Omega}{}^t_\tau(\varphi)$, respectively: $\omega^{(n)}_{ij} \equiv 0$ for $i > n$ or $j > n$. By $G_0(\tau, \varphi)$ and $\overset{(n)}{G}_0(\tau, \varphi)$ we denote the Green functions of the problem of invariant tori for systems (8.1) and (9.1), respectively. Under the conditions of the theorem, they have the form

$$G_0(\tau, \varphi) = \begin{cases} \Omega^0_\tau(\varphi), & \tau \leq 0, \\ 0, & \tau > 0, \end{cases}$$

and

$$\overset{(n)}{G}_0(\tau,\varphi) = \begin{cases} \overset{(n)}{\Omega}{}^0_\tau(\varphi), & \tau \le 0, \\ 0, & \tau > 0. \end{cases}$$

We denote $\overset{(n)}{c}(\varphi) = \{c_1(\varphi), c_2(\varphi), \dots, c_n(\varphi)\}$. Let

$$\begin{aligned} \overset{(n)}{c}(\varphi) &= \{c_1(\varphi_\tau(\varphi)), c_2(\varphi_\tau(\varphi)), \dots, c_n(\varphi_\tau(\varphi))\}, \\ \overset{(n+1)}{c}(\varphi) &= \{c_1(\varphi_\tau(\varphi)), c_2(\varphi_\tau(\varphi)), \dots, c_{n+1}(\varphi_\tau(\varphi))\} \\ &= \{\overset{(n)}{c}(\varphi_\tau(\varphi)), c_{n+1}(\varphi_\tau(\varphi))\}. \end{aligned}$$

Theorem 2.1 implies the equality

$$\lim_{n\to\infty} \overset{(n)}{\Omega}{}^t_\tau(\varphi)\overset{(n)}{c}(\varphi_\tau(\varphi)) = \Omega^t_\tau(\varphi)c(\varphi_\tau(\varphi)), \tag{9.4}$$

where the limit transition is performed coordinatewise.

According to Theorem 8.1 and Remark 7.2, the invariant tori of the system of equations (8.1) and truncated system (9.1) can be represented in the form

$$\mathcal{F} : u(\varphi) = \int_{-\infty}^{0} \Omega^0_\tau(\varphi)c(\varphi_\tau(\varphi))d\tau, \quad \varphi \in \mathcal{F}_m,$$

and

$$\mathcal{F}^{(n)} : \overset{(n)}{u}(\varphi) = \int_{-\infty}^{0} \overset{(n)}{\Omega}{}^0_\tau \overset{(n)}{c}(\varphi_\tau(\varphi))d\tau, \quad \varphi \in \mathcal{F}_m,$$

respectively. In view of (9.4), to prove the theorem, it remains to show that

$$\lim_{n\to\infty} \int_{-\infty}^{0} \overset{(n)}{\Omega}{}^0_\tau \overset{(n)}{c}(\varphi_\tau(\varphi))d\tau = \int_{-\infty}^{0} \lim_{n\to\infty} \overset{(n)}{\Omega}{}^0_\tau \overset{(n)}{c}(\varphi_\tau(\varphi))d\tau. \tag{9.5}$$

To simplify our calculations, we denote the vector functions under the limit signs on the right-hand and left-hand sides of equality (9.4) by

$$z(\tau,\varphi) = \{z_1(\tau,\varphi), z_2(\tau,\varphi), \dots\}$$

and

$$\overset{(n)}{z}(\tau,\varphi) = \{\overset{(n)}{z}_1(\tau,\varphi),\ldots,\overset{(n)}{z}_n(\tau,\varphi)\},$$

respectively.

In this case, the expression

$$\overset{(n)}{z}(\tau,\varphi) = \{\overset{(n)}{z}_1(\tau,\varphi),\ldots,\overset{(n)}{z}_n(\tau,\varphi),\overset{(n)}{z}_{n+1}(\tau,\varphi),\ldots\}$$

is regarded as the vector $\{\overset{(n)}{z}_1(\tau,\varphi),\ldots,\overset{(n)}{z}_n(\tau,\varphi),0,0,0,\ldots\}$.

Let A be an arbitrarily small negative number. According to the theorem on the limit transition under the sign of Lebesgue integral, we have

$$\lim_{n\to\infty}\int_A^0 \overset{(n)}{z}_i(\tau,\varphi)\,d\tau = \int_A^0 \lim_{n\to\infty}\overset{(n)}{z}_i(\tau,\varphi)\,d\tau = \int_A^0 z_i(\tau,\varphi)\,d\tau. \tag{9.6}$$

Denote

$$F(n,A) = \int_A^0 \overset{(n)}{z}_i(\tau,\varphi)\,d\tau.$$

Since

$$\left|\int_{-\infty}^0 \overset{(n)}{z}_i(\tau,\varphi)\,d\tau - \int_A^0 \overset{(n)}{z}_i(\tau,\varphi)\,d\tau\right|$$

$$\le \left|\int_{-\infty}^A \overset{(n)}{z}_i(\tau,\varphi)\,d\tau\right| \le \int_{-\infty}^A cN\exp\{\gamma\tau\}\,d\tau = \frac{cN}{\gamma}\exp\{\gamma A\},$$

in view of equality (9.6) we conclude that

$$\lim_{n\to\infty} F(n,A) = \lim_{n\to\infty}\int_A^0 \overset{(n)}{z}_i(\tau,\varphi)\,d\tau = \int_A^0 z_i(\tau,\varphi)\,d\tau$$

and

$$\lim_{A\to-\infty} F(n,A) = \lim_{A\to-\infty}\int_A^0 \overset{(n)}{z}_i(\tau,\varphi)\,d\tau = \int_{-\infty}^0 \overset{(n)}{z}_i(\tau,\varphi)\,d\tau$$

uniformly in n.

Further, according to the theorem on the permutation of limit transitions, we obtain

$$\lim_{n\to\infty}\lim_{A\to-\infty} F(n,A) = \lim_{A\to-\infty}\lim_{n\to\infty} F(n,A),$$

which yields relation (9.5).

We now consider a subset $C^l_{\mathrm{Lip}}(\mathcal{F}_m)$ of the set $C^l(\mathcal{F}_m)$ whose elements satisfy the Lipschitz condition in φ together with their derivatives with respect to φ up to the lth order inclusive.

Theorem 9.2. *Assume that the system of equations (8.1) satisfies the conditions of Theorem 9.1, the functions $a(\varphi)$, $P(\varphi)$, and $c(\varphi)$ belong to the set $C^1_{\mathrm{Lip}}(\mathcal{F}_m)$, and inequality (8.8) is true.*

Then the invariant torus $\mathcal{F}$ of the system of equations (8.1) belongs to the space $C^1(\mathcal{F}_m)$.

Proof. All conditions of Theorem 9.1 are satisfied. Hence, the invariant torus $\mathcal{F}: x = u(\varphi)$, $\varphi \in \mathcal{F}_m$, of the system of equations (8.1) can be represented as the coordinatewise limit of the sequence of invariant tori $\mathcal{F}^n$: $\overset{(n)}{x}\overset{(n)}{u}(\varphi)$, $\varphi \in \mathcal{F}_m$, for systems of equations of the form (9.1):

$$\lim_{n\to\infty} \overset{(n)}{u}(\varphi) = u(\varphi). \tag{9.7}$$

Consider a sequence of vector functions

$$\frac{\partial \overset{(1)}{u}(\varphi)}{\partial \varphi_i}, \frac{\partial \overset{(2)}{u}(\varphi)}{\partial \varphi_i}, \ldots, \frac{\partial \overset{(n)}{u}(\varphi)}{\partial \varphi_i}, \ldots, \quad i \in \overline{1,m}, \tag{9.8}$$

and the sequence of their sth coordinates

$$\frac{\partial \overset{(1)}{u}_s(\varphi)}{\partial \varphi_i}, \frac{\partial \overset{(2)}{u}_s(\varphi)}{\partial \varphi_i}, \ldots, \frac{\partial \overset{(n)}{u}_s(\varphi)}{\partial \varphi_i}, \ldots, \quad s = 1, 2, \ldots . \tag{9.9}$$

If sequence (9.8) is uniformly bounded and equicontinuous in norm, then the same is true for sequence (9.9) for any $s = 1, 2, \ldots$. By virtue of equality (9.7), this yields the existence of $\dfrac{\partial u_s(\varphi)}{\partial \varphi_i}$ for any natural s and, hence, the existence of $\dfrac{\partial u(\varphi)}{\partial \varphi_i}$.

Thus, to prove the theorem, it is necessary to show that sequence (9.8) is uniformly bounded and equicontinuous.

In [MSK2], it is shown that, under the conditions of Theorem 9.2, the partial derivative of the matrizant $\overset{(n)}{\Omega}{}^0_\tau(\varphi)$ of the truncated system of equations (9.2) with respect to φ_i, $i = \overline{1, m}$, is continuous and satisfies the inequality

$$\left\| \frac{\partial \overset{(n)}{\Omega}{}^0_\tau(\varphi)}{\partial \varphi_i} \right\| \leq K_1 \exp\{-(\gamma - \alpha)\,|\,\tau\,|\,\}, \quad \tau \in R^1.$$

Moreover, the solution $\varphi_t(\varphi)$ of the first equation in system (8.1) satisfies the inequality

$$\left\| \frac{\partial \varphi_t(\varphi)}{\partial \varphi_i} \right\| \leq K_2 \exp\{\alpha\,|\,t\,|\,\},$$

where K_1 and K_2 are positive constants independent of n, τ, and φ.

It is clear that if $\|c(\varphi)\| \leq c$ and $\left\| \frac{\partial c(\varphi)}{\partial \varphi_i} \right\| \leq R$, then, for any natural n,

$$\|\overset{(n)}{c}\| \leq c \qquad \text{and} \qquad \left\| \frac{\partial \overset{(n)}{c}(\varphi)}{\partial \varphi_i} \right\| \leq R.$$

Similarly, for any n, we have

$$\left\| \overset{(n)}{\Omega}{}^t_\tau(\varphi) \right\| \leq N \exp\{-\gamma(t - \tau)\,\} \quad \text{for} \quad \tau \leq t.$$

The uniform boundedness of sequence (9.8) follows from the inequality

$$\left\| \frac{\partial \overset{(n)}{u}(\varphi)}{\partial \varphi_i} \right\|$$

$$= \left\| \int_{-\infty}^{0} \Big\{ \frac{\partial \overset{(n)}{\Omega}{}^0_\tau(\varphi)}{\partial \varphi_i} \overset{(n)}{c}(\varphi_\tau(\varphi)) + \overset{(n)}{\Omega}{}^0_\tau(\varphi) \frac{\partial \overset{(n)}{c}(\varphi_\tau(\varphi))}{\partial \varphi_i} \Big\} d\tau \right\|$$

$$= \left\| \int_{-\infty}^{0} \Big\{ \frac{\partial \overset{(n)}{\Omega}{}^0_\tau(\varphi)}{\partial \varphi_i} \overset{(n)}{c}(\varphi_\tau(\varphi)) + \overset{(n)}{\Omega}{}^0_\tau(\varphi) \sum_{j=1}^{m} \frac{\partial \overset{(n)}{c}(\varphi_\tau(\varphi))}{\partial \varphi_{\tau j}(\varphi)} \frac{\partial \varphi_{\tau j}(\varphi)}{\partial \varphi_i} \Big\} d\tau \right\|$$

$$\leq \int_{-\infty}^{0} K_1 c \exp\{(\gamma - \alpha)\tau\} d\tau + \int_{-\infty}^{0} N m R K_2 \exp\{(\gamma - \alpha)\tau\} d\tau$$

$$= \frac{K_1 c + N K_2 R m}{\gamma - \alpha}.$$

Prior to proving the equicontinuity of this sequence, we present the following auxiliary statements:

Lemma 9.1. *The inequality*

$$\|X\| = \left\| \frac{\partial \varphi_{t_1}(\varphi)}{\partial \varphi_i} - \frac{\partial \varphi_{t_1}(\bar{\varphi})}{\partial \varphi_i} \right\| \leq \Psi \|\varphi - \bar{\varphi}\|^{\frac{\nu}{\nu+1}} \exp\Big\{\Big(\alpha + \frac{\nu\alpha}{\nu+1}\Big)\,|\,t_1\,|\Big\},$$

where $\Psi = \text{const} > 0$ *is independent of* t, φ, *and* $\bar{\varphi}$, *and* ν *is an arbitrary positive constant, is true under the conditions of Theorem 9.2.*

Proof. It is easy to see that the difference X is a solution of the equation

$$\frac{dX}{dt} = \Big(\frac{\partial a(\varphi_{t_1}(\varphi))}{\partial \varphi_{t_1}(\varphi)}\Big) X + f(t_1), \tag{9.10}$$

where the matrix $\left(\dfrac{\partial a}{\partial \varphi}\right)$ and the vector function f have the form

$$\Big(\frac{\partial a}{\partial \varphi}\Big) = \begin{pmatrix} \dfrac{\partial a_1(\varphi_{t_1}(\varphi))}{\partial \varphi_{t_1 1}(\varphi)} & \dfrac{\partial a_1(\varphi_{t_1}(\varphi))}{\partial \varphi_{t_1 2}(\varphi)} & \cdots & \dfrac{\partial a_1(\varphi_{t_1}(\varphi))}{\partial \varphi_{t_1 m}(\varphi)} \\ \cdots & \cdots & \cdots & \cdots \\ \dfrac{\partial a_m(\varphi_{t_1}(\varphi))}{\partial \varphi_{t_1 1}(\varphi)} & \dfrac{\partial a_m(\varphi_{t_1}(\varphi))}{\partial \varphi_{t_1 2}(\varphi)} & \cdots & \dfrac{\partial a_m(\varphi_{t_1}(\varphi))}{\partial \varphi_{t_1 m}(\varphi)} \end{pmatrix},$$

$$f(t_1) = \sum_{j=1}^{m} \Big(\frac{\partial a(\varphi_{t_1}(\varphi))}{\partial \varphi_{t_1 j}(\varphi)} - \frac{\partial a(\varphi_{t_1}(\bar{\varphi}))}{\partial \varphi_{t_1 j}(\bar{\varphi})} \Big) \frac{\partial \varphi_{t_1 j}(\bar{\varphi})}{\partial \varphi_i}.$$

We denote the matrizant of the system of equations (9.10) by $U_\tau^{t_1}(\varphi)$. By using condition (8.8) in Theorem 9.2, we obtain

$$\|U_\tau^{t_1}(\varphi)\| \leq M \exp\{\alpha\,|\,t_1 - \tau\,|\,\}, \quad \tau,\ t_1 \in R^1, \quad M = \text{const} < \infty.$$

In view of the fact that $X(0) = 0$, we conclude that

$$X = U_0^{t_1} 0 + U_0^{t_1}(\varphi) \int_0^{t_1} U_s^0(\varphi) \sum_{j=1}^{m} \Big(\frac{\partial a(\varphi_s(\varphi))}{\partial \varphi_{sj}(\varphi)} - \frac{\partial a(\varphi_s(\bar{\varphi}))}{\partial \varphi_{sj}(\bar{\varphi})} \Big) \frac{\partial \varphi_{sj}(\bar{\varphi}))}{\partial \varphi_i}\, ds\,.$$

Since $\dfrac{\partial a(\varphi)}{\partial \varphi_i} \in C^0_{\text{Lip}}(\mathcal{F}_m)$, $i = \overline{1,m}$, one can find constants A^0 and η independent of φ and such that $\left\|\dfrac{\partial a(\varphi)}{\partial \varphi_i}\right\| \leq A^0$ and

$$\Big\|\frac{\partial a(\varphi)}{\partial \varphi_i} - \frac{\partial a(\bar{\varphi})}{\partial \varphi_i}\Big\| \leq \eta \|\varphi - \bar{\varphi}\|, \quad i = \overline{1,m}.$$

This yields

$$\Big\|\frac{\partial a(\varphi)}{\partial \varphi_i} - \frac{\partial a(\bar\varphi)}{\partial \varphi_i}\Big\| \le (2A^0\eta^\nu)^{\frac{1}{\nu+1}}\|\varphi - \bar\varphi\|^{\frac{\nu}{\nu+1}}, \quad \nu = \text{const} > 0,$$

and, hence,

$$\|X\| \le \int_0^{t_1} M\exp\{\alpha\,|\,t_1 - s\,|\,\} m(2A^0\eta^\nu)^{\frac{1}{\nu+1}} \exp\Big\{\frac{\alpha\nu}{\nu+1}\,|\,s\,|\Big\}$$
$$\times\|\varphi - \bar\varphi\|^{\frac{\nu}{\nu+1}} K_2 \exp\{\alpha\,|\,s\,|\,\}\,ds.$$

For $\Psi = \dfrac{mMK_2(2A^0\eta^\nu)^{\frac{1}{\nu+1}}(\nu+1)}{\alpha\nu}$, this implies the statement of Lemma 9.1.

Lemma 9.2. *The inequality*

$$\|G\| = \Big\|\frac{\partial \overset{(n)}{\Omega}{}^0_\tau(\varphi)}{\partial\varphi_i} - \frac{\partial \overset{(n)}{\Omega}{}^0_\tau(\bar\varphi)}{\partial\varphi_i}\Big\|$$
$$\le \Psi_0\|\varphi - \bar\varphi\|^{\frac{\nu}{\nu+1}} \exp\Big\{\Big(\gamma - \alpha - \frac{\alpha\nu}{\nu+1}\Big)\tau\Big\},$$

where Ψ_0 is a constant independent of n, τ, and φ, and $\tau \le 0$, holds under the conditions of Theorem 9.2.

Proof. By using the equality

$$\frac{d}{dt}\Big(\overset{(n)}{\Omega}{}^t_\tau(\varphi) - \overset{(n)}{\Omega}{}^t_\tau(\bar\varphi)\Big) = \overset{(n)}{P}(\varphi_t(\varphi))\Big(\overset{(n)}{\Omega}{}^t_\tau(\varphi) - \overset{(n)}{\Omega}{}^t_\tau(\bar\varphi)\Big)$$
$$+\Big(\overset{(n)}{P}(\varphi_t(\varphi)) - \overset{(n)}{P}(\varphi_t(\bar\varphi))\Big)\overset{(n)}{\Omega}{}^t_\tau(\bar\varphi)$$

and the properties of the matrizant established in Section 1.4, we obtain

$$\overset{(n)}{\Omega}{}^0_\tau(\varphi) - \overset{(n)}{\Omega}{}^0_\tau(\bar\varphi) = \int_\tau^0 \overset{(n)}{\Omega}{}^0_{t_1}(\varphi)\Big[\overset{(n)}{P}(\varphi_{t_1}(\varphi)) - \overset{(n)}{P}(\varphi_{t_1}(\bar\varphi))\Big]\overset{(n)}{\Omega}{}^{t_1}_\tau(\bar\varphi)\,dt_1. \quad (9.11)$$

We set

$$\|P(\varphi) - P(\bar\varphi)\| \le \beta\|\varphi - \bar\varphi\|, \quad \|P(\varphi)\| \le p^0,$$

where β and p^0 are positive constants independent of φ.

It is clear that

$$\|\overset{(n)}{P}(\varphi_{t_1}(\varphi)) - \overset{(n)}{P}(\varphi_{t_1}(\bar\varphi))\|$$

$$\leq \left(2p^0\beta\right)^{\frac{1}{\nu+1}} \|\varphi - \bar\varphi\|^{\frac{\nu}{\nu+1}} \exp\Big\{\frac{\alpha\nu}{\nu+1}\,|\,t_1\,|\Big\}, \qquad \nu = \text{const} > 0\,.$$

This yields

$$\|\overset{(n)}{\Omega}{}^0_\tau(\varphi) - \overset{(n)}{\Omega}{}^0_\tau(\bar\varphi)\| \leq \Psi_1 \|\varphi - \bar\varphi\|^{\frac{\nu}{\nu+1}} \exp\Big\{(\gamma - \frac{\alpha\nu}{\nu+1})\tau\Big\}, \qquad \tau \leq 0,$$

where

$$\Psi_1 = N^2 (2p^0\beta)^{\frac{1}{\nu+1}} \frac{\nu+1}{\alpha\nu} = \text{const.}$$

We assume that φ and $\bar\varphi$ differ only by their ith coordinates, divide both sides of equality (9.11) by $\bar\varphi_i - \varphi_i$, and pass to the limit as $\bar\varphi_i \to \varphi_i$. This gives

$$\frac{\partial \overset{(n)}{\Omega}{}^0_\tau(\varphi)}{\partial \varphi_i} = \lim_{\bar\varphi_i \to \varphi_i} \frac{\overset{(n)}{\Omega}{}^0_\tau(\bar\varphi) - \overset{(n)}{\Omega}{}^0_\tau(\varphi)}{\bar\varphi_i - \varphi_i}$$

$$= \int_\tau^0 \overset{(n)}{\Omega}{}^0_{t_1}(\varphi) \sum_{j=1}^m \frac{\partial \overset{(n)}{P}(\varphi_{t_1}(\varphi))}{\partial \varphi_{t_1 j}(\varphi)} \frac{\partial \varphi_{t_1 j}(\varphi)}{\partial \varphi_i} \overset{(n)}{\Omega}{}^{t_1}_\tau(\varphi)\, dt_1.$$

After obvious transformations, we arrive at the following representation:

$$G = \int_\tau^0 \Big(\overset{(n)}{\Omega}{}^0_{t_1}(\varphi) - \overset{(n)}{\Omega}{}^0_{t_1}(\bar\varphi)\Big) \sum_{j=1}^m \frac{\partial \overset{(n)}{P}(\varphi_{t_1}(\varphi))}{\partial \varphi_{t_1 j}(\varphi)} \frac{\partial \varphi_{t_1 j}(\varphi)}{\partial \varphi_i} \overset{(n)}{\Omega}{}^{t_1}_\tau(\varphi)\, dt_1$$

$$+ \int_\tau^0 \overset{(n)}{\Omega}{}^0_{t_1}(\bar\varphi) \sum_{j=1}^m \left\{ \frac{\partial \overset{(n)}{P}(\varphi_{t_1}(\varphi))}{\partial \varphi_{t_1 j}(\varphi)} \frac{\partial \varphi_{t_1 j}(\varphi)}{\partial \varphi_i} \overset{(n)}{\Omega}{}^{t_1}_\tau(\varphi) \right.$$

$$\left. - \frac{\partial \overset{(n)}{P}(\varphi_{t_1}(\bar\varphi))}{\partial \varphi_{t_1 j}(\bar\varphi)} \frac{\partial \varphi_{t_1 j}(\bar\varphi)}{\partial \varphi_i} \overset{(n)}{\Omega}{}^{t_1}_\tau(\bar\varphi) \right\} dt_1.$$

We denote the integrals in this equality by J_1 and J_2, respectively, and assume that

$$\Big\|\frac{\partial P(\varphi)}{\partial \varphi_i}\Big\| \leq P_0 = \text{const} < \infty.$$

For J_1, we obtain

$$\|J_1\| \le \int_\tau^0 \Psi_1 \|\varphi - \bar{\varphi}\|^{\frac{\nu}{\nu+1}} m P_0 K_2 N$$

$$\times \exp\Big\{(\gamma - \frac{\alpha\nu}{\nu+1})t_1 - \alpha t_1 - \gamma(t_1 - \tau)\Big\} dt_1$$

$$\le \Psi_2 \|\varphi - \bar{\varphi}\|^{\frac{\nu}{\nu+1}} \exp\Big\{(\gamma - \alpha - \frac{\alpha\nu}{\nu+1})\tau\Big\},$$

$$\Psi_2 = \Psi_1 m P_0 K_2 N \frac{1}{\alpha + \frac{\alpha\nu}{\nu+1}}, \quad \tau \le 0.$$

Similarly,

$$\|\overset{(n)}{\Omega}{}^t_\tau(\varphi) - \overset{(n)}{\Omega}{}^t_\tau(\bar{\varphi})\|$$

$$\le \int_\tau^t \|\overset{(n)}{\Omega}{}^t_{t_1}(\varphi)\| \|\overset{(n)}{P}(\varphi_{t_1}(\varphi)) - \overset{(n)}{P}(\varphi_{t_1}(\bar{\varphi}))\| \|\overset{(n)}{\Omega}{}^{t_1}_\tau(\bar{\varphi})\| dt_1$$

$$\le N^2 (2p^0\beta)^{\frac{1}{\nu+1}} \|\varphi - \bar{\varphi}\|^{\frac{\nu}{\nu+1}}$$

$$\times \int_\tau^t \exp\Big\{-\gamma(t - t_1) - \gamma(t_1 - \tau) - \frac{\alpha\nu}{\nu+1} t_1\Big\} dt_1$$

$$\le N^2 (2p^0\beta)^{\frac{1}{\nu+1}} \frac{\nu+1}{\alpha\nu} \|\varphi - \bar{\varphi}\|^{\frac{\nu}{\nu+1}} \exp\Big\{-\gamma t + \Big(\gamma - \frac{\alpha\nu}{\nu+1}\Big)\tau\Big\}$$

$$= \Psi_1 \|\varphi - \bar{\varphi}\|^{\frac{\nu}{\nu+1}} \exp\Big\{-\gamma t + \Big(\gamma - \frac{\alpha\nu}{\nu+1}\Big)\tau\Big\}$$

for $\tau \le t \le 0$. Let us now estimate the difference $\mathcal{L}$. For this purpose, we represent it in the form

$$\mathcal{L} = \frac{\partial \overset{(n)}{P}(\varphi_{t_1}(\varphi))}{\partial \varphi_{t_1 j}(\varphi)} \frac{\partial \varphi_{t_1 j}(\varphi)}{\partial \varphi_i} \overset{(n)}{\Omega}{}^{t_1}_\tau(\varphi)$$

$$- \frac{\partial \overset{(n)}{P}(\varphi_{t_1}(\bar{\varphi}))}{\partial \varphi_{t_1 j}(\bar{\varphi})} \frac{\partial \varphi_{t_1 j}(\bar{\varphi})}{\partial \varphi_i} \overset{(n)}{\Omega}{}^{t_1}_\tau(\bar{\varphi})$$

$$= \frac{\partial \overset{(n)}{P}(\varphi_{t_1}(\varphi))}{\partial \varphi_{t_1 j}(\varphi)} \frac{\partial \varphi_{t_1 j}(\varphi)}{\partial \varphi_i} \Big(\overset{(n)}{\Omega}{}^{t_1}_{\tau}(\varphi) - \overset{(n)}{\Omega}{}^{t_1}_{\tau}(\bar{\varphi}) \Big)$$

$$+ \frac{\partial \overset{(n)}{P}(\varphi_{t_1}(\varphi))}{\partial \varphi_{t_1 j}(\varphi)} \Big(\frac{\partial \varphi_{t_1 j}(\varphi)}{\partial \varphi_i} - \frac{\partial \varphi_{t_1 j}(\bar{\varphi})}{\partial \varphi_i} \Big) \overset{(n)}{\Omega}{}^{t_1}_{\tau}(\bar{\varphi})$$

$$+ \Big(\frac{\partial \overset{(n)}{P}(\varphi_{t_1}(\varphi))}{\partial \varphi_{t_1 j}(\varphi)} - \frac{\partial \overset{(n)}{P}(\varphi_{t_1}(\bar{\varphi}))}{\partial \varphi_{t_1 j}(\bar{\varphi})} \Big) \frac{\partial \varphi_{t_1 j}(\bar{\varphi})}{\partial \varphi_i} \overset{(n)}{\Omega}{}^{t_1}_{\tau}(\bar{\varphi}).$$

We denote the terms on the right-hand side of this equality in the order of their appearance by J_3, J_4, and J_5.

In view of Lemma 9.1, for $\tau \leq t_1 \leq 0$ we obtain

$$\|J_3\| \leq P_0 K_2 \Psi_1 \|\varphi - \bar{\varphi}\|^{\frac{\nu}{\nu+1}} \exp\Big\{ -\alpha t_1 + \Big(\gamma - \frac{\alpha\nu}{\nu+1} \Big)\tau - \gamma t_1 \Big\},$$

$$\|J_4\| \leq P_0 N \Psi \|\varphi - \bar{\varphi}\|^{\frac{\nu}{\nu+1}} \exp\Big\{ -\gamma(t_1 - \tau) - \Big(\alpha + \frac{\nu\alpha}{\nu+1} \Big) t_1 \Big\},$$

$$\|J_5\| \leq \Psi_3 K_2 N \|\varphi - \bar{\varphi}\|^{\frac{\nu}{\nu+1}} \exp\Big\{ \Big(-\alpha - \frac{\alpha\nu}{\nu+1} \Big) t_1 - \gamma(t_1 - \tau) \Big\},$$

where $\Psi_3 = (2P_0\zeta^\nu)^{\frac{1}{\nu+1}} = \text{const} < \infty$ and $\zeta > 0$ is the Lipschitz constant for the matrices $\dfrac{\partial P(\varphi)}{\partial \varphi_i}$, $i = \overline{1, m}$.

By using these representations, we get

$$\|\mathcal{L}\| \leq \|J_3\| + \|J_4\| + \|J_5\|.$$

This guarantees the validity of the inequality

$$\begin{aligned} \|J_2\| \ &\leq \ \int_\tau^0 Nm \exp\{\gamma t_1\} (\|J_3\| + \|J_4\| + \|J_5\|) dt_1 \\ &\leq \ (\Psi_4 + \Psi_5 + \Psi_6) \|\varphi - \bar{\varphi}\|^{\frac{\nu}{\nu+1}} \exp\Big\{ \Big(\gamma - \alpha - \frac{\alpha\nu}{\nu+1} \Big) \tau \Big\}, \end{aligned}$$

where

$$\Psi_4 = \frac{mNP_0K_2\Psi_1}{\alpha}, \quad \Psi_5 = \frac{mN^2P_0\Psi}{\alpha + \frac{\alpha\nu}{\nu+1}}, \quad \text{and} \quad \Psi_6 = \frac{mN^2\Psi_3K_2}{\alpha + \frac{\alpha\nu}{\nu+1}}$$

are constants independent of n, φ, and τ.

Thus, we get

$$\begin{aligned} \|G\| &\le \|J_1\| + \|J_2\| \\ &\le \Psi_0 \|\varphi - \bar{\varphi}\|^{\frac{\nu}{\nu+1}} \exp\Big\{\Big(\gamma - \alpha - \frac{\alpha\nu}{\nu+1}\Big)\tau\Big\}, \end{aligned}$$

where $\Psi_0 = \Psi_2 + \Psi_4 + \Psi_5 + \Psi_6$, which proves Lemma 9.2.

We now prove that sequence (9.8) is equicontinuous. Indeed,

$$\begin{aligned}
&\Big\| \frac{\partial \overset{(n)}{u}(\varphi)}{\partial \varphi_i} - \frac{\partial \overset{(n)}{u}(\bar{\varphi})}{\partial \varphi_i} \Big\| \\
&= \Big\| \frac{\partial}{\partial \varphi_i} \int_{-\infty}^{0} \overset{(n)}{\Omega}{}^0_\tau(\varphi) \overset{(n)}{c}(\varphi_\tau(\varphi))\, d\tau - \frac{\partial}{\partial \varphi_i} \int_{-\infty}^{0} \overset{(n)}{\Omega}{}^0_\tau(\bar{\varphi}) \overset{(n)}{c}(\varphi_\tau(\bar{\varphi}))\, d\tau \Big\| \\
&= \Big\| \int_{-\infty}^{0} \frac{\partial \overset{(n)}{\Omega}{}^0_\tau(\varphi)}{\partial \varphi_i} \overset{(n)}{c}(\varphi_\tau(\varphi))\, d\tau \\
&\qquad + \int_{-\infty}^{0} \overset{(n)}{\Omega}{}^0_\tau(\varphi) \sum_{j=1}^{m} \frac{\partial \overset{(n)}{c}(\varphi_\tau(\varphi))}{\partial \varphi_{\tau j}(\varphi)} \frac{\partial \varphi_{\tau j}(\varphi)}{\partial \varphi_i}\, d\tau \\
&\qquad - \int_{-\infty}^{0} \frac{\partial \overset{(n)}{\Omega}{}^0_\tau(\bar{\varphi})}{\partial \varphi_i} \overset{(n)}{c}(\varphi_\tau(\bar{\varphi}))\, d\tau \\
&\qquad - \int_{-\infty}^{0} \overset{(n)}{\Omega}{}^0_\tau(\bar{\varphi}) \sum_{j=1}^{m} \frac{\partial \overset{(n)}{c}(\varphi_\tau(\bar{\varphi}))}{\partial \varphi_{\tau j}(\bar{\varphi})} \frac{\partial \varphi_{\tau j}(\bar{\varphi})}{\partial \varphi_i}\, d\tau \Big\| \\
&= \Big\| \int_{-\infty}^{0} \frac{\partial \overset{(n)}{\Omega}{}^0_\tau(\varphi)}{\partial \varphi_i} \overset{(n)}{c}(\varphi_\tau(\varphi))\, d\tau + \int_{-\infty}^{0} \frac{\partial \overset{(n)}{\Omega}{}^0_\tau(\varphi)}{\partial \varphi_i} \overset{(n)}{c}(\varphi_\tau(\bar{\varphi}))\, d\tau \\
&\qquad - \int_{-\infty}^{0} \frac{\partial \overset{(n)}{\Omega}{}^0_\tau(\varphi)}{\partial \varphi_i} \overset{(n)}{c}(\varphi_\tau(\bar{\varphi}))\, d\tau - \int_{-\infty}^{0} \frac{\partial \overset{(n)}{\Omega}{}^0_\tau(\bar{\varphi})}{\partial \varphi_i} \overset{(n)}{c}(\varphi_\tau(\bar{\varphi}))\, d\tau
\end{aligned}$$

$$+\int_{-\infty}^{0}\overset{(n)}{\Omega}{}_{\tau}^{0}(\varphi)\sum_{j=1}^{m}\frac{\partial\overset{(n)}{c}(\varphi_\tau(\varphi))}{\partial\varphi_{\tau j}(\varphi)}\frac{\partial\varphi_{\tau j}(\varphi)}{\partial\varphi_i}\,d\tau$$

$$+\int_{-\infty}^{0}\overset{(n)}{\Omega}{}_{\tau}^{0}(\varphi)\sum_{j=1}^{m}\frac{\partial\overset{(n)}{c}(\varphi_\tau(\bar\varphi))}{\partial\varphi_{\tau j}(\bar\varphi)}\frac{\partial\varphi_{\tau j}(\bar\varphi)}{\partial\varphi_i}\,d\tau$$

$$-\int_{-\infty}^{0}\overset{(n)}{\Omega}{}_{\tau}^{0}(\varphi)\sum_{j=1}^{m}\frac{\partial\overset{(n)}{c}(\varphi_\tau(\bar\varphi))}{\partial\varphi_{\tau j}(\bar\varphi)}\frac{\partial\varphi_{\tau j}(\bar\varphi)}{\partial\varphi_i}\,d\tau$$

$$-\int_{-\infty}^{0}\overset{(n)}{\Omega}{}_{\tau}^{0}(\bar\varphi)\sum_{j=1}^{m}\frac{\partial\overset{(n)}{c}(\varphi_\tau(\bar\varphi))}{\partial\varphi_{\tau j}(\bar\varphi)}\frac{\partial\varphi_{\tau j}(\bar\varphi)}{\partial\varphi_i}\,d\tau\Big\|$$

$$\leq\int_{-\infty}^{0}\left\|\frac{\partial\overset{(n)}{\Omega}{}_{\tau}^{0}(\varphi)}{\partial\varphi_i}\right\|\|\overset{(n)}{c}(\varphi_\tau(\varphi))-\overset{(n)}{c}(\varphi_\tau(\bar\varphi))\|\,d\tau \tag{9.12}$$

$$+\int_{-\infty}^{0}\|\overset{(n)}{c}(\varphi_\tau(\bar\varphi))\|\left\|\frac{\partial\overset{(n)}{\Omega}{}_{\tau}^{0}(\varphi)}{\partial\varphi_i}-\frac{\partial\overset{(n)}{\Omega}{}_{\tau}^{0}(\bar\varphi)}{\partial\varphi_i}\right\|d\tau \tag{9.13}$$

$$+\int_{-\infty}^{0}\|\overset{(n)}{\Omega}{}_{\tau}^{0}(\varphi)\|\sum_{j=1}^{m}\Big\|\frac{\partial\overset{(n)}{c}(\varphi_\tau(\varphi))}{\partial\varphi_{\tau j}(\varphi)}\frac{\partial\varphi_{\tau j}(\varphi)}{\partial\varphi_i}-\frac{\partial\overset{(n)}{c}(\varphi_\tau(\bar\varphi))}{\partial\varphi_{\tau j}(\bar\varphi)}\frac{\partial\varphi_{\tau j}(\bar\varphi)}{\partial\varphi_i}\Big\|\,d\tau \tag{9.14}$$

$$+\int_{-\infty}^{0}\sum_{j=1}^{m}\left\|\frac{\partial\overset{(n)}{c}(\varphi_\tau(\bar\varphi))}{\partial\varphi_{\tau j}(\bar\varphi)}\frac{\partial\varphi_{\tau j}(\bar\varphi)}{\partial\varphi_i}\right\|\|\overset{(n)}{\Omega}{}_{\tau}^{0}(\varphi)-\overset{(n)}{\Omega}{}_{\tau}^{0}(\bar\varphi)\|\,d\tau. \tag{9.15}$$

We now separately estimate each term on the right-hand side of the last inequality. For this purpose, we denote these terms by J_6, J_7, J_8, and J_9 (in the order of their appearance).

The required estimates for integrals (9.12), (9.13), and (9.15) are obvious. Indeed, let $\|c(\varphi)-c(\bar\varphi)\|\leq\xi\|\varphi-\bar\varphi\|$. Then

$$J_6 \le \int_{-\infty}^{0} K_1 \exp\{(\gamma-\alpha)\tau\}(2c\xi^\nu)^{\frac{1}{\nu+1}} \|\varphi-\bar\varphi\|^{\frac{\nu}{\nu+1}} \exp\Big\{-\frac{\alpha\nu}{\nu+1}\tau\Big\}\, d\tau$$

$$\le Z_1 \|\varphi-\bar\varphi\|^{\frac{\nu}{\nu+1}},$$

where

$$Z_1 = \frac{K_1(2c\xi^\nu)^{\frac{1}{\nu+1}}}{\gamma-\alpha-\alpha\nu/(\nu+1)} = \text{const} > 0,$$

and

$$J_7 \le \int_{-\infty}^{0} c\Psi_0 \|\varphi-\bar\varphi\|^{\frac{\nu}{\nu+1}} \exp\Big\{\Big(\gamma-\alpha-\frac{\alpha\nu}{\nu+1}\Big)\tau\Big\}\, d\tau$$

$$\le Z_2 \|\varphi-\bar\varphi\|^{\frac{\nu}{\nu+1}},$$

where

$$Z_2 = c\Psi_0 \frac{1}{\gamma-\alpha-\alpha\nu/(\nu+1)} = \text{const} > 0$$

and $\nu > 0$ is chosen from the inequality

$$\gamma - \Big(1+\frac{\nu}{\nu+1}\Big)\alpha > 0.$$

Further,

$$J_9 \le \int_{-\infty}^{0} mRK_2\Psi_1 \|\varphi-\bar\varphi\|^{\frac{\nu}{\nu+1}} \exp\Big\{-\alpha\tau + \Big(\gamma-\frac{\alpha\nu}{\nu+1}\Big)\tau\Big\}\, d\tau$$

$$\le Z_3 \|\varphi-\bar\varphi\|^{\frac{\nu}{\nu+1}},$$

where

$$Z_3 = \frac{mRK_2\Psi_1}{\gamma-\alpha-\alpha\nu/(\nu+1)} = \text{const} > 0.$$

It is clear that the constants Z_1, Z_2, and Z_3 are independent of τ, φ, and the order of truncation n.

It remains to estimate integral (9.14). We have

$$
\begin{aligned}
L_1 &= \frac{\partial \overset{(n)}{c}(\varphi_\tau(\varphi))}{\partial \varphi_{\tau j}(\varphi)} \frac{\partial \varphi_{\tau j}(\varphi)}{\partial \varphi_i} - \frac{\partial \overset{(n)}{c}(\varphi_\tau(\bar\varphi))}{\partial \varphi_{\tau j}(\bar\varphi)} \frac{\partial \varphi_{\tau j}(\bar\varphi)}{\partial \varphi_i} \\
&= \frac{\partial \overset{(n)}{c}(\varphi_\tau(\varphi))}{\partial \varphi_{\tau j}(\varphi)} \Big(\frac{\partial \varphi_{\tau j}(\varphi)}{\partial \varphi_i} - \frac{\partial \varphi_{\tau j}(\bar\varphi)}{\partial \varphi_i}\Big) \\
&\qquad + \Big(\frac{\partial \overset{(n)}{c}(\varphi_\tau(\varphi))}{\partial \varphi_{\tau j}(\varphi)} - \frac{\partial \overset{(n)}{c}(\varphi_{\tau j}(\bar\varphi))}{\partial \varphi_{\tau j}(\bar\varphi)}\Big) \frac{\partial \varphi_{\tau j}(\bar\varphi)}{\partial \varphi_i}.
\end{aligned}
$$

Assume that $\Big\|\dfrac{\partial c(\varphi)}{\partial \varphi_i} - \dfrac{\partial c(\bar\varphi)}{\partial \varphi_i}\Big\| \le \delta \|\varphi - \bar\varphi\|$. By using Lemma 9.1 and the inequality

$$
\begin{aligned}
&\Big\|\frac{\partial \overset{(n)}{c}(\varphi_\tau(\varphi))}{\partial \varphi_{\tau j}(\varphi)} - \frac{\partial \overset{(n)}{c}(\varphi_{\tau j}(\bar\varphi))}{\partial \varphi_{\tau j}(\bar\varphi)}\Big\| \\
&\qquad \le \Big(2R\delta^\nu\Big)^{\frac{1}{\nu+1}} \|\varphi - \bar\varphi\|^{\frac{\nu}{\nu+1}} \exp\Big\{-\frac{\alpha\nu}{\nu+1}\tau\Big\}, \qquad \tau \le 0,
\end{aligned}
$$

we obtain

$$
\|L_1\| \le \Big(\, R\Psi + K_2(2R\delta^\nu)^{\frac{1}{\nu+1}}\Big)\|\varphi - \bar\varphi\|^{\frac{\nu}{\nu+1}} \exp\Big\{-\Big(\alpha + \frac{\alpha\nu}{\nu+1}\Big)\tau\Big\}.
$$

This yields

$$
J_8 \le \int_{-\infty}^{0} mN\|L_1\| \exp\{\gamma\tau\}\, d\tau \le Z_4 \|\varphi - \bar\varphi\|^{\frac{\nu}{\nu+1}},
$$

where

$$
Z_4 = \frac{mNR\Psi + mNK_2(2R\delta^\nu)^{\frac{1}{\nu+1}}}{\gamma - \alpha - \alpha\nu/(\nu+1)} = \text{const} > 0
$$

is independent of n, τ, and φ.

In view of the estimates for integrals (9.12)–(9.15), we conclude that

$$
\Big\|\frac{\partial \overset{(n)}{u}(\varphi)}{\partial \varphi_i} - \frac{\partial \overset{(n)}{u}(\bar\varphi)}{\partial \varphi_i}\Big\| \le \Phi \|\varphi - \bar\varphi\|^{\frac{\nu}{\nu+1}},
$$

where $\Phi = \sum\limits_{s=1}^{4} Z_s$ is a positive constant independent of n, φ, and τ.

We now establish conditions under which the invariant torus $\mathcal{F}$ of the system of equations (8.1) is l times differentiable. By $D^s_\varphi(\Phi(\varphi))$ we denote any sth derivative of the function $\Phi(\varphi)$ with respect to $\varphi = (\varphi_1, \ldots, \varphi_m)$. The following assertion is true:

Theorem 9.3. *Assume that the system of equations (8.1) satisfies the conditions of Theorem 9.2 and the functions $a(\varphi)$, $P(\varphi)$, and $c(\varphi)$ belong to $C^l_{\mathrm{Lip}}(\mathcal{F}_m)$. If, in addition,*

$$\frac{\gamma}{l} > \alpha = \max_{\varphi \in \mathcal{F}_m} \max_{\|\eta = 1\|} \left\| \frac{\partial a(\varphi)}{\partial \varphi} \eta \right\|, \quad \eta \in R^m,$$

then the invariant torus $\mathcal{F}\colon x = u(\varphi)$, $\varphi \in \mathcal{F}_m$, of system (8.1) belongs to the space $C^l(\mathcal{F}_m)$.

Proof. According to [MSK2], under the conditions of the theorem, the functions $\varphi_t(\varphi)$ and $\overset{(n)}{\Omega}{}^0_\tau(\varphi)$ satisfy the following inequalities (for $\tau \leq 0$):

$$\|D^s_\varphi \varphi_\tau(\varphi)\| \leq K \exp\{-s\alpha\tau\},$$

$$\|D^s_\varphi \overset{(n)}{\Omega}{}^0_\tau(\varphi)\| \leq K \exp\{(\gamma - \alpha s)\tau\}, \quad K = \mathrm{const} > 0,$$

$$\|D^s_\varphi(\overset{(n)}{\Omega}{}^0_\tau(\varphi) - \overset{(n)}{\Omega}{}^0_\tau(\bar{\varphi}))\|$$
$$\leq K\|\varphi - \bar{\varphi}\|^{\frac{1}{z+1}} \exp\Big\{\Big(\gamma - s\alpha - \frac{\alpha}{z+1}\Big)\tau\Big\}, \quad s = \overline{0, l},$$

$$\Big\|D^s_\varphi\Big(\frac{\partial \varphi_\tau(\varphi)}{\partial \varphi_i} - \frac{\partial \varphi_\tau(\bar{\varphi})}{\partial \varphi_i}\Big)\Big\| \leq K\|\varphi - \bar{\varphi}\|^{\frac{1}{z+1}} \exp\Big\{-\Big(s\alpha + \frac{\alpha}{z+1}\Big)\tau\Big\}, \quad (9.16)$$

$s = \overline{0, l-1}$, where φ, $\bar{\varphi} \in \mathcal{F}_m$, and z is a positive constant such that

$$\gamma > \Big(l + \frac{1}{z+1}\Big)\alpha.$$

To prove the theorem, it suffices to show that, for all $s = \overline{1, l}$, the sequence

$$D^s_\varphi \overset{(1)}{u}(\varphi),\ D^s_\varphi \overset{(2)}{u}(\varphi), \ldots, D^s_\varphi \overset{(n)}{u}(\varphi), \ldots \quad (9.17)$$

is uniformly bounded and equicontinuous. This can be proved by induction.

Indeed, let $s = 1$. In this case, uniform boundedness is established just as in the proof of Theorem 9.2. The property of equicontinuity is also established as in the proof of Theorem 9.2 with the only difference that

$$J_7 \leq \int\limits_{-\infty}^{0} Kc\exp\Big\{\Big(\gamma - \alpha - \frac{\alpha}{z+1}\Big)\tau\Big\}\|\varphi - \bar{\varphi}\|^{\frac{1}{z+1}}\, d\tau \leq K_3\|\varphi - \bar{\varphi}\|^{\frac{1}{z+1}}. \quad (9.18)$$

In view of (9.18), we obtain

$$\Big\|D_\varphi^1(\overset{(n)}{u}(\varphi) - \overset{(n)}{u}(\bar{\varphi}))\Big\| \leq \sum_{i=6}^{9} J_i \leq K_4\Big\{\|\varphi - \bar{\varphi}\|^{\frac{\nu}{\nu+1}} + \|\varphi - \bar{\varphi}\|^{\frac{1}{z+1}}\Big\},$$

where $\nu > 0$ is chosen from the inequality

$$\gamma > \Big(1 + \frac{\nu}{\nu+1}\Big)\alpha.$$

For $s = 1$, the required assertion is proved

Let us now assume that sequence (9.17) is uniformly bounded and equicontinuous for $s \leq l - 1$.

Then, for $D_\varphi^l \overset{(n)}{u}(\varphi)$, we can write

$$D_\varphi^l \overset{(n)}{u}(\varphi) = \int\limits_{-\infty}^{0} \mathcal{L}(\varphi, \tau)\, d\tau,$$

where $\mathcal{L}(\varphi, \tau)$ is the sum of finitely many terms of the form

$$D_\varphi^s \overset{(n)}{\Omega}{}_\tau^0(\varphi) D_\varphi^{l-s} \overset{(n)}{c}(\varphi_\tau(\varphi)),$$

and $s = 0, 1, \ldots, l$.

In turn, the quantity $D_\varphi^l \overset{(n)}{c}(\varphi_\tau(\varphi))$ admits a representation in the form of finitely many terms of the type

$$D^s_{\varphi_\tau(\varphi)}\Big(\overset{(n)}{c}(\varphi_\tau(\varphi))\Big)\frac{\partial^{\lambda_1}\varphi_{\tau k_1}(\varphi)}{\partial\varphi_1^{\lambda_1}} \cdots \frac{\partial^{\lambda_m}\varphi_{\tau k_m}(\varphi)}{\partial\varphi_m^{\lambda_m}}, \quad (9.19)$$

where $1 \leq s \leq l$, $\sum\limits_{i=1}^{m} \lambda_i = l$, $\lambda_i \in \{1, 2, \ldots, l\}$, $k_i \in \overline{1, m}$, and $i = 1, 2, \ldots, m$. Then

$$\|D_\varphi^l \overset{(n)}{c}(\varphi_\tau(\varphi))\| \leq K_4 \exp\{-\lambda_1\alpha\tau - \ldots - \lambda_m\alpha\tau\} = K_4 \exp\{-l\alpha\tau\}$$

for $\tau \leq 0$, whence it follows that

$$\|D^l_\varphi \overset{(n)}{u}(\varphi)\| \leq K_5 \int\limits_{-\infty}^{0} \exp\{(\gamma - \alpha s)\tau - (l-s)\alpha\tau\}\, d\tau = \frac{K_5}{\gamma - l\alpha}.$$

This proves that sequence (9.17) is uniformly bounded for $s = l$.

We have

$$D^l_\varphi\Big(\overset{(n)}{u}(\varphi) - \overset{(n)}{u}(\bar\varphi)\Big) = \int\limits_{-\infty}^{0} (\mathcal{L}(\varphi,\tau) - \mathcal{L}(\bar\varphi,\tau))\, d\tau,$$

where the integrand is the sum of finitely many terms of the form

$$D^s_\varphi\Big(\overset{(n)}{\Omega}{}^0_\tau(\varphi) - \overset{(n)}{\Omega}{}^0_\tau(\bar\varphi)\Big) D^{l-s}_\varphi\Big(\overset{(n)}{c}(\varphi_\tau(\varphi))\Big)$$
$$+ D^s_\varphi\Big(\overset{(n)}{\Omega}{}^0_\tau(\varphi)\Big) D^{l-s}_\varphi\Big(\overset{(n)}{c}(\varphi_\tau(\varphi)) - \overset{(n)}{c}(\varphi_\tau(\bar\varphi))\Big).$$

Using expression (9.19), we conclude that $D^l_\varphi\left(\overset{(n)}{c}(\varphi_\tau(\varphi)) - \overset{(n)}{c}(\varphi_\tau(\bar\varphi))\right)$ admits a representation in the form of finitely many differences of the form

$$\begin{aligned}
Z_s &= D^s_{\varphi_\tau(\varphi)}\Big(\overset{(n)}{c}(\varphi_\tau(\varphi))\Big)\frac{\partial^{\lambda_1}\varphi_{\tau k_1}(\varphi)}{\partial\varphi_1^{\lambda_1}}\cdots\frac{\partial^{\lambda_m}\varphi_{\tau k_m}(\varphi)}{\partial\varphi_m^{\lambda_m}} \\
&\quad - D^s_{\varphi_\tau(\varphi)}\Big(\overset{(n)}{c}(\varphi_\tau(\bar\varphi))\Big)\frac{\partial^{\lambda_1}\varphi_{\tau k_1}(\bar\varphi)}{\partial\varphi_1^{\lambda_1}}\cdots\frac{\partial^{\lambda_m}\varphi_{\tau k_m}(\bar\varphi)}{\partial\varphi_m^{\lambda_m}} \\
&= D^s_{\varphi_\tau(\varphi)}\Big(\overset{(n)}{c}(\varphi_\tau(\varphi)) - \overset{(n)}{c}(\varphi_\tau(\bar\varphi))\Big)\frac{\partial^{\lambda_1}\varphi_{\tau k_1}(\varphi)}{\partial\varphi_1^{\lambda_1}}\cdots\frac{\partial^{\lambda_m}\varphi_{\tau k_m}(\varphi)}{\partial\varphi_m^{\lambda_m}} \\
&\quad + D^s_{\varphi_\tau(\varphi)}\Big(\overset{(n)}{c}(\varphi_\tau(\bar\varphi))\Big) Z_s^{(1)},
\end{aligned}$$

where

$$\begin{aligned}
Z_s^{(1)} &= \frac{\partial^{\lambda_1}\varphi_{\tau k_1}(\varphi)}{\partial\varphi_1^{\lambda_1}}\cdots\frac{\partial^{\lambda_m}\varphi_{\tau k_m}(\varphi)}{\partial\varphi_m^{\lambda_m}} - \frac{\partial^{\lambda_1}\varphi_{\tau k_1}(\bar\varphi)}{\partial\varphi_1^{\lambda_1}}\cdots\frac{\partial^{\lambda_m}\varphi_{\tau k_m}(\bar\varphi)}{\partial\varphi_m^{\lambda_m}} \\
&= \Big(\frac{\partial^{\lambda_1}\varphi_{\tau k_1}(\varphi)}{\partial\varphi_1^{\lambda_1}} - \frac{\partial^{\lambda_1}\varphi_{\tau k_1}(\bar\varphi)}{\partial\varphi_1^{\lambda_1}}\Big)\frac{\partial^{\lambda_2}\varphi_{\tau k_2}(\varphi)}{\partial\varphi_2^{\lambda_2}}\cdots\frac{\partial^{\lambda_m}\varphi_{\tau k_m}(\varphi)}{\partial\varphi_m^{\lambda_m}}
\end{aligned}$$

$$+\frac{\partial^{\lambda_1}\varphi_{\tau k_1}(\bar\varphi)}{\partial\varphi_1^{\lambda_1}}\Big(\frac{\partial^{\lambda_2}\varphi_{\tau k_2}(\varphi)}{\partial\varphi_2^{\lambda_2}}-\frac{\partial^{\lambda_2}\varphi_{\tau k_2}(\bar\varphi)}{\partial\varphi_2^{\lambda_2}}\Big)$$

$$\times\frac{\partial^{\lambda_3}\varphi_{\tau k_3}(\varphi)}{\partial\varphi_3^{\lambda_3}}\cdots\frac{\partial^{\lambda_m}\varphi_{\tau k_m}(\varphi)}{\partial\varphi_m^{\lambda_m}}+\ldots+\frac{\partial^{\lambda_1}\varphi_{\tau k_1}(\bar\varphi)}{\partial\varphi_1^{\lambda_1}}\cdots$$

$$\times\frac{\partial^{\lambda_{m-1}}\varphi_{\tau k_{m-1}}(\bar\varphi)}{\partial\varphi_{m-1}^{\lambda_{m-1}}}\Big(\frac{\partial^{\lambda_m}\varphi_{\tau k_m}(\varphi)}{\partial\varphi_m^{\lambda_m}}-\frac{\partial^{\lambda_m}\varphi_{\tau k_m}(\bar\varphi)}{\partial\varphi_m^{\lambda_m}}\Big).$$

By using inequalities (9.16), we obtain

$$\Big\|Z_s^{(1)}\Big\|\le K_6\|\varphi-\bar\varphi\|^{\frac{1}{z+1}}\exp\Big\{-\Big(l\alpha+\frac{\alpha}{z+1}\Big)\tau\Big\}.$$

Consequently,

$$\Big\|D_\varphi^l\Big(\overset{(n)}{c}(\varphi_\tau(\varphi))-\overset{(n)}{c}(\varphi_\tau(\bar\varphi))\Big)\Big\|$$

$$\le K_6\Big\{\exp\Big\{-\Big(l+\frac{\nu}{\nu+1}\Big)\alpha\tau\Big\}\|\varphi-\bar\varphi\|^{\frac{\nu}{\nu+1}}$$

$$+\exp\Big\{-\Big(l+\frac{1}{z+1}\Big)\Big\}\|\varphi-\bar\varphi\|^{\frac{1}{z+1}}\Big\}.$$

Further, we get

$$\Big\|\mathcal{L}(\varphi,\tau)-\mathcal{L}(\bar\varphi,\tau)\Big\|$$

$$\le K_7\Big\{\|\varphi-\bar\varphi\|^{\frac{1}{z+1}}\exp\Big\{\Big(\gamma-s\alpha-\frac{\alpha}{z+1}\Big)\tau-(l-s)\alpha\tau\Big\}$$

$$+\exp\{(\gamma-\alpha s)\tau\}\Big[\exp\Big\{-\Big(l-s+\frac{\nu}{\nu+1}\Big)\alpha\tau\Big\}\|\varphi-\bar\varphi\|^{\frac{\nu}{\nu+1}}$$

$$+\exp\Big\{-\Big(l-s+\frac{1}{z+1}\Big)d\tau\Big\}\|\varphi-\bar\varphi\|^{\frac{1}{z+1}}\Big]\Big\}$$

$$\le K_8\Big[\|\varphi-\bar\varphi\|^{\frac{1}{z+1}}\exp\Big\{\Big(\gamma-l\alpha-\frac{\alpha}{z+1}\Big)\tau\Big\}$$

$$+\|\varphi-\bar\varphi\|^{\frac{\nu}{\nu+1}}\exp\Big\{\Big(\gamma-l\alpha-\frac{\nu\alpha}{\nu+1}\Big)\tau\Big\}\Big],\quad \tau\le 0,$$

and, hence,

$$\Big\|D_\varphi^l\Big(\overset{(n)}{u}(\varphi)-\overset{(n)}{u}(\bar\varphi)\Big)\Big\|\le K_9\Big\{\|\varphi-\bar\varphi\|^{\frac{1}{z+1}}+\|\varphi-\bar\varphi\|^{\frac{\nu}{\nu+1}}\Big\}.$$

This means that sequence (9.17) is equicontinuous for $s = l$.

Theorem 9.3 is proved.

Remark 9.1. Instead of the Lipschitz condition, for the validity of a similar statement, one can require that the lth derivatives must be continuous with a modulus of continuity $\mu(\sigma)$ such that

$$F(u) = \int_d^u \frac{d\sigma}{\mu(\sigma)} \underset{u\to+0}{\longrightarrow} -\infty, \quad \sigma \in [0, d], \quad d > 0.$$

The method used in this section for the truncation of the original system of equations can be applied to the investigation of the problem of differentiability of the invariant torus with respect to the parameter ρ appearing in this system.

Consider a system of equations of the form

$$\frac{d\varphi}{dt} = a(\varphi), \quad \frac{dx}{dt} = P(\varphi, \rho)x + c(\varphi, \rho), \tag{9.20}$$

where $x \in \mathfrak{M}$, $\varphi \in R^m$, $\rho \in [\rho_1, \rho_2] \subset R^1$, and both the vector functions $a(\varphi)$ and $c(\varphi, \rho)$ and the infinite matrix $P(\varphi, \rho)$ are 2π-periodic in φ_i, $i = \overline{1, m}$, and continuous in φ.

By $C^l_{\text{Lip}}(\rho)$ we denote the set of functions and matrices $\Phi(\varphi, \rho)$ bounded in norm and satisfying the Lipschitz condition together with their partial derivatives with respect to ρ up to the lth order inclusive.

Theorem 9.4. *Assume that the system of equations (9.20) satisfies the following conditions:*

(i) $a(\varphi) \in C^0_{\text{Lip}}(\mathcal{F}_m)$ *and* $P(\varphi, \rho)$, $c(\varphi, \rho) \in C^0(\mathcal{F}_m) \cap C^1_{\text{Lip}}(\rho)$;

(ii) $\sum\limits_{j=n+1} \max\limits_{\varphi,\rho} |p_{sj}(\varphi, \rho)| \leq K\varepsilon(n)$, *where* $\varepsilon(n) \to 0$ *as* $n \to \infty$ *uniformly in* ρ *and the constant* K *does not depend on* ρ;

(iii) $\left\| \overset{(n)}{\Omega}{}^0_\tau(\varphi, \rho) \right\| \leq K_1 \exp\{\gamma\tau\}$, $\tau \leq 0$, *where the constants* K *and* γ *are independent of* n *and* ρ.

Then the invariant torus $\mathcal{F}: x = u(\varphi, \rho)$, $\varphi \in \mathcal{F}_m$, *of the system of equations (9.20) belongs to the space* $C^1_{\text{Lip}}(\rho)$.

Proof. By using Theorem 9.1, we represent the invariant torus $\mathcal{F}: x = u(\varphi, \rho)$ of the system of equations (9.20) as the weak limit

$$u(\varphi,\rho) = \lim_{n\to\infty} \overset{(n)}{u}(\varphi,\rho),$$

where $\mathcal{F}^{(n)}: \overset{(n)}{x}\overset{(n)}{u}(\varphi,\rho)$ is the invariant torus truncated to the nth order for the system of equations corresponding to system (9.20).

To prove the theorem, it suffices to show that the sequence

$$\frac{\partial \overset{(1)}{u}(\varphi,\rho)}{\partial\rho}, \frac{\partial \overset{(2)}{u}(\varphi,\rho)}{\partial\rho}, \ldots, \frac{\partial \overset{(n)}{u}(\varphi,\rho)}{\partial\rho}, \ldots \tag{9.21}$$

is uniformly bounded and equicontinuous.

In the analyzed case, equality (9.11) takes the form

$$\overset{(n)}{\Omega}{}^0_\tau(\varphi,\rho) - \overset{(n)}{\Omega}{}^0_\tau(\varphi,\bar\rho)$$

$$= \int_\tau^0 \overset{(n)}{\Omega}{}^0_{t_1}(\varphi,\rho)\Big[\overset{(n)}{P}(\varphi_{t_1}(\varphi),\rho) - \overset{(n)}{P}(\varphi_{t_1}(\varphi),\bar\rho)\Big]\overset{(n)}{\Omega}{}^{t_1}_\tau(\varphi,\bar\rho)\,dt_1, \tag{9.22}$$

whence we easily obtain the representation

$$\frac{\partial \overset{(n)}{\Omega}{}^0_\tau(\varphi,\rho)}{\partial\rho} = \int_\tau^0 \overset{(n)}{\Omega}{}^0_{t_1}(\varphi,\rho)\frac{\partial \overset{(n)}{P}(\varphi_{t_1}(\varphi),\rho)}{\partial\rho}\overset{(n)}{\Omega}{}^{t_1}_\tau(\varphi,\rho)\,dt_1. \tag{9.23}$$

This yields

$$\Big\|\frac{\partial \overset{(n)}{\Omega}{}^0_\tau(\varphi,\rho)}{\partial\rho}\Big\| \le K_2(-\tau)\exp\{\gamma\tau\}, \quad \tau\le 0. \tag{9.24}$$

This inequality and the representation

$$\frac{\partial}{\partial\rho}\overset{(n)}{u}(\varphi,\rho) = \int_{-\infty}^0 \Big\{\frac{\partial \overset{(n)}{\Omega}{}^0_\tau(\varphi,\rho)}{\partial\rho}\overset{(n)}{c}(\varphi_\tau(\varphi),\rho)$$

$$+ \overset{(n)}{\Omega}{}^0_\tau(\varphi,\rho)\frac{\partial \overset{(n)}{c}\varphi_\tau(\varphi,\rho)}{\partial\rho}\Big\}\,d\tau \tag{9.25}$$

imply the uniform boundedness of sequence (9.21):

$$\Big\|\frac{\partial}{\partial\rho}\overset{(n)}{u}(\varphi,\rho)\Big\| \leq K_3\int\limits_{-\infty}^{0}(-\tau)\exp\{\gamma\tau\}d\tau + K_4\int\limits_{-\infty}^{0}\exp\{\gamma\tau\}d\tau$$

$$\leq K_5\Big(1+\frac{1}{\gamma^2}\Big).$$

Relation (9.25) implies that

$$\frac{\partial\overset{(n)}{u}(\varphi,\rho)}{\partial\rho} - \frac{\partial\overset{(n)}{u}(\varphi,\bar\rho)}{\partial\rho}$$

$$= \int\limits_{-\infty}^{0}\Big\{\frac{\partial\overset{(n)}{\Omega}{}^0_\tau(\varphi,\rho)}{\partial\rho}\Big(\overset{(n)}{c}(\varphi_\tau(\varphi),\rho) - \overset{(n)}{c}(\varphi_\tau(\varphi),\bar\rho)\Big)$$

$$+\Big(\frac{\partial\overset{(n)}{\Omega}{}^0_\tau(\varphi,\rho)}{\partial\rho} - \frac{\partial\overset{(n)}{\Omega}{}^0_\tau(\varphi,\bar\rho)}{\partial\rho}\Big)\overset{(n)}{c}(\varphi_\tau(\varphi),\bar\rho)$$

$$+\overset{(n)}{\Omega}{}^0_\tau(\varphi,\rho)\Big(\frac{\partial\overset{(n)}{c}(\varphi_\tau(\varphi),\rho)}{\partial\rho} - \frac{\partial\overset{(n)}{c}(\varphi_\tau(\varphi),\bar\rho)}{\partial\rho}\Big)$$

$$+\Big(\overset{(n)}{\Omega}{}^0_\tau(\varphi,\rho)\rho - \overset{(n)}{\Omega}{}^0_\tau(\varphi,\bar\rho)\Big)\frac{\partial\overset{(n)}{c}(\varphi_\tau(\varphi),\bar\rho)}{\partial\rho}\Big\}\,d\tau.$$

By using the conditions of the theorem, we obtain

$$\max_{\varphi\in\mathcal{F}_m;\rho,\bar\rho\in[\rho_1,\rho_2]}\Big\{\|\overset{(n)}{P}(\varphi,\rho) - \overset{(n)}{P}(\varphi,\bar\rho)\|,\ \|\overset{(n)}{c}(\varphi,\rho) - \overset{(n)}{c}(\varphi,\bar\rho)\|,$$

$$\Big\|\frac{\partial\overset{(n)}{P}(\varphi,\rho)}{\partial\rho} - \frac{\partial\overset{(n)}{P}(\varphi,\bar\rho)}{\partial\rho}\Big\|,\ \Big\|\frac{\partial\overset{(n)}{c}(\varphi,\rho)}{\partial\rho} - \frac{\partial\overset{(n)}{c}(\varphi,\bar\rho)}{\partial\rho}\Big\|\Big\}$$

$$\leq K_6\,|\,\rho-\bar\rho\,|,$$

where K_6 is a positive constant. Moreover, relation (9.22) implies that

$$\|\overset{(n)}{\Omega}{}^0_\tau(\varphi,\rho) - \overset{(n)}{\Omega}{}^0_\tau(\varphi,\bar\rho)\| \leq K_7\int\limits_{\tau}^{0}|\,\rho-\bar\rho\,|\exp\{\gamma,\tau\}dt_1$$

$$\leq K_7\,|\,\rho-\bar\rho\,|\,(-\tau)\exp\{\gamma\tau\}.$$

Let us estimate the difference

$$\frac{\partial \overset{(n)}{\Omega}{}^{0}_{\tau}(\varphi,\rho)}{\partial\rho} - \frac{\partial \overset{(n)}{\Omega}{}^{0}_{\tau}(\varphi,\bar\rho)}{\partial\rho}.$$

In view of relation (9.23), we can rewrite this difference in the form

$$\frac{\partial \overset{(n)}{\Omega}{}^{0}_{\tau}(\varphi,\rho)}{\partial\rho} - \frac{\partial \overset{(n)}{\Omega}{}^{0}_{\tau}(\varphi,\bar\rho)}{\partial\rho}$$

$$= \int\limits_{\tau}^{0} \Big\{ \Big(\overset{(n)}{\Omega}{}^{0}_{t_1}(\varphi,\rho) - \overset{(n)}{\Omega}{}^{0}_{t_1}(\varphi,\bar\rho) \Big) \frac{\partial \overset{(n)}{P}(\varphi_{t_1}(\varphi),\bar\rho)}{\partial\rho} \overset{(n)}{\Omega}{}^{t_1}_{\tau}(\varphi,\bar\rho)$$

$$+ \overset{(n)}{\Omega}{}^{0}_{t_1}(\varphi,\rho) \Big(\frac{\partial \overset{(n)}{P}(\varphi_{t_1}(\varphi),\rho)}{\partial\rho} - \frac{\partial \overset{(n)}{P}(\varphi_{t_1}(\varphi),\bar\rho)}{\partial\rho} \Big) \overset{(n)}{\Omega}{}^{t_1}_{\tau}(\varphi,\bar\rho)$$

$$+ \overset{(n)}{\Omega}{}^{0}_{t_1}(\varphi,\rho) \frac{\partial \overset{(n)}{P}(\varphi_{t_1}(\varphi),\rho)}{\partial\rho} \Big(\overset{(n)}{\Omega}{}^{t_1}_{\tau}(\varphi,\rho) - \overset{(n)}{\Omega}{}^{t_1}_{\tau}(\varphi,\bar\rho) \Big) \Big\}\, dt_1.$$

By using the properties of the matrizant, we obtain

$$\overset{(n)}{\Omega}{}^{t_1}_{\tau}(\varphi,\rho) - \overset{(n)}{\Omega}{}^{t_1}_{\tau}(\varphi,\bar\rho)$$

$$= \int\limits_{\tau}^{t_1} \overset{(n)}{\Omega}{}^{t_1}_{s}(\varphi,\rho) \Big(\overset{(n)}{P}(\varphi_s(\varphi),\rho) - \overset{(n)}{P}(\varphi_s(\varphi),\bar\rho) \Big) \overset{(n)}{\Omega}{}^{s}_{\tau}(\varphi,\bar\rho)\, ds$$

and, hence,

$$\| \overset{(n)}{\Omega}{}^{t_1}_{\tau}(\varphi,\rho) - \overset{(n)}{\Omega}{}^{t_1}_{\tau}(\varphi,\bar\rho) \| \le K_8\, |\rho-\bar\rho|\, (t_1-\tau) \exp\{-\gamma(t_1-\tau)\}$$

for $\tau \le t_1$. Further, we get

$$\Big\| \frac{\partial \overset{(n)}{\Omega}{}^{0}_{\tau}(\varphi,\rho)}{\partial\rho} - \frac{\partial \overset{(n)}{\Omega}{}^{0}_{\tau}(\varphi,\bar\rho)}{\partial\rho} \Big\|$$

$$\le K_9\, |\rho-\bar\rho| \int\limits_{\tau}^{0} (-t_1) \Big(\exp\{\gamma t_1 - \gamma(t_1-\tau)\}$$

$$+ \exp\{\gamma t_1 - \gamma(t_1 - \tau)\} + (t_1 - \tau)\exp\{-\gamma(t_1 - \tau) + \gamma t_1\}\Big) dt_1$$

$$\leq K_{10} \,|\, \rho - \bar{\rho} \,|\, (-\tau)(1 - \tau)\exp\{\gamma\tau\}.$$

Finally, by using this inequality, we arrive at the estimate

$$\left\| \frac{\partial \overset{(n)}{u}(\varphi,\rho)}{\partial \rho} - \frac{\partial \overset{(n)}{u}(\varphi,\bar{\rho})}{\partial \rho} \right\|$$

$$\leq K_{11} \,|\, \rho - \bar{\rho} \,|\Big\{\frac{1}{\gamma} - 3\int_{\tau}^{0} \tau \exp\{\gamma\tau\}\, d\tau + \int_{-\infty}^{0} \tau^2 \exp\{\gamma\tau\} d\tau\Big\}$$

$$= K_{11}\left(\frac{1}{\gamma} + \frac{3}{\gamma^2} + \frac{2}{\gamma^3}\right) |\, \rho - \bar{\rho} \,| = K_{12} \,|\, \rho - \bar{\rho} \,|,$$

where K_{12} is a constant independent of n and ρ.

Theorem 9.4 is proved.

2.10. The Case of Infinitely Many Angular Variables

We consider a system of differential equations of the form

$$\frac{d\varphi}{dt} = a(\varphi), \qquad \frac{dx}{dt} = P(\varphi)x + c(\varphi), \tag{10.1}$$

where $x \in \mathfrak{M}$, $\varphi = (\varphi_1, \varphi_2, \ldots) \in \mathfrak{M}$, and the components of the vector functions $a(\varphi)$ and $c(\varphi)$ and the infinite matrix $P(\varphi)$ are functions 2π-periodic in φ_i, $i = 1, 2, \ldots$, and continuous in φ. Assume that

$$\sum_{j=1}^{\infty} \sup_{\varphi} |p_{sj}(\varphi)| \leq p^0 = \text{const} < \infty, \quad s = 1, 2, \ldots . \tag{10.2}$$

Then the matrix $P(\varphi)$ is a linear operator in the space $\mathfrak{M}$ of bounded numerical sequences mapping a vector $x \in \mathfrak{M}$ into a vector $P(\varphi)$, $x \in \mathfrak{M}$.

We say that a function $f(x)$ satisfies the strengthened Cauchy–Lipschitz conditions with respect to φ with a coefficient $\varepsilon(m)$ if

$$\Big\| f(\varphi_1, \ldots, \varphi_m,\, \varphi_{m+1}, \ldots) - f(\varphi_1, \ldots, \varphi_m,\, \varphi''_{m+1}, \ldots) \Big\|$$

$$\leq \varepsilon(m) \sup\Big\{ |\varphi'_{m+1} - \varphi''_{m+1}|, |\varphi'_{m+2} - \varphi''_{m+2}|, \ldots \Big\},$$

where $\varepsilon(m) \to 0$ as $m \to \infty$ and

$$(\varphi_1, \ldots, \varphi_m, \varphi'_{m+1}, \varphi''_{m+2}, \ldots) \quad \text{and} \quad (\varphi_1, \ldots, \varphi_m, \varphi''_{m+1}, \varphi''_{m+2}, \ldots)$$

are arbitrary points from $\mathfrak{M}$ whose first coordinates coincide.

The set of bounded vector functions and matrices whose components are 2π-periodic in φ_i, $i = 1, 2, \ldots$, and continuous in $\varphi = (\varphi_1, \varphi_2, \ldots)$ is denoted by $C^0(\mathcal{F}_\infty)$. The set of elements from $C^0(\mathcal{F}_\infty)$ satisfying the Lipschitz condition with respect to φ is denoted by $C^0_{\text{Lip}}(\varphi_\infty)$. Here, $\mathcal{F}_\infty$ is a countably dimensional torus.

In the set $C^0_{\text{Lip}}(\mathcal{F}_\infty)$, we select a subset of elements satisfying the strengthened Cauchy–Lipschitz conditions with respect to φ and denote this subset by $\widehat{C}^0_{\text{Lip}}(\mathcal{F}_\infty)$.

Let $a(\varphi) \in C^0_{\text{Lip}}(\mathcal{F}_\infty)$. According to Theorem 1.3, the first equation in system (10.1) possesses a unique solution $\varphi = \varphi_t(\varphi)$ with initial data 0, $\varphi \in \mathfrak{M}$. Moreover, this solution is defined for all $t \in R^1$.

Consider an equation

$$\frac{dx}{dt} = P(\varphi_t(\varphi))x. \tag{10.3}$$

By virtue of (10.2), its right-hand side satisfies condition $(\widehat{P})$ from Section 1.3. Therefore, the matrizant $\Omega^t_\tau(\varphi)$ of equation (10.3) exists.

We define the Green function of the problem of invariant torus for the system of equations (10.1) by setting

$$G_0(\tau, \varphi) = \begin{cases} \Omega^0_\tau(\varphi) C(\varphi_\tau(\varphi)), & \tau \leq 0, \\ \Omega^0_\tau(\varphi) \Big[C(\varphi_\tau(\varphi)) - E \Big], & \tau > 0, \end{cases} \tag{10.4}$$

and assuming that the matrix $C(\varphi)$ belongs to $C^0(\mathcal{F}_m) \cap C_\varphi$ and is such that

$$\|G_0(\tau, \varphi)\| \leq K \exp\{-\gamma|\tau|\}, \quad \tau \in R^1. \tag{10.5}$$

Then the invariant torus $\mathcal{F}\colon x = u(\varphi)$, $\varphi \in \mathcal{F}_\infty$, of the system of equations (10.1) is specified by the function

$$u(\varphi) = \int\limits_{-\infty}^{\infty} G_0(\tau, \varphi) c(\varphi_\tau(\varphi))\, d\tau. \tag{10.6}$$

Theorem 10.1. *Assume that the system of equations (10.1) satisfies the following conditions:*

(i) $a(\varphi), P(\varphi), c(\varphi) \in C^0_{\mathrm{Lip}}(\mathcal{F}_\infty)$ *and* $\|a(\varphi) - a(\bar{\varphi})\| \le \alpha\|\varphi - \bar{\varphi}\|$;

(ii) inequality (10.2) is true;

(iii) the Green function of the problem of invariant torus $G_0(\tau, \varphi)$ *exists and satisfies inequality (10.5).*

Then the system of equations (10.1) possesses the invariant torus $\mathcal{F}$ *admitting representation (10.6) and such that*

$$\|u(\varphi) - u(\bar{\varphi})\| \le \|\varphi - \varphi\|^{\frac{\nu}{2(\nu+1)}},$$

where $\nu > 0$ *satisfies the condition*

$$\frac{\gamma}{\alpha} > \frac{\nu}{\nu + 1}$$

and K_1 *is a positive constant.*

The proof of Theorem 10.1 is similar to the proof of Theorem 8.2. Denote

$$\overset{(m)}{\varphi} = (\varphi_1, \varphi_2, \ldots, \varphi_m, 0, 0, \ldots),$$

$$\varphi_t\big(\overset{(m)}{\varphi}\big) = \{\varphi_1(t, \varphi_1, \ldots, \varphi_m, 0, 0, \ldots), \varphi_2(t, \varphi_1, \ldots, \varphi_m, 0, 0, \ldots), \ldots\},$$

$$\overline{\varphi}_t(\varphi) = \{\varphi_1(t, \varphi), \ldots, \varphi_m(t, \varphi), 0, 0, \ldots\},$$

$$\overline{\overline{\varphi}}_t = \{0, 0, \ldots, 0, \varphi_{m+1}(t, \varphi), \varphi_{m+2}(t, \varphi), \ldots\}.$$

Let us now study the possibility of reducing the problem of invariant torus for the system of equations (10.1) to the finite-dimensional case. First, we prove some auxiliary statements.

Parallel with (10.1), we consider an equation of the form

$$\frac{dx}{dt} = P\Big(\varphi_t\big(\overset{(m)}{\varphi}\big)\Big)x + c\Big(\varphi_t\big(\overset{(m)}{\varphi}\big)\Big), \tag{10.7}$$

where $\varphi_t\big(\overset{(m)}{\varphi}\big)$ is the solution of the first equation in (10.1) such that $\varphi_0\big(\overset{(m)}{\varphi}\big) = \overset{(m)}{\varphi}$. By $\Omega^t_\tau\big(\overset{(m)}{\varphi}\big)$ we denote the matrizant of the homogeneous system corresponding to (10.7).

Lemma 10.1. *Assume that the system of equations (10.7) satisfies the following conditions:*

(i) $a(\varphi), P(\varphi) \in \widehat{C}^0_{\mathrm{Lip}}(\mathcal{F}_\infty)$ *with coefficients* $\alpha(m)$ *and* $\varepsilon(m)$, *respectively;*

(ii) inequality (10.2) is satisfied;

(iii) $\left\|\Omega^0_\tau(\varphi)\right\| \leq N\exp\{\gamma\tau\}$, $\tau \leq 0$, *where* N *and* γ *are positive constants;*

(iv) $\gamma > \alpha(0)$.

Then $\lim\limits_{m\to\infty}\left\|\Omega^0_\tau\left(\overset{(m)}{\varphi}\right) - \Omega^0_\tau(\varphi)\right\| = 0$ *uniformly with respect to* $\varphi \in \mathcal{F}_\infty$ *and* $\tau \in R^- = (-\infty, 0]$.

Proof. First, we estimate the following quantity:

$$V(t) = \left\|\,\overline{\varphi_t(\varphi) - \varphi_t\left(\overset{(m)}{\varphi}\right)}\,\right\|, \quad t \in R^1.$$

For $t > 0$, we have

$$\begin{aligned}
V(t) &= \left\|\int\limits_0^t \overline{\left(a(\varphi_s(\varphi)) - a\left(\varphi_s\left(\overset{(m)}{\varphi}\right)\right)\right)}ds\right\| \\
&\leq \int\limits_0^t \left\|a(\varphi_s(\varphi)) - a\left(\varphi_s\left(\overset{(m)}{\varphi}\right)\right)\right\|ds \\
&\leq \int\limits_0^t \Big\{\left\|a(\varphi_s(\varphi)) - a\left(\overline{\varphi}_s(\varphi) + \overline{\overline{\varphi}}_s\left(\overset{(m)}{\varphi}\right)\right)\right\| \\
&\qquad\qquad + \left\|a\left(\overline{\varphi}_s(\varphi + \overline{\overline{\varphi}}_s\left(\overset{(m)}{\varphi}\right)\right) - a\left(\varphi_s\left(\overset{(m)}{\varphi}\right)\right)\right\|\Big\} \\
&\leq \int\limits_0^t \left\{\alpha(m)\left\|\varphi_s(\varphi) - \varphi_s\left(\overset{(m)}{\varphi}\right)\right\| + \alpha(0)\left\|\,\overline{\varphi_s(\varphi) - \varphi_s\left(\overset{(m)}{\varphi}\right)}\,\right\|\right\}ds.
\end{aligned}$$

According to the Gronwall–Bellman lemma,

$$\left\|\varphi_t(\varphi) - \varphi_t\left(\overset{(m)}{\varphi}\right)\right\| \leq \exp\{\alpha(0)|t|\}\Delta_m\varphi,$$

where $t \in R^1$ and $\Delta_m\varphi = \sup\{|\varphi_{m+1}|, |\varphi_{m+2}|, \ldots\}$.

Hence, for $V(t)$, we obtain

$$V(t) \leq \frac{\alpha(m)}{\alpha(0)} \Delta_m \varphi \exp\{\alpha(0)t\} + \int_0^t \alpha(0) V(s)\, ds, \quad t > 0, \tag{10.8}$$

where, without loss of generality, we can set $\Delta_m \varphi \leq 2\pi$.

In this case, inequality (10.8) takes the form

$$V(t) \leq \frac{\alpha(m)}{\alpha(0)} 2\pi \exp\{\alpha(0)t\} + \int_0^t \alpha(0) V(s)\, ds, \quad t > 0.$$

We solve this inequality to obtain

$$V(t) \leq 2\pi \frac{\alpha(m)}{\alpha(0)} [1 + \alpha(0)t] \exp\{\alpha(0)t\}, \quad t > 0.$$

Similarly, for $t < 0$, we conclude that

$$V(t) \leq 2\pi \frac{\alpha(m)}{\alpha(0)} [1 + \alpha(0)|t|] \exp\{\alpha(0)t\}, \quad t \in R^1, \tag{10.9}$$

Then

$$\begin{aligned}
&\left\| P(\varphi_t(\varphi)) - P\Big(\varphi_t\Big(\overset{(m)}{\varphi}\Big)\Big) \right\| \\
&\quad \leq \left\| P(\varphi_t(\varphi)) - P\Big(\overline{\varphi}_t(\varphi) + \overline{\overline{\varphi}}_t\Big(\overset{(m)}{\varphi}\Big)\Big) \right\| \\
&\qquad\qquad + \left\| P\Big(\overline{\varphi}_t(\varphi) + \overline{\overline{\varphi}}_t\Big(\overset{(m)}{\varphi}\Big)\Big) - P\Big(\varphi_t\Big(\overset{(m)}{\varphi}\Big)\Big) \right\| \\
&\quad \leq \varepsilon(m) \exp\{-\alpha(0)t\} 2\pi + \varepsilon(0) \left\| \varphi_t(\varphi) - \Big(\varphi_t\Big(\overset{(m)}{\varphi}\Big)\Big) \right\| \\
&\quad \leq 2\pi \Big\{ \varepsilon(m) + \frac{\varepsilon(0)\alpha(m)}{\alpha(0)} - \varepsilon(0)\alpha(m)t \Big\} \exp\{-\alpha(0)t\} \quad \text{at} \quad t \leq 0.
\end{aligned}$$

In this case, for $\tau \leq 0$ and $\varphi \in \mathcal{F}_\infty$, we can write

$$\begin{aligned}
&\left\| \Omega_\tau^0(\varphi) - \Omega_\tau^0\Big(\overset{(m)}{\varphi}\Big) \right\| \\
&\quad \leq \int_\tau^0 N^2 \exp\{\gamma s - \gamma(s - \tau)\} \\
&\qquad\qquad \times 2\pi \Big\{ \varepsilon(m) + \frac{\varepsilon(0)\alpha(m)}{\alpha(0)} - \varepsilon(0)\alpha(m)t \Big\} \exp\{-\alpha(0)s\} ds
\end{aligned}$$

$$\leq 2\pi N^2\Big[\frac{1}{\alpha(0)}\Big(\frac{\varepsilon(0)\alpha(m)}{\alpha(0)} + \varepsilon(m)\Big)\exp\{(\gamma-\alpha(0))\tau\}$$

$$+\varepsilon(0)\alpha(m)\exp\{\gamma\tau\}\int_{\tau}^{0}\exp\{-\alpha(0)s\}(-s)ds\Big]$$

$$\leq \frac{2\pi N^2}{\alpha(0)}\Big[\varepsilon(0)\alpha(m) + \varepsilon(m) + \varepsilon(0)\alpha(m)\exp\{\gamma\tau\}$$

$$\times\Big((-\tau)\exp\{-\alpha(0)\tau\} + \frac{1}{\alpha(0)}\Big)\Big]$$

$$\leq \frac{2\pi N^2}{\alpha(0)}\Big[\varepsilon(0)\alpha(m)\Big(1+\frac{1}{\alpha(0)}+\frac{1}{(\gamma-\alpha(0))e}\Big)+\varepsilon(m)$$

$$= \eta(m) \to 0$$

as $m \to \infty$ uniformly in $\varphi \in \mathcal{F}_\infty$ and $\tau \in R^-$.

By $u(\varphi)$ and $u\big(\overset{(m)}{\varphi}\big)$ we denote functions of the form (10.6) specifying the invariant tori of the systems of equations (10.1) and (10.7), respectively.

Theorem 10.2. *Assume that the conditions of Lemma 10.1 are satisfied and* $c(\varphi) \in \widehat{C}^0_{\mathrm{Lip}}(\mathcal{F}_\infty)$. *Then*

$$\lim_{m\to\infty}\Big\|u(\varphi) - u\big(\overset{(m)}{\varphi}\big)\Big\| = 0$$

uniformly in $\varphi \in \mathcal{F}_\infty$.

Proof. The estimates

$$\Big\|\Omega^0_\tau(\varphi) - \Omega^0_\tau\big(\overset{(m)}{\varphi}\big)\Big\| \leq \eta(m),$$

$$\Big\|\Omega^0_\tau(\varphi) - \Omega^0_\tau\big(\overset{(m)}{\varphi}\big)\Big\| \leq 2N\exp\{\gamma\tau\}, \quad \tau \in R^-,$$

yield

$$\Big\|\Omega^0_\tau(\varphi) - \Omega^0_\tau\big(\overset{(m)}{\varphi}\big)\Big\| \leq \big(2N\eta(m)\big)^{\frac{1}{2}}\exp\Big\{\frac{1}{2}\gamma\tau\Big\}. \qquad (10.10)$$

We denote a constant for which $c(\varphi) \in \widehat{C}^0_{\mathrm{Lip}}(\mathcal{F}_\infty)$ by $\beta(m)$. Further, in view of inequality (10.9), we obtain

$$\left\|c(\varphi_t(\varphi)) - c\Big(\varphi_t\Big(\overset{(m)}{\varphi}\Big)\Big)\right\|$$

$$\le \beta(m)\exp\{-\alpha(0)t\}\Delta_m\varphi + \beta(0)\overline{\left\|\varphi_t(\varphi) - \varphi_t\Big(\overset{(m)}{\varphi}\Big)\right\|}$$

$$\le 2\pi\Big\{\beta(m) + \frac{\beta(0)\alpha(m)}{\alpha(0)} - \beta(0)\alpha(m)t\Big\}\exp\{-\alpha(0)t\}, \quad t \le 0.$$

Finally, by using this inequality and inequality (10.10) and setting $\|c(\varphi)\| \le c^0 = \text{const} < \infty$, we get

$$\left\|u(\varphi) - u\Big(\overset{(m)}{\varphi}\Big)\right\| \le \int_{-\infty}^{0} \|\Omega_s^0(\varphi)\| \left\|c(\varphi_s(\varphi)) - c\Big(\varphi_s\Big(\overset{(m)}{\varphi}\Big)\Big)\right\| ds$$

$$+ \int_{-\infty}^{0} \left\|\Omega_s^0(\varphi) - \Omega_s^0\Big(\overset{(m)}{\varphi}\Big)\right\| \left\|c\Big(\varphi_s\Big(\overset{(m)}{\varphi}\Big)\Big)\right\| ds$$

$$\le \int_{-\infty}^{0} N\exp\{\gamma s\}2\pi$$

$$\times\Big\{\beta(m) + \frac{\beta(0)\alpha(m)}{\alpha(0)} - \beta(0)\alpha(m)s\Big\}\exp\{-\alpha(0)s\}ds$$

$$+ \int_{-\infty}^{0} c^0\big(2N\eta(m)\big)^{\frac{1}{2}}\exp\Big\{\frac{s\gamma}{2}\Big\}ds$$

$$\le \frac{2\pi N}{\gamma - \alpha(0)}\Big\{\beta(m) + \beta(0)\alpha(m)\Big(\frac{1}{\alpha(0)} + \frac{1}{\gamma - \alpha(0)}\Big)\Big\}$$

$$+ \frac{2c^0}{\gamma}\big(2N\eta(m)\big)^{\frac{1}{2}} = \zeta(m),$$

where $\zeta(m) \to 0$ as $m \to \infty$ uniformly in $\varphi \in \mathcal{F}_\infty$.

Theorem 10.2 is proved.

Parallel with equation (10.7), we consider the equation

$$\frac{dx}{dt} = P\Big(\overline{\varphi}_t\Big(\overset{(m)}{\varphi}\Big)\Big)x + c\Big(\overline{\varphi}_t\Big(\overset{(m)}{\varphi}\Big)\Big). \tag{10.11}$$

By $\widehat{\Omega}^t_\tau\left(\overset{(m)}{\varphi}\right)$ we denote the matrizant of the equation

$$\frac{dx}{dt} = P\left(\overline{\varphi}_t\left(\overset{(m)}{\varphi}\right)\right)x. \tag{10.11}$$

A function of the form (10.6) specifying the invariant torus of equation (10.11) is denoted by $\widehat{u}\left(\overset{(m)}{\varphi}\right)$.

Theorem 10.3. *Assume that the conditions of Theorem 10.2 are satisfied and, moreover,*

$$\left\|\widehat{\Omega}^0_\tau\left(\overset{(m)}{\varphi}\right)\right\| \le N\exp\{\gamma\tau\}, \quad \tau \le 0, \quad m = 1, 2, 3, \dots,$$

where N is a positive constant. Then

$$\lim_{m\to\infty}\left\|\widehat{u}\left(\overset{(m)}{\varphi}\right) - u(\varphi)\right\| = 0$$

uniformly in $\varphi \in \mathcal{F}_\infty$.

Proof. In view of Theorem 10.2, it suffices to show that

$$\lim_{m\to\infty}\left\|\widehat{u}\left(\overset{(m)}{\varphi}\right) - u\left(\overset{(m)}{\varphi}\right)\right\| = 0.$$

By setting $\|a(\varphi)\| \le \alpha^0 = \text{const} < \infty$, for $\varphi \in \mathcal{F}_\infty$ and $t \le 0$ we obtain

$$\|\varphi_t(\varphi)\| \le 2\pi - \alpha^0 t. \tag{10.12}$$

Then

$$\left\|P\left(\varphi_s\left(\overset{(m)}{\varphi}\right)\right) - P\left(\overline{\varphi}_s\left(\overset{(m)}{\varphi}\right)\right)\right\| \le \varepsilon(m)\left\|\varphi_s\left(\overset{(m)}{\varphi}\right)\right\| \le \varepsilon(m)(2\pi - \alpha^0 s),$$

whence it follows that

$$\begin{aligned}
&\left\|\Omega^0_\tau\left(\overset{(m)}{\varphi}\right) - \widehat{\Omega}^0_\tau\left(\overset{(m)}{\varphi}\right)\right\| \\
&\le \int_\tau^0 \left\|\Omega^0_s\left(\overset{(m)}{\varphi}\right)\right\|\left\|P\left(\varphi_s\left(\overset{(m)}{\varphi}\right)\right) - P\left(\overline{\varphi}_s\left(\overset{(m)}{\varphi}\right)\right)\right\|\left\|\widehat{\Omega}^s_\tau\left(\overset{(m)}{\varphi}\right)\right\|ds \\
&\le \int_\tau^0 N^2\exp\{\gamma\tau\}\varepsilon(m)(2\pi - \alpha^0 s)ds \\
&= \varepsilon(m)N^2\left\{2\pi(-\tau) + \frac{\alpha^0\tau^2}{2}\right\}\exp\{\gamma\tau\}.
\end{aligned}$$

By using estimate (10.12), we get

$$\begin{aligned}
\left\| u\Big(\overset{(m)}{\varphi}\Big) - \widehat{u}\Big(\overset{(m)}{\varphi}\Big)\right\| \le & \int_{-\infty}^{0} \Big\{ \Big\|\Omega_s^0\Big(\overset{(m)}{\varphi}\Big)\Big\| \Big\| c\Big(\varphi_s\Big(\overset{(m)}{\varphi}\Big)\Big) - c\Big(\overline{\varphi}_s\Big(\overset{(m)}{\varphi}\Big)\Big)\Big\| \\
& + \Big\|\Omega_s^0\Big(\overset{(m)}{\varphi}\Big) - \widehat{\Omega}_s^0\Big(\overset{(m)}{\varphi}\Big)\Big\| \Big\| c\Big(\overline{\varphi}_s\Big(\overset{(m)}{\varphi}\Big)\Big)\Big\| \Big\} ds \\
\le & \int_{-\infty}^{0} N \exp\{\gamma s\} \beta(m)(2\pi - \alpha^0 s)\, ds \\
& + \int_{-\infty}^{0} \varepsilon(m) N^2 c^0 \Big\{ 2\pi(-s) + \frac{\alpha^0 s^2}{2} \Big\} \exp\{\gamma s\}\, ds \\
\le & \left(2\pi + \frac{\alpha^0}{\gamma}\right) \frac{N}{\gamma} \Big\{ \beta(m) + \frac{c^0 N}{\gamma} \varepsilon(m) \Big\} = \zeta^0(m),
\end{aligned}$$

where $\zeta^0(m) \to 0$ as $m \to \infty$. Theorem 10.3 is proved.

Consider the equation

$$\frac{dx}{dt} = P\left(\overset{(m)}{\varphi_t}\left(\overset{(m)}{\varphi}\right)\right) x + c\left(\overset{(m)}{\varphi_t}\left(\overset{(m)}{\varphi}\right)\right), \tag{10.13}$$

where $\overset{(m)}{\varphi_t}\left(\overset{(m)}{\varphi}\right)$ is the solution of the finite-dimensional system of equations

$$\frac{d\overline{\varphi}}{dt} = \overline{a}(\overline{\varphi}) \tag{10.14}$$

such that $\overset{(m)}{\varphi_0}\left(\overset{(m)}{\varphi}\right) = \overset{(m)}{\varphi}$. The matrizant of the equation

$$\frac{dx}{dt} = P\left(\overset{(m)}{\varphi_t}\left(\overset{(m)}{\varphi}\right)\right) x$$

is denoted by $\widehat{\widehat{\Omega}}{}_\tau^t\left(\overset{(m)}{\varphi}\right)$, and a function of the form (10.6) specifying the invariant torus of equation (10.13) is denoted by $\widehat{\widehat{u}}\left(\overset{(m)}{\varphi}\right)$.

Theorem 10.4. *Assume that the conditions of Theorem 10.3 are satisfied and*

$$\left\|\widehat{\widehat{\Omega}}{}_\tau^0\left(\overset{(m)}{\varphi}\right)\right\| \le N \exp\{\gamma\tau\}, \quad \tau \le 0, \quad m = 1, 2, \ldots .$$

Then

$$\lim_{m\to\infty}\Big\|\widehat{\widehat{u}}\Big(\overset{(m)}{\varphi}\Big) - u(\varphi)\Big\| = 0$$

uniformly in $\varphi \in \mathcal{F}_\infty$.

Proof. It suffices to show that

$$\lim_{m\to\infty}\Big\|\widehat{\widehat{u}}\Big(\overset{(m)}{\varphi}\Big) - \widehat{u}\Big(\overset{(m)}{\varphi}\Big)\Big\| = 0$$

and use Theorem 10.3.

For $t < 0$, we have

$$\begin{aligned} V^0(t) &= \Big\|\overline{\varphi}_t(\varphi) - \overset{(m)}{\varphi_t}\Big(\overset{(m)}{\varphi}\Big)\Big\| \\ &\le \Big\|\varphi_t(\varphi) - \overset{(m)}{\varphi_t}\Big(\overset{(m)}{\varphi}\Big)\Big\| \\ &\le \int_t^0 \Big\|a(\varphi_s(\varphi)) - a\big(\overline{\varphi}_s(\varphi)\big)\Big\| ds + \int_t^0 \Big\|a\big(\overline{\varphi}_s(\varphi)\big) - a\Big(\overset{(m)}{\varphi_s}\Big(\overset{(m)}{\varphi}\Big)\Big)\Big\| ds \\ &\le \int_t^0 \alpha(m)\|\varphi_s(\varphi)\| ds + \int_t^0 \alpha(0)\Big\|\overline{\varphi}_s(\varphi) - \overset{(m)}{\varphi_s}\Big(\overset{(m)}{\varphi}\Big)\Big\|. \end{aligned}$$

By using estimate (10.12), we obtain

$$V^0(t) \le \alpha(m)\Big(-2\pi t + \frac{\alpha^0}{2}t^2\Big) + \int_t^0 \alpha(0) V^0(s)\, ds, \quad t \le 0.$$

We now solve this inequality with respect to $V^0(t)$. This gives

$$\begin{aligned} V^0(t) &\le V_1^0(t) \\ &= \frac{\alpha(m)}{\alpha(0)}\Big\{-2\pi + \alpha^0 t - \frac{\alpha^0}{\alpha(0)} + \Big[2\pi + \frac{\alpha^0}{\alpha(0)}\Big]\exp\{-\alpha(0)t\}\Big\}, \quad t \le 0, \end{aligned}$$

where $V_1^0(t) = 0$ for $t = 0$ and $V_1^0(t) > 0$ for $t < 0$. In this case,

$$\begin{aligned} &\Big\|\overline{\varphi}_t\Big(\overset{(m)}{\varphi}\Big) - \overset{(m)}{\varphi_t}\Big(\overset{(m)}{\varphi}\Big)\Big\| \\ &\qquad \le \overline{\Big\|\varphi_t(\varphi) - \varphi_t\Big(\overset{(m)}{\varphi}\Big)\Big\|} + \Big\|\overline{\varphi}_t(\varphi) - \overset{(m)}{\varphi_t}\Big(\overset{(m)}{\varphi}\Big)\Big\| \end{aligned}$$

$$\leq \frac{\alpha(m)}{\alpha(0)}\Big[2\pi(1-\alpha(0)t)\exp\{-\alpha(0)t\} + \alpha^0 t$$

$$+ \Big(2\pi + \frac{\alpha^0}{\alpha(0)}\Big)\exp\{-\alpha(0)t\}\Big], \quad t \leq 0. \tag{10.15}$$

Then

$$\Big\| c\Big(\varphi_t\Big(\overset{(m)}{\varphi}\Big)\Big) - c\Big(\overset{(m)}{\varphi_t}\Big(\overset{(m)}{\varphi}\Big)\Big)\Big\|$$

$$\leq \beta(0)\Big\|\varphi_t\Big(\overset{(m)}{\varphi}\Big) - \overset{(m)}{\varphi_t}\Big(\overset{(m)}{\varphi}\Big)\Big\|$$

$$\leq \alpha(m)\Big[P_1(-t)\exp\{-\alpha(0)t\} + P_2 t + P_3\exp\{-\alpha(0)t\}\Big],$$

where

$$P_1 = 2\pi\beta(0), \qquad P_2 = \frac{\alpha^0\beta(0)}{\alpha(0)}, \quad \text{and} \quad P_3 = \frac{\beta(0)}{\alpha(0)}\Big(4\pi + \frac{\alpha^0}{\alpha(0)}\Big)$$

are positive constants. Since $t \leq 0$, we have

$$\Big\| c\Big(\overline{\varphi}_t\Big(\overset{(m)}{\varphi}\Big)\Big) - c\Big(\overset{(m)}{\varphi_t}\Big(\overset{(m)}{\varphi}\Big)\Big)\Big\| \leq \alpha(m)[-P_1 t + P_3]\exp\{-\alpha(0)t\}.$$

Further, we get

$$\Big\|\widehat{\Omega}^0_\tau\Big(\overset{(m)}{\varphi}\Big) - \widehat{\widehat{\Omega}}{}^0_\tau\Big(\overset{(m)}{\varphi}\Big)\Big\|$$

$$\leq \int_\tau^0 N^2\exp\{\gamma\tau\}\varepsilon(0)\Big\|\overline{\varphi}_s\Big(\overset{(m)}{\varphi}\Big) - \overset{(m)}{\varphi_s}\Big(\overset{(m)}{\varphi}\Big)\Big\| ds$$

$$\leq \frac{N^2\varepsilon(0)\alpha(m)}{\beta(0)}\exp\{\gamma\tau\}\int_\tau^0 [-P_1 s + P_3]\exp\Big\{-\alpha(0)s\Big\} ds$$

$$\leq \frac{N^2\varepsilon(0)}{\beta(0)}\alpha(m)\exp\{\gamma\tau\}$$

$$\times \int_\tau^0 \Big\{P_1(-s)\exp\{-\alpha(0)s\} + P_3\exp\{-\alpha(0)s\}\Big\}$$

$$\leq \alpha(m)\frac{N^2\varepsilon(0)}{\beta(0)}\left\{\frac{1}{\alpha(0)}[P_3 - P_1]\exp\left\{(\gamma-\alpha(0))\tau\right\}\right.$$
$$\left.+\frac{P_1}{\alpha^2(0)}\exp\{\gamma\tau\}\right\}.$$

This yields the following estimates:

$$\left\|\widehat{u}\Big(\overset{(m)}{\varphi}\Big) - \widehat{\widehat{u}}\Big(\overset{(m)}{\varphi}\Big)\right\|$$
$$\leq \int_{-\infty}^{0}\left\|\widehat{\Omega}_s^0\Big(\overset{(m)}{\varphi}\Big)\right\|\left\|c\Big(\overline{\varphi}_s\Big(\overset{(m)}{\varphi}\Big)\Big) - c\Big(\overset{(m)}{\varphi_s}\Big(\overset{(m)}{\varphi}\Big)\Big)\right\|ds$$
$$+\int_{-\infty}^{0}\left\|\widehat{\Omega}_s^0\Big(\overset{(m)}{\varphi}\Big)\right\| - \left\|\widehat{\widehat{\Omega}}{}_s^0\Big(\overset{(m)}{\varphi}\Big)\right\|\left\|c\Big(\overset{(m)}{\varphi_s}\Big(\overset{(m)}{\varphi}\Big)\Big)\right\|ds$$
$$\leq \int_{-\infty}^{0} N\exp\{\gamma s\}\alpha(m)$$
$$\times\Big\{P_1(-s)\exp\{-\alpha(0)s\} + P_3\exp\{-\alpha(0)s\}\Big\}ds$$
$$+\int_{-\infty}^{0}\alpha(m)\frac{c^0N^2\varepsilon(0)}{\beta(0)}$$
$$\times\left[\left(\frac{P_3}{\alpha(0)} - \frac{sP_1}{\alpha(0)}\right)\exp\left\{(\gamma-\alpha(0))s\right\} + \frac{P_1}{\alpha^2(0)}\exp\{\gamma s\}\right]ds$$
$$\leq \alpha(m)\kappa,$$

where $\kappa = \text{const} > 0$ and $\alpha(m) \to 0$ as $m \to \infty$.

Theorem 10.4 is proved.

Let us now present the principal result of this section. Parallel with the system of equations (10.1), we consider a finite-dimensional system of the form

$$\frac{d\overline{\varphi}}{dt} = \overline{a}(\overline{\varphi}), \qquad \frac{d\overset{(n)}{x}}{dt} = \overset{(n)}{P}(\overline{\varphi})\overset{(n)}{x} + \overset{(n)}{c}(\overline{\varphi}), \tag{10.16}$$

obtained by truncation with respect to φ up to the mth order and with respect to x up to the nth order inclusive. The invariant torus of this system of equations is denoted by $\widehat{\widehat{\mathcal{F}}}: h = \widehat{\widehat{u}}_n\left(\overset{(m)}{\varphi}\right)$, $\overset{(m)}{\varphi} \in \mathcal{F}_m$, and the matrizant of the equation

$$\frac{d\overset{(n)}{x}}{dt} = \overset{(n)}{P}\left(\overset{(m)}{\varphi_t}\left(\overset{(m)}{\varphi}\right)\right)\overset{(n)}{x} \tag{10.17}$$

is denoted by $\widehat{\widehat{\Omega}}{}^t_\tau\left(\overset{(m)}{\varphi}\right)_n$. Here, $\overset{(m)}{\varphi_t}\left(\overset{(m)}{\varphi}\right)$ is the solution of the first equation in system (10.16) such that $\overset{(m)}{\varphi_0}\left(\overset{(m)}{\varphi}\right) = \overset{(m)}{\varphi}$, $\overset{(n)}{x} = (x_1, x_2, \ldots, x_n)$, $\overset{(n)}{c} = (c_1, c_2, \ldots, c_n)$, and $\overset{(n)}{P}(\varphi) = [p_{ij}(\varphi)]_{i,j=1}^n$.

Theorem 10.5. *Assume that the system of equations (10.1) satisfies the following conditions:*

(i) $a(\varphi)$, $P(\varphi)$, $c(\varphi) \in \widehat{C}^0_{\text{Lip}}(\mathcal{F}_\infty)$; *for the function* $a(\varphi)$*, this is true with a constant* $\alpha(m)$;

(ii) $\sum\limits_{j=n+1}^{\infty} \sup\limits_{\varphi} \|p_{sj}(\varphi)\| \leq K\delta(n)$, $s = 1, 2, \ldots$, *where* $K = \text{const} < \infty$ *and* $\delta(n) \to 0$ *as* $n \to \infty$;

(iii) the matrizant of the system of equations (10.17) satisfies the inequality

$$\left\|\widehat{\widehat{\Omega}}{}^0_\tau\left(\overset{(m)}{\varphi}\right)_n\right\| \leq N \exp\{\gamma t\}, \quad \tau \leq 0,$$

where N *and* γ *are positive constants independent of* n, m, *and* φ;

(iv) $\gamma > \alpha(0)$.

Then the invariant torus $\mathcal{F}: x = u(\varphi)$, $\varphi \in \mathcal{F}_\infty$, *of system (10.1) can be represented in the form of repeated limit as follows:*

$$u(\varphi) = \lim_{m\to\infty} \lim_{n\to\infty} \widehat{\widehat{u}}_n\left(\overset{(m)}{\varphi}\right), \tag{10.18}$$

where the outer limit is understood in the strong sense, the inner limit is understood in the weak sense, and, moreover, $u(\varphi) \in \widehat{C}^0_{\text{Lip}}(\mathcal{F}_\infty)$.

Proof. It is easy to see that Theorem 9.1 is true under the conditions of Theorem 10.5. Therefore, the function $\widehat{\widehat{u}}\binom{(m)}{\varphi}$ can be represented in the form of the following coordinatewise limit:

$$\widehat{\widehat{u}}\binom{(m)}{\varphi} = \lim_{n\to\infty} \widehat{\widehat{u}}_n\binom{(m)}{\varphi}.$$

It remains to use Theorem 10.4 and complete the proof of equality (10.18). For this purpose, it suffices to show that

$$\|\Omega_\tau^0(\varphi)\| \le N\exp\{\gamma\tau\}, \qquad \left\|\Omega_\tau^0\binom{(m)}{\varphi}\right\| \le N\exp\{\gamma\tau\},$$

$$\left\|\widehat{\Omega}_\tau^0\binom{(m)}{\varphi}\right\| \le N\exp\{\gamma\tau\}, \qquad \left\|\widehat{\widehat{\Omega}}_\tau^0\binom{(m)}{\varphi}\right\| \le N\exp\{\gamma\tau\},$$

where $\tau \le 0$. We denote these inequalities in the order of appearance by O_1, O_2, O_3, and O_4.

According to Theorem 9.1, the limit

$$\lim_{n\to\infty} \widehat{\widehat{\Omega}}_\tau^t\binom{(m)}{\varphi}_n = \widehat{\widehat{\Omega}}_\tau^t\binom{(m)}{\varphi}$$

exists in the weak sense (see 9.3) and, hence, the inequality O_4 is true.

We set $p_0 = K\delta(0)$, i.e., $\|P(\varphi)\| \le p^0$. By using inequality (10.15), for $\tau < 0$ we can write

$$\left\|\widehat{\Omega}_\tau^0\binom{(m)}{\varphi} - \widehat{\widehat{\Omega}}_\tau^0\binom{(m)}{\varphi}\right\|$$

$$\le \int_\tau^0 \left\|\widehat{\Omega}_s^0\binom{(m)}{\varphi}\right\| \left\|P\left(\overline{\varphi}_s\binom{(m)}{\varphi}\right) - P\left(\overset{(m)}{\varphi_s}\binom{(m)}{\varphi}\right)\right\| \left\|\widehat{\widehat{\Omega}}_\tau^s\binom{(m)}{\varphi}\right\| ds$$

$$\le \int_\tau^0 \exp\{-p^0 s\}\varepsilon(0)\left\|\overline{\varphi}_t\binom{(m)}{\varphi} - \overset{(m)}{\varphi_t}\binom{(m)}{\varphi}\right\| N\exp\{-\gamma(s-\tau)\}ds$$

$$\le \alpha(m)\frac{N\varepsilon(0)}{\beta(0)} \int_\tau^0 \exp\{-p^0 s - \gamma s + \gamma\tau\}[P_1(-s) + P_3]\exp\{-\alpha(0)s\}ds$$

$$\le \alpha(m) f(\tau),$$

where $f(\tau) > 0$ for $\tau < 0$ is a continuous function and $\alpha(m) \to 0$ as $m \to \infty$. This estimate implies that

$$\Big\|\widehat{\widehat{\Omega}}{}^0_\tau\Big(\overset{(m)}{\varphi}\Big)\Big\|\underset{m\to\infty}{\longrightarrow}\Big\|\widehat{\Omega}^0_\tau\Big(\overset{(m)}{\varphi}\Big)\Big\|$$

for any $\tau \leq 0$, and, hence, the inequality O_3 is satisfied.

The inequalities O_2 and O_1 are proved by analogy.

Equality (10.18) is proved.

Further, everywhere in the proofs of Lemma 10.1 and Theorem 10.2, instead of $\overset{(m)}{\varphi} = (\varphi_1, \ldots, \varphi_m, 0, 0, \ldots)$, we set

$$\tilde{\varphi} = (\varphi_1, \ldots, \varphi_m, \varphi'_{m+1}, \varphi'_{m+2}, \ldots).$$

As a result, we arrive at inequality (10.9) with $\max\{|\varphi_{m+1} - \varphi'_{m+1}|, \ldots\} = \|\overline{\overline{\varphi - \tilde{\varphi}}}\|$ instead of 2π. This yields

$$\|u(\varphi) - u(\widetilde{\varphi})\| \leq \widetilde{\zeta}(m)\|\varphi - \widetilde{\varphi}\|,$$

where $\widetilde{\zeta}(m)$ is a positive constant such that $\widetilde{\zeta}(m) \to 0$ as $m \to \infty$. This means that $u(\varphi) \in \widehat{C}^0_{\text{Lip}}(\mathcal{F}_\infty)$.

Finally, we note that the proof of Theorem 9.1 remains valid in the case where $\varphi \in \mathfrak{M}$ and Theorem 4.10 is true for $x \in R^n$, i.e., $x = \overset{(n)}{x}$. This yields the commutativity of the limit.

Remark 10.1. In the case where $a(\varphi) = w = (w_1, w_2, \ldots) \in \mathfrak{M}$, Theorem 10.5 gives a representation of the invariant torus for a system of equations with almost periodic coefficients (the case of infinite frequency basis).

At the end of the section, we consider a system of equations of the form (10.1) dependent on a parameter p, namely,

$$\frac{d\varphi}{dt} = a(\varphi, p), \qquad \frac{dx}{dt} = P(\varphi, p) + c(\varphi, p), \tag{10.19}$$

where $p \in [p_1, p_2] \subset R^1$ is a positive parameter and $\varphi, x \in \mathfrak{M}$.

By $w(z)$ we denote a continuous scalar function nondecreasing on the segment $[0, p_2 - p_1]$ and such that $w(0) = 0$.

Theorem 10.6. *Assume that the system of equations (10.19) satisfies the following conditions for every $p \in [p_1, p_2]$:*

(i) $\sum\limits_{j=1}^{\infty} \sup\limits_{\varphi, p} |p_{sj}(\varphi, p)| \leq p^0 = \text{const} < \infty, \quad s = 1, 2, \ldots;$

(ii) *the inequality*

$$\max\left\{\|a(\varphi,p)-a(\varphi',p')\|,\ \|P(\varphi,p)-P(\varphi',p')\|,\ \|c(\varphi,p)-c(\varphi',p')\|\right\}$$
$$\leq \alpha_1\|\varphi-\varphi'\|+\alpha_2\omega(|p-p'|),$$

where α_1 *and* α_2 *are positive constants, holds for all* $\varphi,\varphi' \in \mathfrak{M}$ *and* $p' \in [p_1,p_2]$;

(iii) *the matrizant of the equation*

$$\frac{dx}{dt} = P(\varphi_t^p(\varphi),p)x$$

satisfies the inequality

$$\|\Omega_\tau^0(\varphi,p)\| \leq N\exp\{\gamma,\tau\}, \quad \tau \leq 0,$$

where N *and* γ *are positive constants independent of* φ, p, *and* τ.

Then the invariant torus $\mathcal{F}\colon x = u(\varphi,p)$, $\varphi \in \mathcal{F}_\infty$, *of system (10.19) is continuous with respect to the collection of variables* φ *and* p.

Proof. In the analyzed case, we have the following estimate similar to (7.18):

$$\left\|\varphi_t^p(\varphi)-\varphi_t^{p'}(\varphi')\right\| \leq \left[\|\varphi-\varphi'\|+\frac{\alpha_2}{\alpha_1}\omega(|p-p'|)\right]\exp\{-\alpha_1 t\}, \quad t \leq 0.$$

By using this inequality, we obtain

$$\left\|P(\varphi_t^p(\varphi),p)-P(\varphi_t^{p'}(\varphi'),p')\right\|$$
$$\leq (2p^0)^{\frac{\nu}{\nu+1}}\left[\alpha_1\|\varphi-\varphi'\|+2\alpha_2\omega(|p-p'|)\right]^{\frac{1}{\nu+1}}\exp\left\{-\frac{\alpha_1 t}{\nu+1}\right\},$$

$$\left\|c(\varphi_t^p(\varphi),p)-c(\varphi_t^{p'}(\varphi'),p')\right\|$$
$$\leq (2c^0)^{\frac{\nu}{\nu+1}}\left[\alpha_1\|\varphi-\varphi'\|+2\alpha_2\omega(|p-p'|)\right]^{\frac{1}{\nu+1}}\exp\left\{-\frac{\alpha_1 t}{\nu+1}\right\},$$

where ν is an arbitrary positive real number. Then

$$\left\|\Omega_\tau^0(\varphi,p) - \Omega_\tau^0(\varphi',p')\right\|$$

$$\leq (2p^0)^{\frac{\nu}{\nu+1}} \frac{N^2}{\alpha_1/(\nu+1)} \left[\alpha_1\|\varphi-\varphi'\| + 2\alpha_2\omega(|p-p'|)\right]^{\frac{1}{\nu+1}}$$

$$\times \exp\left\{\left(\gamma - \frac{\alpha_1}{\nu+1}\right)\tau\right\}.$$

If we now choose $\nu > 0$ from the inequality $\gamma - \dfrac{\alpha_1}{\nu+1} > 0$, then we get

$$\|u(\varphi,p) - u(\varphi',p')\|$$

$$\leq \int_{-\infty}^{0} \left\|\Omega_\tau^0(\varphi,p)c(\varphi_\tau^p(\varphi),p) - \Omega_\tau^0(\varphi',p')c(\varphi_\tau^{p'}(\varphi'),p')\right\| d\tau$$

$$\leq \frac{N}{\gamma - \alpha_1/(\nu+1)} \left\{\alpha_1\|\varphi-\varphi'\| + 2\alpha_0\omega(|p-p'|)\right\}^{\frac{1}{\nu+1}}$$

$$\times \left[(2c^0)^{\frac{\nu}{\nu+1}} + \frac{c^0 N}{\alpha_1/(\nu+1)}(2p^0)^{\frac{\nu}{\nu+1}}\right],$$

which proves Theorem 10.6.

2.11. Theorem on Convergence of the Sequence of Invariant Tori

Parallel with the system of equations (8.1), we consider a countable sequence of systems

$$\frac{d\varphi}{dt} = a(\varphi), \quad \frac{dx}{dt} = P^{(k)}(\varphi)\,x + c^{(k)}(\varphi), \quad k = 1, 2, \ldots, \tag{11.1}$$

whose right-hand sides satisfy conditions (i)-(iii) of Theorem 8.1 for any natural k.

We now present sufficient conditions under which the sequence of invariant tori of systems (11.1) is convergent and determines the invariant torus of the original system (8.1). We assume that each system in (11.1) has the Green function

$$G_0^{(k)}(\tau,\varphi) = \begin{cases} \overset{(k)}{\Omega}{}_\tau^0(\varphi)\, C^{(k)}(\varphi_\tau(\varphi)), & \tau \leq 0, \\ \overset{(k)}{\Omega}{}_\tau^0(\varphi)\, [C^{(k)}(\varphi_\tau(\varphi)) - E], & \tau > 0, \end{cases}$$

where $\overset{(k)}{\Omega}{}_\tau^t(\varphi)$ is the matrizant of the homogeneous system corresponding to (11.1).

We set

$$\overset{(k)}{\Omega}{}^0_\tau(\varphi) = [\omega^{(k)}_{ij}(\tau,\varphi)]^\infty_{i,j=1}, \quad G^{(k)}_0(\tau,\varphi) = [g^{(k)}_{ij}(\tau,\varphi)]^\infty_{i,j=1},$$

and denote the invariant torus of system (11.1) by $\mathcal{F}^{(k)}\colon x = u^{(k)}(\varphi), \varphi \in \mathcal{F}_m$.

The answer to the question posed is given by the following theorem:

Theorem 11.1. *Assume that the following conditions are satisfied:*

(i) $\|G^{(k)}_0(\tau,\varphi)\| \le N\exp\{-\gamma\,|\,\tau\,|\,\}$, *where the constants* $N > 0$ *and* $\gamma > 0$ *are independent of* k;

(ii) in the weak sense,

$$\overset{(k)}{\Omega}{}^0_\tau(\varphi) \to \Omega^0_\tau(\varphi), \quad C^{(k)}(\varphi) \to C(\varphi) \in C^0(\mathcal{F}_m) \cap C_\varphi,$$

$$\textit{and} \quad c^{(k)}(\varphi) \to c(\varphi) \quad \textit{as} \quad k \to \infty,$$

where $\Omega^t_\tau(\varphi)$ *is the matrizant of system (8.3);*

(iii) the matrices $C^{(k)}(\varphi)$ *and functions* $c^{(k)}$ *are jointly bounded and the series*

$$\sum_{j=1}^{\infty} \left|\omega^{(k)}_{ij}(\tau,\varphi)\right| \quad \textit{and} \quad \sum_{j=1}^{\infty} \left|g^{(k)}_{ij}(\tau,\varphi)\right|, \quad i = 1,2,\ldots,$$

converge uniformly in k.

Then

$$u_j(\varphi) = \lim_{k\to\infty} u^{(k)}_j(\varphi), \quad j = 1,2,\ldots,$$

uniformly in φ, *and* $\mathcal{F}\colon x = u(\varphi)$, $\varphi \in \mathcal{F}_m$, *is the invariant torus of system (8.1).*

Proof. If the inequalities

$$\|\Omega^0_\tau(\varphi)\, C(\varphi_\tau(\varphi))\| \le N\exp\{\gamma\tau\}, \quad \tau \le 0,$$

$$\|\Omega^0_\tau(\varphi)\, [C(\varphi_\tau(\varphi)) - E]\| \le N\exp\{-\gamma\tau\}, \quad \tau > 0, \tag{11.2}$$

are satisfied, then it suffices to use the matrix

$$G_0(\tau,\varphi) = \begin{cases} \Omega^0_\tau(\varphi)\, C(\varphi_\tau(\varphi)), & \tau \le 0, \\ \Omega^0_\tau(\varphi)\, [C(\varphi_\tau(\varphi)) \, - \, E], & \tau > 0, \end{cases}$$

as the Green function of the problem of invariant torus for the system of equations (8.1). Consider the case $\tau \leq 0$. Let

$$\|C^{(k)}(\varphi)\| \leq C_0, \quad \|c^{(k)}(\varphi)\| \leq c^0,$$

$$G_0(\tau,\varphi) = [g_{ij}(\tau,\varphi)]_{i,j=1}^{\infty}, \quad \text{and} \quad C^{(k)}(\varphi) = [c_{ij}^{(k)}(\varphi)]_{i,j=1}^{\infty}.$$

We now write the equality

$$\lim_{k\to\infty} g_{ij}^{(k)}(\tau,\varphi) = \lim_{k\to\infty} \sum_{s=1}^{\infty} \omega_{is}^{(k)}(\tau,\varphi)\, c_{sj}^{(k)}(\varphi_\tau(\varphi)), \quad i,j = 1,2,\ldots .$$

The series $\sum_{s=1}^{\infty} \omega_{is}^{(k)}(\tau,\varphi)\, c_{sj}^{(k)}(\varphi_\tau(\varphi))$ converge uniformly in k because they are majorized by the series

$$\sum_{s=1}^{\infty} \left|\omega_{is}^{(k)}(\tau,\varphi)\right| C_0, \quad i = 1,2,\ldots,$$

which converge uniformly in k. Then

$$\lim_{k\to\infty} g_{ij}^{(k)}(\tau,\varphi) = \sum_{s=1}^{\infty} \lim_{k\to\infty} \left(\omega_{is}^{(k)}(\tau,\varphi)\, c_{sj}^{(k)}(\varphi_\tau(\varphi))\right) = g_{ij}(\tau,\varphi),$$

$$i,j = 1,2,\ldots .$$

This means that, for $\tau \leq 0$, the function $G_0(\tau,\varphi)$ is the coordinatewise limit of the sequence $G_0^{(k)}(\tau,\varphi)$. Thus, the first inequality in (11.2) is true. The case $\tau > 0$ is analyzed by analogy.

It is easy to see that

$$\lim_{k\to\infty} G_0^{(k)}(\tau,\varphi)\, c^{(k)}(\varphi_\tau(\varphi)) = G_0(\tau,\varphi)\, c(\varphi_\tau(\varphi))$$

in the weak sense. In view of the integral representation of the invariant torus, it remains to prove the equality

$$\lim_{k\to\infty} \int_{-\infty}^{\infty} G_0^{(k)}(\tau,\varphi)\, c^{(k)}(\varphi_\tau(\varphi))d\tau$$

$$= \int_{-\infty}^{\infty} \lim_{k\to\infty} \{G_0^{(k)}(\tau,\varphi)\, c^{(k)}(\varphi_\tau(\varphi))\}d\tau. \quad (11.3)$$

Denote

$$G_0^{(k)}(\tau,\varphi)\, c^{(k)}(\varphi_\tau(\varphi)) = z^{(k)}(\tau,\varphi) = \{z_1^{(k)}(\tau,\varphi), z_2^{(k)}(\tau,\varphi),\ldots\},$$

$$G_0(\tau,\varphi)\, c(\varphi_\tau(\varphi)) = z(\tau,\varphi) = \{z_1(\tau,\varphi), z_2(\tau,\varphi),\ldots\},$$

where A is an arbitrarily large positive number.

By analogy with the proof of Theorem 9.1, we arrive at the following relation valid for all $i = 1, 2, \ldots$:

$$\lim_{k\to\infty} \int_{-A}^{A} z_i^{(k)}(\tau,\varphi)d\tau = \int_{-A}^{A} \lim_{k\to\infty} z_i^{(k)}(\tau,\varphi)d\tau = \int_{-A}^{A} z_i(\tau,\varphi)d\tau.$$

Denote

$$\int_{-A}^{A} z_i^{(k)}(\tau,\varphi)d\tau = F(k,A).$$

Then

$$\lim_{k\to\infty} F(k,A) = \int_{-A}^{A} z_i(\tau,\varphi)d\tau,$$

$$\lim_{A\to\infty} F(k,A) = \int_{-\infty}^{\infty} z_i^{(k)}(\tau,\varphi)d\tau,$$

where the last limit transition is uniform with respect to k because

$$\left|\int_{-\infty}^{\infty} z_i^{(k)}(\tau,\varphi)d\tau - \int_{-A}^{A} z_i^{(k)}(\tau,\varphi)d\tau\right|$$

$$\leq \left|\int_{-\infty}^{-A} z_i^{(k)}(\tau,\varphi)d\tau\right| + \left|\int_{A}^{\infty} z_i^{(k)}(\tau,\varphi)d\tau\right|$$

$$\leq \int_{-\infty}^{-A} c^0 N \exp\{\gamma\tau\}d\tau + \int_{A}^{\infty} c^0 N \exp\{-\gamma\tau\}d\tau$$

$$\leq \frac{2c^0 N}{\gamma} \exp\{-\gamma A\} < \varepsilon$$

for all $k = 1, 2, \ldots$ whenever $A > \dfrac{1}{\gamma} \ln \dfrac{\varepsilon\gamma}{2c^0 N}$.

In this case, according to the theorem on changing the order of limit transitions, we obtain

$$\lim_{k\to\infty} \lim_{A\to\infty} F(k,A) = \lim_{A\to\infty} \lim_{k\to\infty} F(k,A),$$

which yields the required relation (11.3).

Remark 11.1. In the proof of Theorem 11.1, the order of system (11.1) is not used. Therefore, each system in (11.1) can be regarded as finite-dimensional and obtained by using one of the procedures of truncation of system (8.1). It is clear that the validity of the conditions of Theorem 11.1 strongly depends on the choice of this procedure. This can be seen, in particular, if we consider, as an example, a block-diagonal system with

$$P(\varphi) = \operatorname{diag}\{P_1(\varphi), P_2(\varphi), \ldots\},$$

for which it is natural to perform truncation without breaking the blocks.

2.12. Invariant Tori of Nonlinear Systems

We consider a nonlinear system of differential equations

$$\frac{d\varphi}{dt} = a(\varphi, h, \mu), \quad \frac{dh}{dt} = P(\varphi, h, \mu)h + c(\varphi, \mu), \tag{12.1}$$

where $\varphi = (\varphi_1, \ldots, \varphi_m) \in R^m$, $h = (h_1, h_2, \ldots) \in \mathfrak{M}$, $a(\varphi, h, \mu)$ and $c(\varphi, \mu)$ are bounded vector functions 2π-periodic in φ_i, $i = \overline{1, m}$, $P(\varphi, h, \mu)$ is an infinite matrix 2π-periodic in φ_i, $\mu \in (0, \mu_0]$ is a positive parameter, and $c(\varphi, 0) \equiv 0$.

We assume that, in a domain D,

$$\|h\| \le d, \quad \varphi \in \mathcal{F}_m, \quad \mu \in (0, \mu_0],$$

$a(\varphi, h, \mu)$, $P(\varphi, h, \mu)$, $c(\varphi, \mu) \in C^l(\mathcal{F}_m)$ for $h \in C^l(\mathcal{F}_m)$, and the following inequalities are true:

$$\|a(\varphi, h, \mu) - a(\bar\varphi, \bar h, \mu)\| \le \sigma(\|\varphi - \bar\varphi\| + \|h - \bar h\|), \tag{12.2}$$

$$\|P(\varphi, h, \mu) - P(\varphi, \bar h, \mu)\| \le \sigma\|h - \bar h\|, \quad \sigma = \text{const} > 0, \tag{12.3}$$

$$\sum_{j=1}^{\infty} \sup_{\varphi, h} |p_{ij}(\varphi, h, \mu)| \le p^0 = \text{const}, \quad i = 1, 2, \ldots, \tag{12.4}$$

where the elements $p_{ij}(\varphi, h, \mu)$, $i, j = 1, 2, \ldots,$ of the matrix P are continuous with respect to φ.

The set $\mathcal{F}: h = u(\varphi, \mu) = \{u_1(\varphi, \mu), u_2(\varphi, \mu), \ldots\}$, $\varphi \in \mathcal{F}_m$, is called the invariant torus of the system of equations (12.1) if there exists $\tilde{\mu} \in (o, \mu_0]$ such that, for any $\mu \in (0, \tilde{\mu}]$ and $u(\varphi, \mu) \in C^0(\mathcal{F}_m)$, the functions $u_i(\varphi_t, \mu)$, $i = 1, 2, \ldots,$ are continuously differentiable with respect to t and

$$\frac{du(\varphi_t, \mu)}{dt} = P(\varphi_t, u(\varphi_t, \mu), \mu)u(\varphi_t, \mu) + c(\varphi_t, \mu) \tag{12.5}$$

for all $t \in R^1$ and $\varphi \in \mathcal{F}_m$, where $\varphi_t = \varphi_t(\varphi, \mu)$ is the solution of the equation

$$\frac{d\varphi}{dt} = a(\varphi, u(\varphi, \mu), \mu) \tag{12.6}$$

satisfying the condition $\varphi_0(\varphi, \mu) = \varphi \in \mathcal{F}_m$.

We study the problem of the existence of the invariant torus for system (12.1). To do this, we use an iterative process linearizing the problem of finding the invariant torus in each step.

The procedure of linearization reduces to the construction of a sequence of invariant tori

$$\mathcal{F}^{(0)}(\mu), \mathcal{F}^{(1)}(\mu), \mathcal{F}^{(2)}(\mu), \ldots, \mathcal{F}^{(i)}(\mu), \ldots \tag{12.7}$$

in which $\mathcal{F}^{(0)}(\mu): h = 0$, $\varphi \in \mathcal{F}_m$, and the $(i+1)$th term is determined as the invariant torus $\mathcal{F}^{i+1}(\mu): h = u^{(i+1)}(\varphi, \mu)$, $\varphi \in \mathcal{F}_m$, of the system of equations

$$\frac{d\varphi}{dt} = a(\varphi, u^{(i)}(\varphi, \mu), \mu),$$

$$\frac{dh}{dt} = P(\varphi, u^{(i)}(\varphi, \mu), \mu)h + c(\varphi, \mu). \tag{12.8}$$

If the function $a(\varphi, u^{(i)}(\varphi, \mu), \mu)$ satisfies the Lipschitz condition with respect to φ, then there exists a unique solution $\varphi_t^i = \varphi_t^i(\varphi, \mu)$ of the first equation in (12.8) satisfying the condition $\varphi_0^i = \varphi \in \mathcal{F}_m$. According to the results of Chapter 1, the homogeneous system corresponding to the system of equations

$$\frac{dh}{dt} = P(\varphi_t^i, u^{(i)}(\varphi_t^i, \mu), \mu)h + c(\varphi_t^i, \mu) \tag{12.9}$$

possesses a matrizant. It is denoted by $\overset{(i)}{\Omega}{}_\tau^t(\varphi, \mu)$. Assume that system (12.8) has the Green function of the problem of invariant torus $G_0^{(i)}(\tau, \varphi, \mu)$ for any

$\mu \in (0, \mu_0]$. Then the torus $\mathcal{F}^{(i+1)}(\mu)$ exists and is given by the function

$$u^{(i+1)} = \int_{-\infty}^{\infty} G_0^{(i)}(\tau, \varphi, \mu) c(\varphi_\tau^i, \mu) d\tau. \tag{12.10}$$

The theorem on the existence of the Green function for each system in (12.9) with $c \equiv 0$ guarantees the possibility of infinite continuation of the iterative process.

We now establish conditions for the convergence of this process.

First, we prove two auxiliary statements.

Lemma 12.1 *Assume that the series* $\sum\limits_{i=1}^{\infty} u_i^{(k)}(t)$ *converges uniformly in t and k to a function* $f^{(k)}(t)$. *If, in addition,* $\lim\limits_{k\to\infty} u_i^{(k)}(t) = u_i(t)$, $i = 1, 2, 3, \ldots$, *uniformly in t, then the series* $\sum\limits_{i=1}^{\infty} u_i(t)$ *and the sequence* $\{f^{(k)}(t)\}$ *also converge uniformly in t and, moreover,*

$$\sum_{i=1}^{\infty} u_i(t) = \lim_{k\to\infty} f^{(k)}(t)\,.$$

Proof. According to the condition of the lemma, for any $\varepsilon > 0$ there exists a number N such that $\left|\sum\limits_{i=n}^{n+m} u_i^{(k)}(t)\right| < \varepsilon$ for all natural k and m and all t, provided that $n > N$. Passing to the limit as $k \to \infty$, we conclude that $\left|\sum\limits_{i=n}^{n+m} u_i(t)\right| \le \varepsilon$ for all t, i.e., the series $\sum\limits_{i=1}^{\infty} u_i(t)$ converges uniformly in t.

The following inequality is true:

$$\left|f^{(k)}(t) - \sum_{i=1}^{\infty} u_i(t)\right|$$

$$\le \left|\sum_{i=1}^{n} u_i^{(k)}(t) - \sum_{i=1}^{n} u_i(t)\right| + \left|\sum_{i=n+1}^{\infty} u_i^{(k)}(t)\right| + \left|\sum_{i=n+1}^{\infty} u_i(t)\right|. \tag{12.11}$$

Let $\varepsilon > 0$ be an arbitrarily small number. Then there exists a number N^0 such that, for $n > N^0$ and all t and k, we have

$$\left|\sum_{i=n+1}^{\infty} u_i^{(k)}(t)\right| \le \frac{\varepsilon}{3} \qquad \text{and} \qquad \left|\sum_{i=n+1}^{\infty} u_i(t)\right| \le \frac{\varepsilon}{3}$$

(as remainders of series convergent uniformly in t and k).

At the same time, $u_i^{(k)}(t) \to u_i(t)$ as $k \to \infty$ uniformly in t. Hence, for $\dfrac{\varepsilon}{3n} > 0$, there exists a number N_i such that, for all $k > N_i$ and all t, we have $\left| u_i^{(k)}(t) - u_i(t) \right| \leq \dfrac{\varepsilon}{3n}$. Then, for $k > \max\limits_{i=\overline{1,n}} N_i = N_k$ and all t,

$$\sum_{i=1}^{n} \left| u_i^{(k)}(t) - u_i(t) \right| \leq \sum_{i=1}^{n} \frac{\varepsilon}{3n} = \frac{\varepsilon}{3}.$$

In relation (12.11), we now fix the number $n \geq N^0$. In this case, the inequality

$$\left| \sum_{i=1}^{\infty} u_i(t) - f^{(k)}(t) \right| \leq \varepsilon$$

is true for all $k > N_k$ and all t, which proves Lemma 12.1.

Lemma 12.2. *Assume that the system of equations (12.8) is such that the sequence (12.7) of invariant tori $\mathcal{F}^{(i)}(\mu)$ is specified by a sequence of functions $u^{(i)}(\varphi, \mu) \in C^1(\mathcal{F}_m)$ bounded in the norm $\| \cdot \|$ by a constant d.*

If, for all $\mu \in (0, \bar{\mu}] \subset (0, \mu_0]$,

$$\|u(\varphi, \mu) - u^{(i)}(\varphi, \mu)\| \to 0 \quad as \quad i \to \infty$$

uniformly in $\varphi \in \mathcal{F}_m$, then the function $u(\varphi, \mu)$, $\varphi \in \mathcal{F}_m$, specifies the invariant torus of the system of equations (12.1) for $\mu \in (0, \bar{\mu}]$.

Proof. Since $\mathcal{F}^{(i+1)}(\mu)$ is the invariant torus of the system of equations (12.8), the following identities are true:

$$\frac{d\varphi_t^i}{dt} = a(\varphi_t^i, u^{(i)}(\varphi_t^i, \mu), \mu),$$

$$\frac{du^{(i+1)}(\varphi_t^i, \mu)}{dt} = P(\varphi_t^i, u^{(i)}(\varphi_t^i, \mu), \mu)u^{(i+1)}(\varphi_t^i, \mu) + c(\varphi_t^i, \mu). \tag{12.12}$$

Let t be an arbitrary finite point of time, let $T_0 = [0, T^0]$ be an arbitrary segment containing this point, and let $\left| T^0 \right|$ be its length. We integrate equalities (12.12) from 0 to t and obtain

$$\varphi_t^i = \varphi_0^i + \int_0^t a(\varphi_s^i, u^{(i)}(\varphi_s^i, \mu), \mu) ds, \tag{12.13}$$

$$u^{(i+1)}(\varphi_t^i, \mu)$$
$$= u^{(i+1)}(\varphi_0^i, \mu)$$
$$+ \int_0^t \Big\{ P(\varphi_s^i, u^{(i)}(\varphi_s^i, \mu), \mu) u^{(i+1)}(\varphi_s^i, \mu) + c(\varphi_s^i, \mu) \Big\} ds. \tag{12.14}$$

Since $\|u^{(i)}(\varphi, \mu)\| \le d$ for all $i = 1, 2, 3, \ldots$, the function $a(\varphi_t^i, u^{(i)}(\varphi_t^i, \mu), \mu)$ is bounded on the segment T_0. We set

$$\|a(\varphi_t^i, u^{(i)}(\varphi_t^i, \mu), \mu)\| \le A = \text{const} < \infty.$$

In this case, the sequence φ_t^i is uniformly bounded and equicontinuous on T_0. Indeed,

$$\|\varphi_t^i\| \le \|\varphi\| + A\,|\,t\,|, \quad \text{where} \quad \varphi \in \mathcal{F}_m,$$

and

$$\|\varphi_t^i - \varphi_{\bar t}^i\| \le \Big\| \int_{\bar t}^t A ds = A\,|\,t - \bar t\,|, \quad t, \bar t \in T_0.$$

For $|\,t - \bar t\,| < \dfrac{\varepsilon}{A}$, we conclude that $\|\varphi_t^i - \varphi_{\bar t}^i\| \le \varepsilon$ uniformly in $t, \bar t \in T_0$. Then the sequence φ_t^i contains a subsequence $\varphi_t^{i_k}$ that converges to a function $\varphi_t = \varphi(t, \mu)$ uniformly in $t \in T_0$.

We now show that

$$\lim_{k \to \infty} a(\varphi_t^{i_k}, u^{(i_k)}(\varphi_t^{i_k}, \mu), \mu) = a(\varphi_t, u(\varphi_t, \mu), \mu) \tag{12.15}$$

uniformly in $t \in T_0$. The following relations are true:

$$\|a(\varphi_t, u(\varphi_t, \mu), \mu) - a(\varphi_t^{i_k}, u^{(i_k)}(\varphi_t^{i_k}, \mu), \mu)\|$$
$$\le \|a(\varphi_t, u(\varphi_t, \mu), \mu) - a(\varphi_t^{i_k}, u(\varphi_t, \mu), \mu)\|$$
$$+ \|a(\varphi_t^{i_k}, u(\varphi_t, \mu), \mu) - a(\varphi_t^{i_k}, u^{(i_k)}(\varphi_t^{i_k}, \mu), \mu)\|$$
$$= S_1 + S_2.$$

Let ε be an arbitrarily small positive number. Since $a(\varphi, h, \mu)$ is a continuous function of φ, there exists $\delta > 0$ such that

$$S_1 \le \frac{\varepsilon}{2}$$

for any φ_t and $\varphi_t^{i_k}$ satisfying the inequality

$$\|\varphi_t - \varphi_t^{i_k}\| < \delta. \tag{12.16}$$

As shown above, there exists a number N_1 such that, for all $k \geq N_1$, this inequality holds uniformly in $t \in T_0$.

Similarly, the fact that $a(\varphi, h, \mu)$ is a continuous function of h implies that there exists $\delta_1 > 0$ such that

$$S_2 \leq \frac{\varepsilon}{2}$$

for all $\bar{u} = \|u(\varphi_t, \mu) - u^{(i_k)}(\varphi_t^{i_k}, \mu)\| < \delta_1$. Further, we denote

$$\|u(\varphi_t, \mu) - u^{(i_k)}(\varphi_t, \mu)\| = S_3 \quad \text{and} \quad \|u^{(i_k)}(\varphi_t, \mu) - u^{(i_k)}(\varphi_t^{i_k}, \mu)\| = S_4.$$

According to the condition of the lemma, there exists a number N_2 such that $S_3 \leq \dfrac{\delta_1}{2}$ for $k \geq N_2$ and all $t \in T_0$.

It is also easy to see that the function $u^{(i_k)}(\varphi, \mu)$ is continuous in φ in the norm $\|\cdot\|$. Indeed, if we denote $u^{(i_k)}(\varphi, \mu) = (u_1^{(i_k)}(\varphi, \mu), u_2^{(i_k)}(\varphi, \mu), \ldots)$, take into account the fact that

$$\left|\frac{du_j^{(i_k)}(\varphi, \mu)}{d\varphi_r}\right| < K_1 = \text{const}, \quad r = \overline{1, m}, \;\; j = 1, 2, \ldots,$$

and use the well-known Lagrange lemma, then we get

$$|u_j^{(i_k)}(\varphi, \mu) - u_j^{(i_k)}(\bar{\varphi}, \mu)| < \xi = \text{const}, \quad j = 1, 2, \ldots,$$

whenever $\|\varphi - \bar{\varphi}\| < \dfrac{\xi}{mK_1}$. This also implies that

$$\|u^{(i_k)}(\varphi, \mu) - u^{(i_k)}(\bar{\varphi}, \mu)\| = \sup_{j=1,2,\ldots} \left\{|u_j^{(i_k)}(\varphi, \mu) - u_j^{(i_k)}(\bar{\varphi}, \mu)|\right\} < \xi.$$

Then there exists $\delta_2 > 0$ such that $S_4 \leq \dfrac{\delta_1}{2}$ whenever

$$\|\varphi_t - \varphi_t^{i_k}\| < \delta_2. \tag{12.17}$$

At the same time, inequality (12.17) holds for all $k \geq N_3$ and $t \in T_0$.

This implies that $\bar{u} \leq S_3 + S_4 \leq \delta_1$ for $k \geq \max\{N_2, N_3\}$ and, hence, $S_1 + S_2 < \varepsilon$ whenever $k \geq \max\{N_1, N_2, N_3\}$, i.e., equality (12.15) is proved.

If we now pass to the limit as $k \to \infty$ in the integrand in equality (12.13) with i replaced by i_k and take into account the fact that $\varphi_0^i = \varphi$, then we get

$$\varphi_t = \varphi + \int_0^t a(\varphi_s, u(\varphi_s, \mu), \mu) ds. \tag{12.18}$$

The fact that $u^{(i_k)}(\varphi, \mu)$ are continuous functions of φ and the formula

$$\lim_{k\to\infty} u^{i_{k+1}}(\varphi_t^{i_{k+1}}, \mu) = u(\varphi_t, \mu)$$

immediately yield the representation

$$\lim_{k\to\infty} u^{(i_{k+1})}(\varphi_t^{i_k}, \mu) = u(\varphi_t, \mu), \tag{12.19}$$

whence it follows that

$$u^{(i_{k+1})}(\varphi_0^i, \mu) \to u(\varphi, \mu) \quad \text{as} \quad k \to \infty. \tag{12.20}$$

It is also clear that

$$c(\varphi_t^{i_k}, \mu) \to c(\varphi_t, \mu) \quad \text{as} \quad k \to \infty. \tag{12.21}$$

Let us now show that the following limit transition takes place in the weak sense uniformly in $t \in T_0$:

$$\lim_{k\to\infty} P(\varphi_t^{i_k}, u^{(i_k)}(\varphi_t^{i_k}, \mu), \mu) u^{(i_{k+1})}(\varphi_t^{i_k}, \mu) = P(\varphi_t, u(\varphi_t, \mu), \mu) u(\varphi_t, \mu). \tag{12.22}$$

We prove this fact for the first coordinates of the vector functions appearing on the left- and right-hand sides of equality (12.22).

Clearly, the series

$$\sum_{r=1}^{\infty} p_{1r}(\varphi_t^{i_k}, u^{(i_k)}(\varphi_t^{i_k}, \mu), \mu) u_r^{(i_{k+1})}(\varphi_t^{i_k}, \mu)$$

converges uniformly in k and $t \in T_0$ because it can be majorized by the following convergent numerical series:

$$\sum_{r=1}^{\infty} \sup_{\varphi, h} |\, p_{1r}(\varphi, h, \mu) \,|\, d.$$

Moreover,

$$p_{1r}(\varphi_t^{i_k}, u^{(i_k)}(\varphi_t^{i_k}, \mu), \mu)u_r^{(i_k+1)}(\varphi_t^{i_k}, \mu) \to p_{1r}(\varphi_t, u(\varphi, \mu), \mu)u_r(\varphi_t, \mu)$$

as $k \to \infty$ uniformly in $t \in T_0$, which follows from the continuity of $p_{1r}(\varphi, h, \mu)$ with respect to φ and h and the fact that $\varphi_t^{i_k} \to \varphi_t$, $u^{(i_k)}(\varphi_t^{i_k}, \mu) \to u(\varphi_t, \mu)$, and $u^{(i_k+1)}(\varphi_t^{i_k}, \mu) \to u(\varphi_t, \mu)$ uniformly as $k \to \infty$. Then, according to Lemma 12.1, we have

$$\sum_{r=1}^{\infty} p_{1r}(\varphi_t^{i_k}, u^{(i_k)}(\varphi_t^{i_k}, \mu), \mu)u_r^{(i_k+1)}(\varphi_t^{i_k}, \mu)$$

$$\to \sum_{r=1}^{\infty} p_{1r}(\varphi_t, u(\varphi_t, \mu), \mu)u_r(\varphi_t, \mu)$$

as $k \to \infty$ uniformly in $t \in T_0$. In this case, the series on the right-hand side also converges uniformly in $t \in T_0$. This proves equality (12.22).

By using equalities (12.19)–(12.22), we pass to the limit as $k \to \infty$ in equality (12.14) with i replaced by i_k. This gives

$$u(\varphi_t, \mu) = u(\varphi, \mu) + \int_0^t \Big\{P(\varphi_s, u(\varphi_s, \mu), \mu)u(\varphi_s, \mu) + c(\varphi_s, \mu)\Big\}ds.$$

Together with equality (12.18), this proves the assertion of Lemma 12.2.

We now introduce the notion of coarseness of the Green function of a linear system of equations. For this purpose, we set

$$|\, f(\varphi)\,|_0 = \sup_{\varphi} \|f(\varphi)\| \quad \text{and} \quad \|f(\varphi)\|_l = \sup_{o \le s \le l} \Big| D_\varphi^s f(\varphi)\Big|_0.$$

The Green function $G_0(\tau, \varphi)$ of the system

$$\frac{d\varphi}{dt} = a(\varphi), \quad \frac{dh}{dt} = P(\varphi)h \tag{12.23}$$

is called coarse if there exists a constant $\delta > 0$ such that, for all $a^1(\varphi) \in C^l(\mathcal{F}_m)$ satisfying the inequality $\|a^1(\varphi)\|_l \le \delta$, the Green function $\overline{G}_0(\tau, \varphi)$ of the system of equations

$$\frac{d\varphi}{dt} = a(\varphi) + a^1(\varphi), \quad \frac{dh}{dt} = P(\varphi)h \tag{12.24}$$

exists and is such that $\overline{G}_0(\tau,\varphi)f(\varphi) \in C^l(\mathcal{F}_m)$ and

$$\|\overline{G}_0(\tau,\varphi)f(\varphi_\tau(\varphi))\|_l \le K\exp\{-\gamma|\tau|\}\|f\|_l, \tag{12.25}$$

where $\varphi_t(\varphi)$ is the solution of the first equation in (12.24), $f(\varphi)$ is an arbitrary function from $C^l(\mathcal{F}_m)$, and K and γ are positive constants independent of δ, φ, and f.

Lemma 12.3. *Assume that the system of equations (12.23) possesses a coarse Green function $G_0(\tau,\varphi)$ and $P(\varphi) \in C_\varphi$. Then one can find $\rho = \rho(\delta) > 0$, $\rho \to 0$ as $\delta \to 0$, such that, for any $a^1(\varphi) \in C^l(\mathcal{F}_m)$ and $P^1(\varphi) \in C^l(\mathcal{F}_m) \cap C_\varphi$ satisfying the inequality*

$$\max\{\|a^1(\varphi)\|_l, \|P^1(\varphi)\|_l\} \le \rho$$

and an arbitrary function $f(\varphi) \in C^l(\mathcal{F}_m)$, the system of equations

$$\frac{d\varphi}{dt} = a(\varphi) + a^1(\varphi), \qquad \frac{dh}{dt} = (P(\varphi) + P^1(\varphi))h + f(\varphi) \tag{12.26}$$

possesses the invariant torus $\mathcal{F}$: $h = u(\varphi)$, $\varphi \in \mathcal{F}_m$, satisfying the condition

$$\|u(\varphi)\|_l \le K_1\|f(\varphi)\|_l, \tag{12.27}$$

where K_1 is an arbitrary constant independent of ρ and f.

Proof. We choose $\rho(\delta) \le \delta$. In this case, system (12.24) possesses the Green function $\overline{G}_0(\tau,\varphi)$ satisfying condition (12.25). By using the operator G defined on functions $f \in C^l(\mathcal{F}_m)$ by the formula

$$Gf(\varphi) = \int_{-\infty}^{\infty} \overline{G}_0(\tau,\varphi)f(\varphi_\tau(\varphi))d\tau\,,$$

we introduce the equation

$$u = GP^1u + Gf. \tag{12.28}$$

For sufficiently small ρ, the norm of the operator GP^1 in the space $C^l(\mathcal{F}_m)$ satisfies the inequality

$$\|GP^1\|_l \le \frac{2K}{\gamma}\|P^1\|_l \le \frac{2K}{\gamma}\rho \le d_0 < 1.$$

Therefore, equation (12.28) possesses a unique solution in $C^l(\mathcal{F}_m)$ given by the formula

$$u(\varphi) = \sum_{k=0}^{\infty}(GP^1)^k Gf$$

and such that

$$\|u(\varphi)\|_l \leq \frac{2K}{\gamma(1-d_0)}\|f(\varphi)\|_l.$$

This inequality yields inequality (12.27). It follows from the definition of the operator G that the function $u(\varphi)$ specifies the required invariant torus of the system of equations (12.26).

By using Lemmas 12.2 and 12.3, we now prove the following theorem on the existence of the invariant torus for system (12.1):

Theorem 12.1. *Assume that the system of equations (12.1) satisfies conditions (12.2)–(12.4) in the domain D, the system*

$$\frac{d\varphi}{dt} = a(\varphi, 0, 0), \quad \frac{dh}{dt} = P(\varphi, 0, 0)h \tag{12.29}$$

possesses a coarse Green function, and the inequality

$$\max\{\|a(\varphi,h,\mu)-a(\varphi,0,0)\|_l,\ \|P(\varphi,h,\mu)-P(\varphi,0,0)\|_l,\ \|c(\varphi,\mu)\|_l\}$$
$$\leq \mathcal{L}_l(d,\mu_0), \tag{12.30}$$

where $\mathcal{L}_l(d,\mu_0) \to 0$ as $d \to 0$ and $\mu_0 \to 0$, holds for $h \in C^l(\mathcal{F}_m)$.

Then there exists $\tilde{\mu} \in (0,\mu_0]$ such that, for all $\mu \in [0,\tilde{\mu}]$, system (12.1) possesses the invariant torus $\mathcal{F}(\mu)$: $h = u(\varphi,\mu) \in C^{l-1}(\mathcal{F}_m)$, $\varphi \in \mathcal{F}_m$, $l \geq 1$, and, moreover, $\|u(\varphi,\mu)\| \to 0$ as $\mu \to 0$.

Proof. We rewrite the system of equations (12.1) in the form

$$\frac{d\varphi}{dt} = a_0(\varphi) + a^1(\varphi,h,\mu),$$

$$\frac{dh}{dt} = [P_0(\varphi) + P^1(\varphi,h,\mu)]h + c(\varphi,\mu)$$

by setting

$$a_0(\varphi) = a(\varphi,0,0), \quad P_0(\varphi) = P[\varphi,0,0],$$
$$a^1(\varphi,h,\mu) = a(\varphi,h,\mu) - a(\varphi,0,0),$$
$$P^1(\varphi,h,\mu) = P(\varphi,h,\mu) - P(\varphi,0,0).$$

To find the invariant torus of this system, we use the iterative process described above. A function $u^{(0)}(\varphi, \mu) \equiv 0$, $\varphi \in \mathcal{T}_m$, is regarded as the zeroth-order iteration of the process. The first approximation is determined from the system of equations

$$\frac{d\varphi}{dt} = a_0(\varphi) + a^1(\varphi, 0, \mu),$$

$$\frac{dh}{dt} = [P_0(\varphi) + P^1(\varphi, 0, \mu)]h + c(\varphi, \mu) \tag{12.31}$$

as its invariant torus $\mathcal{T}^{(1)}: h = u^{(1)}(\varphi, \mu)$, $\varphi \in \mathcal{T}_m$.

We choose $\bar{\mu} \in (0, \mu_0]$ to guarantee the validity of the inequalities

$$K_1\mathcal{L}_l(0, \bar{\mu}) \leq d \quad \text{and} \quad \mathcal{L}_l(K_1\mathcal{L}_l(0, \bar{\mu}), \bar{\mu}) \leq \rho,$$

where K_1 and ρ are the same constants as in Lemma 12.3. Condition (12.30) guarantees the possibility of this choice. Then, according to Lemma 12.3, the invariant torus $\mathcal{T}^{(1)}$ of system (12.31) exists. In this case,

$$\|u^{(1)}(\varphi, \mu)\|_l \leq K_1\mathcal{L}_l(0, \bar{\mu}) \leq d, \quad \mu \in (0, \bar{\mu}].$$

We assume that, for all $i = 1, 2, \ldots, k$, the invariant tori $\mathcal{T}^{(i)}: h = u^{(i)}(\varphi, \mu)$, $\varphi \in \mathcal{T}_m$, $\mu \in (0, \bar{\mu}]$, from sequence (12.7) exist and are such that

$$\|u^{(i)}(\varphi, \mu)\|_l \leq K_1\mathcal{L}_l(0, \bar{\mu}).$$

But then, for all $\mu \in (0, \bar{\mu}]$, we have

$$\max\{\|a^1(\varphi, u^{(k)}(\varphi, \mu), \mu)\|_l \; \|P^1(\varphi, u^{(k)}(\varphi, \mu), \mu)\|_l\}$$
$$\leq \mathcal{L}_l(K_1\mathcal{L}_l(0, \bar{\mu}), \mu) \leq \rho\,,$$

and, hence, Lemma 12.3 can be applied to the system of equations

$$\frac{d\varphi}{dt} = a_0(\varphi) + a^1(\varphi, u^{(k)}(\varphi, \mu), \mu),$$

$$\frac{dh}{dt} = \left[P_0(\varphi) + P^1(\varphi, u^{(k)}(\varphi, \mu), \mu)\right]h + c(\varphi, \mu)\,. \tag{12.32}$$

This means that the invariant torus $\mathcal{T}^{(k+1)}: h = u^{(k+1)}(\varphi, \mu)$, $\varphi \in \mathcal{T}_m$, exists for all $\mu \in (0, \bar{\mu}]$ and, in addition,

$$\|u^{(k+1)}(\varphi, \mu)\|_l \leq K_1\|c(\varphi, \mu)\|_l \leq K_1\mathcal{L}_l(0, \bar{\mu}) \leq d.$$

By induction, we can prove that the elements of sequence (12.7) exist for all $i = 0, 1, 2, \dots$ and are jointly bounded in the norm $\| \cdot \|_l$ by a constant $d > 0$ for all $\mu \in (0, \bar{\mu}]$. It remains to show that sequence (12.7) is convergent and use Lemma 12.2.

Since $u^{(k+1)}(\varphi, \mu) \in C^l(\mathcal{F}_m) \subset C^1(\mathcal{F}_m)$ for any $\mu \in (0, \bar{\mu}]$ and $t \in R^1$, we obtain

$$\sum_{i=1}^{m} \frac{\partial u^{(k+1)}(\varphi, \mu)}{\partial \varphi_i}\Big|_{\varphi=\varphi_t^k} [a_{0_i}(\varphi_t^k) + a_i^1(\varphi_t^k, u^{(k)}(\varphi_t^k, \mu), \mu)]$$
$$= [P_0(\varphi_t^k) + P^1(\varphi_t^k, u^{(k)}(\varphi_t^k, \mu), \mu)] u^{(k+1)}(\varphi_t^k, \mu) + c(\varphi_t^k, \mu), \quad (12.33)$$

where $\varphi_t^k = \varphi_t^k(\varphi, \mu)$ is the solution of the first equation in (12.32) such that $\varphi_0^k = \varphi \in \mathcal{F}_m$.

Denote

$$\sum_{i=1}^{m} \frac{\partial u^{(k+1)}(\varphi, \mu)}{\partial \varphi_i} [a_{0_i}(\varphi) + a_i^1(\varphi, u^{(k)}(\varphi, \mu), \mu)] = \frac{\partial u^{(k+1)}}{\partial \varphi}[a_0 + a^1].$$

It follows from equality (12.33) that

$$\frac{\partial u^{(k+1)}}{\partial \varphi}[a_0(\varphi) + a^1(\varphi, u^{(k)}(\varphi, \mu), \mu)]$$
$$= [P_0(\varphi) + P^1(\varphi, u^{(k)}(\varphi, \mu), \mu)] u^{(k+1)}(\varphi, \mu) + c(\varphi, \mu) \qquad (12.34)$$

for any $\varphi \in \mathcal{F}_m$ and $\mu \in (0, \bar{\mu}]$.

Similarly, we arrive at the formula

$$\frac{\partial u^{(k)}}{\partial \varphi}[a_0(\varphi) + a^1(\varphi, u^{(k-1)}(\varphi, \mu), \mu)]$$
$$= [P_0(\varphi) + P^1(\varphi, u^{(k-1)}(\varphi, \mu), \mu)] u^{(k)}(\varphi, \mu) + c(\varphi, \mu),$$

whence, by using relation (12.34), for the function

$$u^{(k+1)}(\varphi, \mu) - u^{(k)}(\varphi, \mu) = \omega^{(k+1)}(\varphi, \mu),$$

we obtain

$$\begin{aligned}\frac{\partial\omega^{(k+1)}}{\partial\varphi}&[a_0(\varphi)+a^1(\varphi,u^{(k)},\mu)]\\ &=[P_0(\varphi)+P^1(\varphi,u^{(k)},\mu)]\omega^{(k+1)}\\ &\quad+\frac{\partial u^{(k)}}{\partial\varphi}[a^1(\varphi,u^{(k-1)},\mu)-a^1(\varphi,u^{(k)},\mu)]\\ &\quad+[P^1(\varphi,u^{(k)},\mu)-P^1(\varphi,u^{(k-1)},\mu)]u^{(k)}.\end{aligned}$$

This means that the function $h=\omega^{(k+1)}(\varphi,\mu)$ specifies the invariant torus of the system of equations

$$\frac{d\varphi}{dt}=a_0(\varphi)+a^1(\varphi,u^{(k)}(\varphi,\mu),\mu),$$

$$\frac{dh}{dt}=[P_0(\varphi)+P^1(\varphi,u^{(k)},\mu)]h+c_k(\varphi,\mu), \tag{12.35}$$

where

$$\begin{aligned}c_k(\varphi,\mu)\;=\;&\frac{\partial u^{(k)}}{\partial\varphi}[a^1(\varphi,u^{(k-1)},\mu)-a^1(\varphi,u^{(k)},\mu)]\\ &+[P^1(\varphi,u^{(k)},\mu)-P^1(\varphi,u^{(k-1)},\mu)]u^{(k)}.\end{aligned}$$

The system of equations (12.35) has the form of system (12.32) and, therefore,

$$\|\omega^{(k+1)}(\varphi,\mu)\|\le K_1\|c_k(\varphi,\mu)\|.$$

Since

$$\|c_k(\varphi,\mu)\|\le K_1\sigma(m+1)\mathcal{L}_1(0,\bar\mu)\|u^{(k)}(\varphi,\mu)-u^{(k-1)}(\varphi,\mu)\|,$$

we arrive at the estimate

$$\|\omega^{(k+1)}(\varphi,\mu)\|\le\rho_0\|\omega^{(k)}(\varphi,\mu)\|,$$

where $\rho_0=K_1^2\sigma(m+1)\mathcal{L}_1(0,\bar\mu)$. Similarly, we obtain the chain of inequalities

$$\begin{aligned}\|\omega^{(k+1)}(\varphi,\mu)\|\le\rho_0\|\omega^{(k)}(\varphi,\mu)\|\;&\le\;\rho^2\|\omega^{(k-1)}(\varphi,\mu)\|\\ &\le\;\ldots\le\rho_0^k\|\omega^{(1)}(\varphi,\mu)\|.\end{aligned}$$

Since $\mathcal{L}_0(0,\bar{\mu}) \to 0$ as $\bar{\mu} \to 0$, we conclude that there exists $\tilde{\mu} \in (0,\bar{\mu}]$ such that $\rho_0 < 1$. This means that the sequence $u^{(i)}(\varphi,\mu)$, $i = 0,1,2,\ldots,$ converges as $i \to \infty$ uniformly in $\varphi \in \mathcal{F}_m$.

Denote

$$\lim_{i\to\infty} u^{(i)}(\varphi,\mu) = u(\varphi,\mu).$$

Let us show that $u(\varphi,\mu) \in C^{l-1}(\mathcal{F}_m)$ for $l \geq 1$.

Consider the sequence of sth coordinates of the functions $u^{(i)}(\varphi,\mu)$ and the sequence of their partial derivatives with respect to φ_j :

$$\Big\{\frac{\partial u_s^{(i)}(\varphi,\mu)}{\partial\varphi_j}\Big\}, \quad i = 1,2,\ldots, \tag{12.36}$$

where j is a fixed natural number such that $1 \leq j \leq m$.

It is clear that sequence (12.36) is uniformly bounded. Since $u^{(i)}(\varphi,\mu) \in C^2(\mathcal{F}_m)$, by virtue of the Lagrange theorem we arrive at the estimate

$$\Big|\frac{\partial u_s^{(i)}(\varphi,\mu)}{\partial\varphi_j} - \frac{\partial u_s^{(i)}(\bar{\varphi},\mu)}{\partial\varphi_j}\Big| \leq K_1 m\mathcal{L}_2(0,\tilde{\mu})\|\varphi-\bar{\varphi}\|, \quad s = 1,2,\ldots,$$

i.e., sequence (12.36) is equicontinuous in φ. By using the Arzelà theorem, we conclude that there exists a subsequence

$$\Big\{\frac{\partial u_s^{(i_\alpha)}(\varphi,\mu)}{\partial\varphi_j}\Big\}, \quad i_\alpha = 1,2,\ldots,$$

of sequence (12.36) that converges to the derivative $\dfrac{\partial u_s(\varphi,\mu)}{\partial\varphi_j}$ uniformly in φ.

Thus, for $l = 2$, the required assertion is proved.

Continuing the reasoning presented above and taking into account that the sequence $\{D^\rho_\varphi u^{(k)}(\varphi,\mu)\}$ of partial derivatives of the functions $u^{(k)}(\varphi,\mu)$, $k = 1,2,\ldots,$ of order $\rho \leq l-1$ is uniformly bounded and equicontinuous, we can easily show that the functions $u(\varphi,\mu)$ belong to the space $C^{l-1}(\mathcal{F}_m)$.

Remark 12.1. It becomes clear from the proof of Theorem 12.1 that the requirement of the coarseness of the Green function of the corresponding system of equations can be weakened if the right-hand side of the equation for φ is independent of h. In this case, it suffices to demand the existence of the Green function.

Thus, for the system of equations

$$\frac{d\varphi}{dt} = a(\varphi), \quad \frac{dh}{dt} = P(\varphi)h + \mu c(\varphi, h) \tag{12.37}$$

with $P(\varphi) \in C^0(\mathcal{F}_m) \cap C_\varphi$ and $a(\varphi) \in C^0_{\text{Lip}}(\mathcal{F}_m)$, the following statement is true:

Theorem 12.2. *Assume that the function $c(\varphi, h)$ satisfies the Lipschitz condition with respect to h for $\varphi \in \mathcal{F}_m$ and $\|h\| \le d$ and, for $\mu = 0$, the system of equations (12.37) possesses the Green function of the problem of invariant tori $G_0(\tau, \varphi)$ satisfying inequality (8.5).*

Then there exists $\mu_0 > 0$ such that, for all $\mu \in [0, \mu_0]$, system (12.37) has the invariant torus $\mathcal{F}(\mu)$: $h = u(\varphi)$, $\varphi \in \mathcal{F}_m$, such that $\lim\limits_{\mu \to 0} \|u(\varphi, \mu)\| = 0$.

2.13. Exponential Attraction of Motions in a Neighborhood of the Invariant Torus of a System of Equations to Its Motions on the Torus

The Green function studied in the previous section becomes especially simple in the case where the matrizant $\Omega^t_\tau(\varphi)$ of the corresponding linear system satisfies the condition of exponential decay

$$\|\Omega^t_0(\varphi)\| \le K \exp\{-\gamma t\}, \quad t \in R^+ = [0, +\infty). \tag{13.1}$$

This condition leads to the exponential stability of the trivial torus of the linear system and guarantees the attraction of its solutions to this torus.

In the present section, we formulate conditions under which the behavior of solutions of nonlinear countable systems is similar.

First, we consider the quasilinear system of equations (12.37). For this system, we have the following theorem on exponential stability of the invariant torus:

Theorem 13.1. *Assume that the system of equations (12.37) satisfies the conditions of Theorem 12.2 and the matrizant of the corresponding homogeneous system satisfies inequality (13.1). Then, for sufficiently small μ, the trajectory of a solution $h_t = h(h_0, t, \varphi, \mu)$, $h_0 = h(h_0, 0, \varphi, \mu)$, of this system of equations is attracted as $t \to +\infty$ to a trajectory lying on its invariant torus $\mathcal{F}$: $h = u(\varphi, \mu)$, $\varphi \in \mathcal{F}_m$, according to the law*

$$\|h_t - u(\varphi_t(\varphi))\| \le K_1 \|h_0 - u(\varphi, \mu)\| \exp\{-\gamma_1 t\}, \quad t \ge 0, \tag{13.2}$$

where K_1 and γ_1 are positive constants.

Proof. In system (12.37), we perform the change of variables

$$h_t = u(\varphi_t(\varphi), \mu) + \mu x\,(t, x_0, \varphi),$$

where $x_0 = x(0, x_0, \varphi)$. As a result, we arrive at the following system of equations for $x_t = x(t, x_0, \varphi)$:

$$\frac{dx}{dt} = P(\varphi_t(\varphi))x + \{c\,(\varphi_t(\varphi), u(\varphi_t(\varphi), \mu) + \mu x) - c(\varphi_t(\varphi), u(\varphi_t(\varphi), \mu))\}.$$

The solution $x = x_t$ of this system of equations satisfies the identity

$$x_t = \Omega_0^t x_0 + \int_0^t \Omega_s^t \{c(\varphi_s, u(\varphi_s, \mu) + \mu x_s) - c(\varphi_s, u(\varphi_s, \mu))\}ds$$

in the entire segment $t \in [0, T]$ of the existence of the solution.

Setting

$$\|c(\varphi, h) - c(\varphi, \bar{h})\| \le L\|h - \bar{h}\|, \quad L = \text{const} > 0$$

and using inequality (13.1), we obtain

$$\|x_t\| \le K\{\|x_0\| \exp\{-\gamma t\} + \mu L \int_0^t \|x_s\| \exp\{-\gamma(t-s)\}ds\}, \quad t \in [0, T].$$

In view of the Gronwall–Bellman inequality, we conclude that

$$\|x_t\| \le K\|x_0\| \exp\{-(\gamma - KL\mu)t\}, \quad t \in [0, T].$$

For sufficiently small μ, the last inequality implies that $T = +\infty$ and, hence,

$$\|x_t\| \le K\|x_0\| \exp\{-(\gamma - KL\mu)t\}, \quad t \ge 0.$$

Returning to the original variables, we find that

$$\|h_t - u(\varphi_t(\varphi), \mu)\| \le \mu K\|h_0 - u(\varphi, \mu)\| \exp\{-(\gamma - KL\mu)t\}, \quad t \ge 0,$$

which yields inequality (13.2).

Note that, for sufficiently small μ, the torus $\mathcal{F}$: $h = u(\varphi, \mu)$, $\varphi \in \mathcal{F}_m$, belongs to the domain D together with its ρ-neighborhood ($d > \rho$). Thus, if the point h_0 belongs to the indicated neighborhood, then the semitrajectory of the solution h_t for $t \in R^+$ also belongs to this neighborhood.

We now consider a system of equations of the general form

$$\frac{d\varphi}{dt} = a(\varphi, h), \quad \frac{dh}{dt} = P(\varphi, h)h + c(\varphi), \tag{13.3}$$

where $\varphi \in R^m$, $h \in \mathfrak{M}$, and $P(\varphi, h)$, $c(\varphi) \in C^1(\mathcal{F}_m)$ for $h \in C^1(\mathcal{F}_m)$. We define the norm $\|\cdot\|_D$ of an infinite matrix $P(\varphi, h)$ by the formula

$$\|P\|_D = \sup_i \sum_{j=1}^{\infty} \sup_{\varphi h \in D} |\, p_{ij}(\varphi, h)\,|, \quad D : \varphi \in \mathcal{F}_m, \ \|h\| \leq d.$$

The matrix $A(\varphi, h)$ is called Lipschitzian with respect to $\varphi, h \in D$ if, for any $\bar{\varphi}, \bar{h} \in D$,

$$\|A(\varphi, h) - A(\bar{\varphi}, \bar{h})\| \leq K\{\|\varphi - \bar{\varphi}\| + \|h - \bar{h}\|\}, \quad K = \text{const} > 0.$$

By C_D we denote the set of matrices bounded in the norm $\|\cdot\|_D$ and satisfying the Lipschitz condition with respect to φ and h in the domain D.

Let $\mathcal{F}: h = u(\varphi) \in C^1_{\text{Lip}}(\mathcal{F}_m)$, $\varphi \in \mathcal{F}_m$, be the invariant torus of system (13.3) contained in the domain D together with its ρ-neighborhood.

We assume that $a(\varphi, u(\varphi)) \in C^1_{\text{Lip}}(\mathcal{F}_m)$, the vector functions $a(\varphi, h)$ and $P(\varphi, h)h$ are Frechét differentiable with respect to h, and, in addition, the derivatives $\dfrac{\partial a(\varphi, h)}{\partial h}$ and $\dfrac{\partial P(\varphi, h)h}{\partial h}$ belong to the set C_D.

Let $\psi = \psi_t(\psi_0)$, $\psi_0 = \psi_0(\psi_0)$, be a solution of the equation

$$\frac{d\psi}{dt} = a(\psi, u(\psi)), \quad \psi_0 \in \mathcal{F}_m. \tag{13.4}$$

Then $u = u(\psi_t(\psi_0))$ is a trajectory on the torus $\mathcal{F}$. We represent the solution of the system of equations (13.3) in the form $\varphi = \varphi_t(\varphi_0, h_0)$, $h = h_t(\varphi_0, h_0)$, where $\varphi_0(\varphi_0, h_0) = \varphi_0$ and $h_0(\varphi_0, h_0) = h_0$.

We say that the motions of system (13.3) in a neighborhood of the invariant torus $\mathcal{F}$ are attracted (as $t \to +\infty$) to its motions on the torus according to the exponential law if there exist positive constants ρ_0, $\rho_1 \in R^1$ such that, for any solution $\{\varphi, h\}$ of this system of equations satisfying the condition

$$\min_{\varphi \in \mathcal{F}_m} \|\varphi_0 - \varphi\| \leq \rho_0 \quad \text{and} \quad \|h_0 - u(\varphi_0)\| \leq \rho_1, \tag{13.5}$$

one can find a trajectory on the torus $u(\psi_t(\psi_0))$, where ψ_0 depends on h_0 and φ_0, such that the following inequality is satisfied:

$$\|\varphi_t(\varphi_0, h_0) - \psi_t(\psi_0)\| + \|h_t(\varphi_0, h_0) - u(\psi_t(\psi_0))\| \leq K \exp\{-\gamma t\}, \tag{13.6}$$

where K and γ are positive constants and $t \geq 0$.

To solve the posed problem, we represent system (13.3) in a neighborhood of the torus $\mathcal{F}$ in the required form. To do this, we perform the change of variables

$$\varphi = \psi(t) + \mu\theta, \qquad h = u(\psi + \mu\theta) + \mu^2 h_1,$$

where μ is a positive parameter. This change of variables transforms system (13.3) in the neighborhood of the torus $\mathcal{F}$ into

$$\frac{d\theta}{dt} = A^1(\psi, \theta, \mu)\theta + \mu A^2(\psi, \theta, h_1, \mu)h_1,$$

$$\frac{dh_1}{dt} = P^1(\psi, \theta, h_1, \mu)h_1, \tag{13.7}$$

where

$$A^1(\psi, \theta, \mu) = \int_0^1 \frac{\partial a(\psi + \tau\mu\theta, u(\psi + \tau\mu\theta))}{\partial(\psi + \tau\mu\theta)} d\tau,$$

$$A^2(\psi, \theta, h_1, \mu) = \int_0^1 \frac{\partial a(\psi + \mu\theta, u(\psi + \mu\theta) + \mu^2 h_1 \tau)}{\partial(u(\psi + \mu\theta) + \mu^2 h_1 \tau)} d\tau,$$

and

$$P^1(\psi, \theta, h_1, \mu)$$

$$= \int_0^1 \frac{\partial[P(\psi + \mu\theta, u(\psi + \mu\theta) + \tau\mu^2 h_1)\,(u(\psi + \mu\theta) + \tau\mu^2 h_1)]}{\partial(u(\psi + \mu\theta) + \tau\mu^2 h_1)} d\tau$$

$$- \frac{\partial u(\psi + \mu\theta)}{\partial(\psi + \mu\theta)} \int_0^1 \frac{\partial a(\psi + \mu\theta, u(\psi + \mu\theta) + \tau\mu^2 h_1)}{\partial(u(\psi + \mu\theta) + \tau\mu^2 h_1)} d\tau, \tag{13.8}$$

where $\theta \in R^m$, $h_1 \in \mathfrak{M}$, A^1 is an $m \times m$ square matrix, the matrix A^2 has m rows and infinitely many columns, P^1 is an infinite matrix, and the matrix $\dfrac{\partial u(\psi + \mu\theta)}{\partial(\psi + \mu\theta)}$ has m columns and infinitely many rows.

It follows from expressions (13.8) that there exists $0 < \mu_0 < 1$ such that, for

$$\|h_1\| \leq 1 \quad \text{and} \quad 0 < \mu \leq \mu_0 \leq \min\{\sqrt{\rho}, 1\}, \tag{13.9}$$

the matrices A^1, A^2, and P^1 are continuous with respect to the collection of variables θ, h_1, and ψ in the matrix norm $\|\cdot\|$ and satisfy the inequalities

$$\|A^i(\psi, \theta, \mu)\| \leq M_1, \; i = 1, 2, \qquad \|P^1(\psi, \theta, h_1, \mu)\| \leq M_2,$$

$$\|A^1(\psi,\theta,\mu) - A^1(\psi,\bar\theta,\mu)\| \le \mu K_1\|\theta-\bar\theta\|,$$
$$\|A^2(\psi,\theta,h_1,\mu) - A^2(\psi,\bar\theta,\bar h_1,\mu)\| \le \mu K_1(\|\theta-\bar\theta\| + \|h_1-\bar h_1\|),$$
$$\|P^1(\psi,\theta,h_1,\mu) - P^1(\psi,\bar\theta,\bar h_1,\mu)\| \le \mu K_2(\|\theta-\bar\theta\| + \|h_1-\bar h_1\|),$$

where M_1, M_2, K_1, and K_2 are positive constants.

By $\Omega_\tau^t(P(t))$ we denote the matrizant of a homogeneous linear system of equations of normal form whose matrix of coefficients is equal to $P(t)$. We set

$$\|P(t)\|_T = \|P(t)\|_D,$$

where

$$D = T = [a,b] \subset R^1.$$

Lemma 13.1. *Assume that the infinite matrices $P(t)$ and $P_1(t)$ are continuous in t and bounded in the norm $\|\cdot\|_{T_1}$ for any finite segment $T_1 \subset [\tau,+\infty)$ ($T_1 \subset (-\infty,\tau]$). If the inequality*

$$\|\Omega_\tau^t(P)x_0\| \le L\exp\Big\{\int_\tau^t \beta(s)ds\Big\}\|x_0\|,$$

where $L = \text{const} > 0$ and $\beta(t)$ is a function integrable for $t \ge \tau$ ($t \le \tau$), holds for all $t \ge \tau$ ($t \le \tau$) and $x_0 \in \mathfrak{M}$, then

$$\|\Omega_\tau^t(P+P_1)x_0\| \le L\exp\Big\{\int_\tau^t \beta(s)ds + L\Big|\int_\tau^t \|P_1(s)\|\,ds\Big|\Big\}\|x_0\|$$

for any $t \ge \tau$ ($t \le \tau$).

Under the conditions of this lemma, the matrices P and $P+P_1$ satisfy requirement $(\widehat{P})$ from Section 1.3 and, hence, the matrizants $\Omega_\tau^t(P)$ and $\Omega_\tau^t(P+P_1)$ indeed exist. By virtue of Lemma 12.1, the proof of Lemma 13.1 is performed by analogy with the proof of Lemma 1 in [Sam10, Chap. 4, Sec. 4].

Let $\Omega_\tau^t(P_0)$ be the matrizant of the system of equations

$$\frac{dh_1}{dt} = P^1(\psi_t,0,0,0)h_1$$

and let $\Omega_\tau^t(A_0)$ be the matrizant of the system

$$\frac{d\theta}{dt} = \left(\frac{\partial a(\psi_t,u(\psi_t))}{\partial\psi_t}\right)\theta.$$

The behavior of solutions of system (13.7) is described by the following theorem:

Theorem 13.2. *Assume that the system of equations (13.7) is such that the following inequalities hold in domain (13.9):*

$$\|\Omega_t^s(P_0)h_0\| \leq \exp\Big\{-\int_t^s \beta(\psi_\tau)d\tau\Big\}\|h_0\|, \quad h_0 \in \mathfrak{M}, \tag{13.10}$$

$$\|\Omega_s^t(A_0)\theta_0\| \leq \exp\Big\{\int_s^t \alpha(\psi_\tau)d\tau\Big\}\|\theta_0\|, \quad \theta_0 \in R^m, \tag{13.11}$$

where $0 \leq t \leq s$, $\min \beta(\psi) \geq \beta_0 > 0$, $\alpha(\psi) \leq 0$, $\beta(\psi) + \alpha(\psi) > \gamma = \text{const} > 0$, *and* $\alpha(\psi_t)$ *and* $\beta(\psi_t)$ *are integrable functions.*

Then there exists a positive constant $\mu^* \leq \mu_0$ *such that, for all* μ: $0 < \mu \leq \mu^*$, *the system of equations (13.7) possesses a solution exponentially vanishing as* $t \to +\infty$.

Proof. We have

$$\frac{d\theta^n}{dt} = A^1(\psi, \theta^{n-1}, \mu)\theta^n + \mu A^2(\psi, \theta^{n-1}, h^n, \mu)h^n,$$

$$\frac{dh^n}{dt} = P^1(\psi, \theta^{n-1}, h^{n-1}, \mu)h^n, \tag{13.12}$$

provided that $\theta^0 = 0 \in R^m$ and $h^0 = 0 \in \mathfrak{M}$. For $n = 1$, we arrive at the system of equations

$$\frac{d\theta^1}{dt} = A_0(\psi)\theta^1 + \mu A^2(\psi, 0, h^1, \mu)h^1,$$

$$\frac{dh^1}{dt} = P_0(\psi)h^1.$$

As a solution of the last system for $t \geq 0$, we take a vector function (θ^1, h^1) of the form

$$\theta^1 = -\mu \int_t^\infty \Omega_s^t(A_0)\, A^2(\psi, 0, h^1, \mu)h^1 ds,$$

$$h^1 = \Omega_0^t(P_0)h_0^1, \tag{13.13}$$

where $\|h_0^1\| \le 1$. It is clear that, for $t \ge 0$, inequalities (13.10) and (13.11) imply the estimates

$$\|h^1\| \le \exp\Big\{-\int_0^t \beta(\psi)d\tau\Big\}\|h_0^1\| \le 1,$$

$$\begin{aligned}
&\|\Omega_s^t(A_0)A^2(\psi,0,h^1,\mu)h^1\| \\
&\qquad \le \exp\Big\{\int_s^t \alpha(\psi_\tau)d\tau\Big\}\|A^2(\psi,0,h^1,\mu)h^1\| \\
&\qquad \le M_1\exp\Big\{\int_s^t \alpha(\psi_\tau)d\tau - \int_0^s \beta(\psi_\tau)d\tau\Big\}\|h_0^1\| \\
&\qquad = M_1\exp\Big\{\int_0^t \alpha(\psi_\tau)d\tau - \int_0^s (\beta(\psi_\tau)+\alpha(\psi_\tau))d\tau\Big\}\|h_0^1\| \\
&\qquad \le M_1\exp\Big\{-\int_0^s (\beta(\psi_\tau)+\alpha(\psi_\tau))d\tau\Big\}\|h_0^1\| \\
&\qquad < M_1 e^{-\gamma s}\|h_0^1\|,
\end{aligned} \tag{13.14}$$

where $0 \le t \le s$.

This yields

$$\|\theta^1(t)\| < \mu M_1 \int_t^\infty \exp\{-\gamma s\}\|h_0^1\|ds = \frac{\mu M_1}{\gamma}\exp\{-\gamma t\}\|h_0^1\| \tag{13.15}$$

for $t \ge 0$, i.e., $\|\theta^1(t)\| < 1$ for $\mu \le \min\left\{\mu_0, \dfrac{\gamma}{M_1}\right\}$ and $t \ge 0$.

We have the following representations:

$$\begin{aligned}
P_{n-1} &= P^1(\psi,\theta^{n-1},h^{n-1},\mu) \\
&= P_0(\psi) + (P^1(\psi,\theta^{n-1},h^{n-1},\mu) - P_0(\psi)),
\end{aligned}$$

$$A_{n-1} = A^1(\psi,\theta^{n-1},\mu) = A_0(\psi) + (A^1(\psi,\theta^{n-1},\mu) - A_0(\psi)). \tag{13.16}$$

The solution (θ^n, h^n) of the system of equations (13.12) for $n \geq 1$ is given by the formulas

$$\theta^n = -\mu \int_t^\infty \Omega_s^t(A_{n-1}) A^2(\psi, \theta^{n-1}, h^n, \mu) h^n ds,$$

$$h^n = \Omega_0^t(P_{n-1}) h_0^1 \quad (t \geq 0). \tag{13.17}$$

Let $\varepsilon > 0$ be a constant such that $\varepsilon < \min\{\frac{\gamma}{2}; \beta_0\}$ and let

$$\mu_* = \min\Big\{\mu^0; \frac{\varepsilon}{K_1}; \frac{\varepsilon}{2K_2}; \frac{\gamma - 2\varepsilon}{M_1}\Big\}.$$

Assume that, for $n = 2, 3, \ldots, k-1$, solution (13.17) satisfies the inequalities

$$\|\theta^n\| \leq \exp\{-\gamma t + 2\varepsilon t\} \|h_0^1\|, \quad \gamma \geq 2\varepsilon, \ t \geq 0,$$

$$\|h^n\| \leq \exp\Big\{-\int_0^t \beta(\psi_\tau) d\tau + \varepsilon t\Big\} \|h_0^1\|, \quad t \geq 0. \tag{13.18}$$

For $n = 1$, we prove these estimates by using relations (13.14) and (13.15). Let us show that inequalities (13.18) are true for $n = k$. We have

$$\|P^1(\psi, \theta^{k-1}, h^{k-1}, \mu) - P_0(\psi)\| \leq \mu K_2(\|\theta^{k-1}\| + \|h^{k-1}\|) \leq 2\mu K_2 < \varepsilon,$$

$$\|A^1(\psi, \theta^{k-1}, \mu) - A_0(\psi)\| \leq \mu K_1 \|\theta^{k-1}\| \leq \mu K_1 < \varepsilon, \quad t \geq 0, \ \mu \leq \mu_*.$$

By virtue of Lemma 13.1, this yields

$$\|\Omega_s^t(A_{k-1})\zeta\| \leq \exp\Big\{\int_s^t \alpha(\psi_\tau) d\tau + \varepsilon(s-t)\Big\} \|\zeta\|, \quad \zeta \in R^m,$$

$$\|\Omega_s^t(P_{k-1})\eta\| \leq \exp\Big\{-\int_t^s \beta(\psi_\tau) d\tau + \varepsilon(s-t)\Big\} \|\eta\|,$$

$\|\eta| \leq 1$, $\eta \in \mathfrak{M}$, for all $0 \leq t \leq s$.

Further, for $t \geq 0$, we can write

$$\|h^k\| \leq \exp\Big\{-\int_0^t \beta(\psi_\tau) d\tau + \varepsilon t\Big\} \|h_0^1\| \leq 1, \quad \beta_0 > \varepsilon,$$

$$\begin{aligned}
&\|\Omega_s^t(A_{k-1})A^2(\psi,\theta^{k-1},h^k,\mu)h^k\| \\
&\qquad \le M_1\|h^k\|\exp\Big\{\int_s^t \alpha(\psi_\tau)d\tau+\varepsilon(s-t)\Big\} \\
&\qquad \le M_1\exp\Big\{-\int_0^s(\beta(\psi_\tau)+\alpha(\psi_\tau))d\tau+2\varepsilon s\Big\}\|h_0^1\| \\
&\qquad \le M_1\|h_0^1\|\exp\{-\gamma s+2\varepsilon s\},
\end{aligned}$$

whence it follows that

$$\begin{aligned}
\|\theta^k\| &\le \mu M_1\int_t^\infty \exp\{-(\gamma-2\varepsilon)s\}ds\|h_0^1\| \\
&\le \frac{\mu M_1}{\gamma-2\varepsilon}\exp\{-(\gamma-2\varepsilon)t\}\|h_0^1\|\le 1.
\end{aligned}$$

Thus, we have actually proved by induction that the function (θ^n,h^n) is defined for all $n=1,2,3,\dots,$ $t\ge 0$, $\mu\le\mu_*$, and $\|h_0^1\|\le 1$ and, moreover, inequalities (13.18) are satisfied.

In what follows, the initial value h_0^1 is denoted by h_0.

We now prove that the sequences $\{\theta^n\}$ and $\{h^n\}$ converge as $n\to\infty$ uniformly in t, ψ_0, h_0, and μ.

We set

$$r_{n+1}=h^{n+1}-h^n=\Big[\Omega_0^t(P_n)-\Omega_0^t(P_{n-1})\Big]h_0.$$

Then

$$\frac{dr_{n+1}}{dt}=P_nh^{n+1}-P_{n-1}h^n=P_nr_{n+1}+(P_n-P_{n-1})h^n.$$

Since $r_{n+1}(0)=0$, we have

$$r_{n+1}=\int_0^t \Omega_s^t(P_n)(P_n-P_{n-1})\Omega_0^s(P_{n-1})h_0ds.$$

Hence, for $t\ge 0$ and $\mu_*\le\dfrac{1}{2K_2}$, we get

$$\begin{aligned}\|r_{n+1}\| &\leq \int_0^t \exp\Big\{-\int_s^t \beta(\psi_\tau)d\tau + \varepsilon(t-s)\Big\}\mu K_2[\|\theta^n - \theta^{n-1}\| \\ &\quad + \|h^n - h^{n-1}\|]\exp\Big\{-\int_0^s \beta(\psi_\tau)d\tau + \varepsilon s\Big\}\|h_0\|ds \\ &\leq \exp\Big\{-\int_0^t \beta(\psi_\tau)d\tau + \varepsilon t\Big\}[\|\theta^n - \theta^{n-1}\|_0 \\ &\quad + \|h^n - h^{n-1}\|_0]\|h_0\|t\mu K_2, \end{aligned} \tag{13.19}$$

where $\|\cdot\|_0 = \sup\limits_{t\geq 0}\|\cdot\|$.

In view of inequality (13.19), we obtain

$$\|r_{n+1}\| \leq \exp\left\{-\int_0^t \beta(\psi_\tau)d\tau + \varepsilon t\right\}\left(\|\theta^n - \theta^{n-1}\|_0 + \|r_n\|_0\right)t, \quad t \geq 0. \tag{13.20}$$

Let us now estimate the norm of the difference $\xi_{n+1} = \theta^{n+1} - \theta^n$. For $t \geq 0$, we find

$$\begin{aligned}\|\xi_{n+1}\| &\leq \mu\int_t^\infty \left\|\left[\Omega_s^t(A_n) - \Omega_s^t(A_{n-1})\right]A^2(\psi, \theta^n, h^{n+1}, \mu)h^{n+1}\right\| ds \\ &\quad + \mu\int_t^\infty \left\|\Omega_s^t(A_{n-1})A^2(\psi, \theta^n, h^{n+1}, \mu)(h^{n+1} - h^n)\right\| ds \\ &\quad + \mu\int_t^\infty \Big\|\Omega_s^t(A_{n-1})\left[A^2(\psi, \theta^n, h^{n+1}, \mu)\right. \\ &\quad \left. - A^2(\psi, \theta^{n-1}, h^n, \mu)\right]h^n\Big\| ds. \end{aligned} \tag{13.21}$$

Further, we denote $\Delta_n = \Omega_s^t(A_{n-1})\theta_0$ and consider the difference

$$\Delta_{n+1} - \Delta_n = \left[\Omega_s^t(A_n) - \Omega_s^t(A_{n-1})\right]\theta_0.$$

This gives

$$\frac{d(\Delta_{n+1}-\Delta_n)}{dt} = A_n(\Delta_{n+1}-\Delta_n) + (A_n - A_{n-1})\Delta_n,$$

and, in view of the fact that $(\Delta_{n+1}-\Delta_n) = 0$ for $t = s$, we get

$$\Delta_{n+1}-\Delta_n = \int_s^t \Omega_\tau^t(A_n)(A_n - A_{n-1})\Omega_s^\tau(A_{n-1})\theta_0 d\tau.$$

This implies the inequality

$$\begin{aligned}
&\|\Delta_{n+1}-\Delta_n\| \\
&\quad \le \int_t^s \exp\Big\{\int_\tau^t \alpha(\psi_t)dt + \varepsilon(\tau - t)\Big\}\mu K_1\|\theta^n - \theta^{n-1}\| \\
&\qquad \times \exp\Big\{\int_s^\tau \alpha(\psi_t)dt + \varepsilon(s-\tau)\Big\}\|\theta_0\|d\tau \\
&\quad \le \mu K_1\|\theta_0\|\,\|\theta^n - \theta^{n-1}\|_0 \exp\Big\{\int_s^t \alpha(\psi_t)dt + \varepsilon(s-t)\Big\}(s-t) \\
&\quad \le \|\theta^n - \theta^{n-1}\|_0 \exp\Big\{\int_s^t \alpha(\psi_t)dt + \varepsilon(s-t)\Big\}(s-t) \qquad (13.22)
\end{aligned}$$

for $\mu_* \le \dfrac{1}{K_1}$ and $\|\theta_0\| \le 1$.

By using inequalities (13.21) and (13.22), we obtain

$$\begin{aligned}
\|\xi_{n+1}\| \;\le\; & \mu\int_t^\infty \|\theta^n - \theta^{n-1}\|_0 \exp\Big\{\int_s^t \alpha(\psi_t)dt + \varepsilon(s-t)\Big\}(s-t)M_1 \\
& \times \exp\Big\{-\int_0^s \beta(\psi_\tau)d\tau + \varepsilon s\Big\}\|h_0\|ds \\
& + \mu\int_t^\infty \exp\Big\{\int_s^t \alpha(\psi_\tau)d\tau + \varepsilon(s-t)\Big\}M_1\|h^{n+1} - h^n\|ds
\end{aligned}$$

$$+\mu\int_t^\infty \exp\Big\{\int_s^t \alpha(\psi_\tau)d\tau+\varepsilon(s-t)\Big\}K_1(\|\theta^n-\theta^{n-1}\|$$

$$+\|h^{n+1}-h^n\|)\exp\Big\{-\int_0^s \beta(\psi_\tau)d\tau+\varepsilon s\Big\}\|h_0\|ds$$

$$\le \mu M_1\|\xi_n\|_0\|h_0\|J_1+\mu M_1 J_2$$

$$+\mu K_1(\|\xi_n\|_0+\|r_{n+1}\|_0)\|h_0\|J_3, \tag{13.23}$$

where the integrals J_1 and J_3 are such that

$$J_1 \le \int_t^\infty \exp\Big\{-\int_0^s(\beta(\psi_\tau)+\alpha(\psi_\tau))d\tau+2\varepsilon s\Big\}(s-t)ds$$

$$\le \int_t^\infty \exp\{-(\gamma-2\varepsilon)s\}(s-t)ds=\frac{\exp\{-(\gamma-2\varepsilon)t\}}{(\gamma-2\varepsilon)^2},$$

$$J_3 \le \int_t^\infty \exp\{-(\gamma-2\varepsilon)s\}ds=\frac{\exp\{-(\gamma-2\varepsilon)t\}}{(\gamma-2\varepsilon)}$$

for $t\ge 0$. In view of inequality (13.20), for the integral J_2 we can write

$$\|J_2\| \le (\|\xi_n\|_0+\|r_n\|_0)$$

$$\times\int_t^\infty \exp\left\{\int_s^t \alpha(\psi_\tau)d\tau+\varepsilon(s-t)-\int_0^t \beta(\psi_\tau)d\tau+\varepsilon s\right\}sds$$

$$\le (\|\xi_n\|_0+\|r_n\|_0)\int_t^\infty \exp\{-(\gamma-2\varepsilon)s\}sds$$

$$\le (\|\xi_n\|_0+\|r_n\|_0)\frac{1}{\gamma-2\varepsilon}\Big(t+\frac{1}{\gamma-2\varepsilon}\Big)\exp\{-(\gamma-2\varepsilon)t\}$$

for $t\ge 0$. By using representation (13.23), we arrive at the following inequality for ξ_{n+1}:

$$\|\xi_{n+1}\| \leq \frac{\mu}{\gamma - 2\varepsilon}(\|\xi_n\|_0 + \|r_n\|_0)\Big\{M_1 \exp\{-\gamma t + 2\varepsilon t\}\frac{2}{\gamma - 2\varepsilon}$$

$$+ M_1 t \exp\{-(\gamma - 2\varepsilon)t\} + K_1(1+t)\exp\{-(\gamma - 2\varepsilon)t\}\Big\}$$

$$\leq \frac{\mu}{\gamma - 2\varepsilon} \leq (\|\xi_n\|_0 + \|r_n\|_0)\frac{M_1(2e+1) + K_1(e\gamma + 1)}{e(\gamma - 2\varepsilon)}$$

$$= C_1\mu(\|\xi_n\|_0 + \|r_n\|_0),$$

where

$$C_1 = \frac{M_1(2e+1) + K_1(e\gamma+1)}{e(\gamma - 2\varepsilon)^2} = \text{const} > 0.$$

In view of relation (13.20), we get

$$\|\xi_{n+1}\|_0 + \|r_{n+1}\|_0 \leq \mu\Big(C_1 + \frac{K_2}{e(\beta_0 - \varepsilon)}\Big)(\|\xi_n\|_0 + \|r_n\|_0).$$

We now choose sufficiently small μ'_* such that

$$\mu'_*\Big(C_1 + \frac{K_2}{e(\beta_0 - \varepsilon)}\Big) \leq \frac{1}{2}.$$

Then the inequality

$$\|\xi_{n+1}\|_0 + \|r_{n+1}\|_0 \leq \frac{1}{2}(\|\xi_n\|_0 + \|r_n\|_0)$$

means that the sequences $\{\theta^m\}$ and $\{h^n\}$ are convergent uniformly in t, ψ_0, h_0, and μ in the domain D^0:

$$\|h_0\| \leq 1,\ \psi_0 \in \mathfrak{F}_m,\ t \geq 0,\ \mu \leq \mu^* = \min\Big\{\frac{1}{K_1};\ \frac{1}{2K_2};\ \mu_*;\ \mu'_*\Big\}.$$

We set

$$\theta(t) = \lim_{n\to\infty} \theta^n(t) \quad \text{and} \quad h_1(t) = \lim_{n\to\infty} h^n(t). \tag{13.24}$$

Clearly, the limiting functions $\theta(t)$ and $h_1(t)$ satisfy the inequalities

$$\|\theta(t)\| \leq \exp\{-(\gamma - 2\varepsilon)t\}\|h_0\|,$$

$$\|h_1(t)\| \leq \exp\Big\{-\int_0^t \beta(\psi_\tau)d\tau + \varepsilon t\Big\}\|h_0\|. \tag{13.25}$$

It remains to show that the vector function $\{\theta(t), h_1(t)\}$ is a solution of the system of equations (13.17).

It is clear that the following identities are true in the domain D^0:

$$\frac{d\theta^n(t)}{dt} = A^1(\psi_t, \theta^{n-1}(t), \mu)\Theta^n(t) + \mu A^2(\psi_t, \theta^{n-1}(t), h^n(t), \mu)h^n(t),$$

$$\frac{dh^n(t)}{dt} = P^1(\psi_t, \theta^{n-1}(t), h^{n-1}(t), \mu)h^n(t). \tag{13.26}$$

The fact that the $(m \times m)$ matrix A^1 is finite-dimensional immediately implies that

$$\lim_{n\to\infty} A^1(\psi_t, \theta^{n-1}(t), \mu)\theta^n(t) = A^1(\psi_t, \theta(t), \mu)\theta(t)$$

uniformly in t for any finite segment $T_1 \subset [0, +\infty)$.

The limit transition in the second vector equality in (13.26) is performed coordinatewise. Thus, for any ith coordinate ($i = 1, 2, 3, \ldots$), we obtain

$$\lim_{n\to\infty} \frac{dh_i^n(t)}{dt} = \lim_{n\to\infty} \sum_{j=1}^{\infty} p_{ij}^1(\psi_t, \theta^{n-1}(t), h^{n-1}(t), \mu)h_j^n(t)\,.$$

Since the matrix P^1 defined in the domain D^0 is bounded in the norm $\|\cdot\|_{T_1}$ on the segment T_1 and $\|h^n(t)\| \leq 1$, the series on the right-hand side of the last equality converges absolutely and uniformly in $t \in T_1$ and n. Moreover,

$$\lim_{n\to\infty} p_{ij}^1(\psi_t, \theta^{n-1}(t), h^{n-1}(t), \mu)h_j^n(t) = p_{ij}^1(\psi_t, \theta(t), h_1(t), \mu)h_{1j}(t)$$

uniformly in $t \in T_1$, which enables us to use Lemma 12.1 and write the following formula (for $t \geq 0$):

$$\lim_{n\to\infty} \frac{dh^n(t)}{dt} = \frac{d}{dt}\Big(\lim_{n\to\infty} h^n(t)\Big) = P^1(\psi_t, \theta(t), h_1(t), \mu)h_1(t),$$

where the limit is taken coordinatewise.

Similarly, one can easily show that (coordinatewise)

$$\lim_{n\to\infty} A^2(\psi_t, \theta^{n-1}(t), h^n(t), \mu)h^n(t) = A^2(\psi_t, \theta(t), h_1(t), \mu)h_1(t)$$

uniformly in $t \in T_1$.

This means that the vector function $\{\theta(t), h_1(t)\}$ is differentiable with respect to t at any point of the semiaxis $[0, +\infty)$ and solves the system of equations (13.7) in the domain D^0. This completes the proof of Theorem 13.2.

By using the theorem just proved, we can easily clarify the behavior of solutions in the neighborhood of the invariant torus of the system of equations (13.3) as compared with the behavior of solutions of this system on the torus. The following theorem is true:

Theorem 13.3. *Assume that* $a(\varphi, h)$, $P(\varphi, h)$, *and* $c(\varphi)$ *belong to the set* $C^1(\mathcal{F}_m)$ *for* $h \in C^1(\mathcal{F}_m)$ *and the Frechét derivatives*

$$\frac{\partial a(\varphi, h)}{\partial h}, \ \frac{\partial (P(\varphi, h)h)}{\partial h} \in C_D$$

exist. If the torus $\mathcal{F}\colon h = u(\varphi)$, $\varphi \in \mathcal{F}_m$, *belongs to the domain* D *together with its ρ-neighborhood and is an invariant set of system (13.3) and, in addition,* $u(\varphi)$, $a(\varphi, u(\varphi)) \in C^1_{\mathrm{Lip}}(\mathcal{F}_m)$, *then, under conditions (13.10) and (13.11), the motions of system (13.3) in the neighborhood of the torus* $\mathcal{F}$ *are attracted as* $t \to +\infty$ *to its motions on the torus according to the exponential law.*

Proof. Under the conditions of the formulated theorem, Theorem 13.2 is true. This means that, for any fixed $\psi_0 \in \mathcal{F}_m$ and sufficiently small $\mu \leq \mu^*$, there exists a solution

$$\varphi_t = \psi_t + \mu\theta(t),$$
$$h_t = u(\psi_t) + [u(\psi_t + \mu\theta(t)) - u(\psi_t)] + \mu^2 h_1(t) \tag{13.27}$$

of the system of equations (13.3) such that

$$\|\varphi_t - \psi_t\| \leq \mu \exp\{-(\gamma - 2\varepsilon)t\},$$
$$\|h_t - u(\psi_t)\| \leq \mu C_2 \exp\{-(\gamma - 2\varepsilon)t\}, \tag{13.28}$$

where C_2 is a positive constant independent of t.

The second inequality in (13.28) can be obtained as follows: Since $u(\psi) \in C_{\mathrm{Lip}}(\psi)$, there exists $P = \mathrm{const} > 0$ such that

$$\begin{aligned}
\|h_t - u(\psi_t)\| &\leq P\mu\|\theta(t)\| + \mu^2\|h_1(t)\| \\
&\leq \mu\Big(P\exp\{-(\gamma - 2\varepsilon)t\} + \mu\exp\Big\{-\int_o^t \beta(\psi_\tau)d\tau + \varepsilon t\Big\}\Big) \\
&\leq \mu(P + \mu)\exp\{-\gamma_2 t\} \\
&= \mu C_2 \exp\{-\gamma_2 t\},
\end{aligned}$$

where $\gamma_2 = \min\{\gamma - 2\varepsilon,\ b_0 - \varepsilon\}$ and $C_2 = P + \mu$. At the same time, according to the condition of the theorem, we have $\gamma \leq \beta_0$ and, hence, $\gamma - 2\varepsilon < b_0 - \varepsilon$, i.e., $\gamma_2 = \gamma - 2\varepsilon$. Inequalities (13.28) yield (13.6). In this case, the solution (φ_t, h_t) of system (13.3) corresponding to $\psi_0 \in \mathcal{F}_m$ satisfies the inequalities

$$\|\varphi_0 - \psi_0\| \leq \mu \quad \text{and} \quad \|h_0 - u(\varphi_0)\| \leq \mu^2,$$

and, hence, for $\rho_0 = \mu$ and $\rho_1 = \mu^2$, condition (13.5) is satisfied. Moreover, ρ_0 and ρ_1 are independent of $\psi_0 \in \mathcal{F}_m$.

Denote $\theta(0) = \theta(\mu, \psi_0, h_1(0))$. As shown above, the solution (φ_t, h_t) of the system of equations (13.3) for which

$$\varphi_0 = \varphi(0) = \psi_0 + \mu\theta(\mu, \psi_0, h_1(0)),$$

$$h_0 = h(0) = u(\varphi_0) + \mu^2 h_1(0), \quad \psi_0 \in \mathcal{F}_m,$$

satisfies inequality (13.6) for $t \geq 0$.

We fix arbitrary values of φ_0 and h_0 satisfying conditions (13.5). Let us show that, for these φ_0 and h_0 the equation

$$\varphi_0 = \psi_0 + \mu\theta\Big(\mu, \psi_0, \frac{h_0 - u(\varphi_0)}{\mu^2}\Big) \tag{13.29}$$

possesses a solution $\psi_0(\varphi_0, h_0)$. This will complete the proof of Theorem 13.3.

We introduce an unknown quantity $v = \varphi_0 - \psi_0$ and rewrite equation (13.29) in the form

$$v = \mu\theta\Big(\mu, \varphi_0 - v, \frac{h_0 - u(\varphi_0)}{\mu^2}\Big). \tag{13.30}$$

Since $\left\|\mu\theta(\mu, \varphi_0 - v, \dfrac{h_0 - u(\varphi_0)}{\mu^2}\right\| \leq \mu$, the right-hand side of equation (13.30) specifies an operator that maps the ball $\|v\| \leq \mu$ into itself. It remains to show that this operator is continuous with respect to v and use the Brouwer fixed-point theorem.

To do this, by virtue of (13.24), it suffices to prove that the functions $\theta^n(t, \psi_0)$ given by relations (13.17) with $n = 1, 2, 3, \ldots$ are continuous with respect to ψ_0.

The function $\theta^1(t, \psi_0)$ is given by relations (13.13). By virtue of the theorem on continuous dependence of solutions on the initial conditions, $\psi = \psi_t(\psi_0)$ is a continuous function of ψ_0. The matrices $A_0(\psi_t)$ and $P_0(\psi_t)$ are continuous in ψ_t and, therefore, in ψ_0. For $h^1(t, \psi_0)$, the following inequality is satisfied:

$$\|h_t^1(\psi_0) - h_t^1(\bar{\psi}_0)\|$$

$$\leq M_2 \int_0^t \|h_s^1(\psi_0) - h_s^1(\bar{\psi}_0)\| ds + \int_0^t \|P_0(s, \psi_0) - P_0(s, \psi_0)\| ds. \quad (13.31)$$

Further, by using the fact that the matrix $P_0(s, \psi_0)$ is continuous with respect to ψ_0, the Gronwall–Bellman lemma, and inequality (13.31), we conclude that the function $h_t^1(\psi_0)$ is continuous with respect to ψ_0 uniformly in t for any finite interval $[0, T_0] \subset R^1$. Then the matrix $A^2(\psi, 0, h_1, \mu)$ is also continuous with respect to ψ_0. Hence, the integrand in (13.13) is also continuous with respect to ψ_0 in any finite segment of variation of t and, therefore, the function $\theta^1(t, \psi_0)$ is also continuous with respect to ψ_0 because the indicated integral is uniformly convergent.

By induction, using the same reasoning as above, we prove that $\theta^n(t, \psi_0)$ are continuous functions of ψ_0 for any natural n, which completes the proof of Theorem 13.3.

Remark 13.1. Theorem 13.3 remains valid if conditions (13.10) and (13.11) are replaced by the inequalities

$$\|\Omega_t^s(P_0)h_0\| \leq L_1 \exp\Big\{-\int_t^s \beta(\psi_\tau)d\tau\Big\}\|h_0\|, \quad h_0 \in \mathfrak{M},$$

$$\|\Omega_s^t(A_0)\theta_0\| \leq L_2 \exp\Big\{\int_s^t \alpha(\psi_\tau)d\tau\Big\}\|\theta_0\|, \quad \theta_0 \in R^m,$$

where L_1 and L_2 are constants greater than one.

More precise laws of attraction to the invariant torus of system (13.3) than the law discussed in this section can be obtained in the general case by using the results presented in [Sam12, Sam13] for the case of finite-dimensional systems.

3. REDUCIBILITY OF LINEAR SYSTEMS

In the present chapter, we study the problem of reducibility of countable linear systems of differential equations, i.e., the possibility of mapping the trajectories of their solutions onto the trajectories of systems with simpler structure. For finite-dimensional systems with periodic coefficients, one can especially mention the well-known results of Floquet and Lyapunov establishing the admissibility of this mapping [CoL]. Linear systems with quasiperiodic coefficients are of great interest for applications. Numerous works (see, e.g., [And, Arn3, Art, Bar2, Bel, Gel, MiS2]) are devoted to the problem of reducibility of these systems and construction of their solutions. At present, the problem of reducibility of almost periodic linear systems is of serious practical significance. In this direction, we can mention the works [Bli1], [Bli2], [Fil], and [Par] devoted to the solution of the problem of reducibility for fairly narrow classes of almost periodic systems. We also refer the reader to the works dealing with the reducibility of the original system of equations to a block- diagonal form [Byl1, Mil, MSK2].

In the present chapter, we formulate sufficient conditions for the reducibility (in the Lyapunov sense) of a periodic countable system, study the possibility of reducing the problem of reducibility of an almost periodic system of equations to the case of a system with quasiperiodic coefficients, construct a process with accelerated convergence of iterations reducing a quasiperiodic system with unbounded right-hand side, and analyze the problem of decomposition of a countable system of equations.

3.14. Erugin and Floquet–Lyapunov Theorems

The theory of reducibility of finite-dimensional linear systems is well developed (see, e.g., [CoL, Pon]). In the present section, we generalize some results of this theory to the case of countable systems.

In the space $\mathfrak{M}$, we consider a differential equation

$$\frac{dx}{dt} = A(t)x, \tag{14.1}$$

where $x \in \mathfrak{M}$ and $A(t) = [a_{ij}(t)]_{i,j=1}^{\infty}$ is an infinite matrix.

We assume that the functions $a_{ij}(t)$ $(i,\ j = 1, 2, \ldots)$ are continuous in t in the interval $T = R^1$ and, moreover,

$$\|A(t)\|_{T_1} \leq \Psi_{T_1} = \text{const} < \infty \tag{14.2}$$

in any finite segment $T_1 = [a,\ b] \subset T$.

Then, as shown in Chapter 1, equation (14.1) possesses a unique solution with given initial conditions $x_0 \in \mathfrak{M}$ and $t_0 \in T$ defined for all $t \in T$.

Equation (14.1) is called reducible on T if there exist an infinite matrix $\Phi(t)$ and a differential equation

$$\frac{dy}{dt} = Py, \quad y \in \mathfrak{M}, \ \ \|P\| < \infty \tag{14.3}$$

with constant matrix P such that any solution $x = x(t)$ of equation (14.1) can be expressed via the solution $y = y(t)$ of equation (14.3) by the formula $x(t) = \Phi(t)y(t)$ valid for $t \in T$.

A matrix solution $X(t)$ of equation (14.1) is called a fundamental matrix if any vector solution of this equation can be represented in the form $X(t)c$, where $c \in \mathfrak{M}$ is a constant vector. Thus, the matrizant of equation (14.1) is its fundamental matrix.

A matrix $L(t)$ is called a Lyapunov matrix if it is continuously differentiable and invertible on T and, in addition,

$$T_1 \subset T \ \max\Big\{\|L(t)\|_{T_1},\ \|L^{-1}(t)\|_{T_1},\ \Big\|\frac{d}{dt}L(t)\Big\|_{T_1}\Big\} < \infty$$

on every finite interval.

We say that equation (14.1) is reduced to the form (14.3) in the Lyapunov sense if the matrix $\Phi(t)$ realizing reducibility is a Lyapunov matrix.

The set of all infinite constant matrices bounded in the norm $\|\cdot\|$ is denoted by $\mathcal{B}$. It is easy to see that the set $\mathcal{B}$ forms a Banach space over the field of real numbers and the following statements are true:

1. If $P \in \mathcal{B}$, then $e^P = E + P + \dfrac{P^2}{2!} + \ldots + \dfrac{P^k}{k!} + \ldots$ exists and belongs to $\mathcal{B}$.

2. The matrices P and e^P are commutative in the sense of multiplication.

3. The matrix e^{-P} is inverse to e^P.

4. The derivative $(e^{Pt})' = e^{Pt} \cdot P$.

5. If $P_k \in \mathcal{B}$ $(k = 1, 2, \ldots)$, $t \in T$, then

$$\prod_{k=1}^{s} P_k(t) = P_1(t)\Big(\prod_{k=2}^{s} P_k(t)\Big).$$

The following theorem is an analog of the Erugin theorem and establishes the possibility of reducing equation (14.1) to equation (14.3) with the help of the Lyapunov reducing matrix:

Theorem 14.1. *Equation (14.1) is reducible to equation (14.3) in the Lyapunov sense if and only if a fundamental matrix $X(t)$ of the original equation admits a representation $X(t) = L(t)\exp\{Pt\}$, where $L(t)$ is a Lyapunov matrix.*

Proof. The following assertion is true: If the matrices $Z(t) = [z_{ij}(t)]_{i,j=1}^{\infty}$ and $G(t) = [g_{ij}(t)]_{i,j=1}^{\infty}$ are differentiable on the segment T_1 and, in addition, both the matrices and their derivatives are bounded on T_1 in the norm $\|\cdot\|_{T_1}$, then

$$\frac{d}{dt}(Z(t)G(t)) = \frac{d}{dt}(Z(t))G(t) + Z(t)\frac{d}{dt}(G(t)), \quad t \in T_1.$$

To prove this assertion, it suffices to show that, for all $i, j = 1, 2, \ldots$,

$$\Big(\sum_{s=1}^{\infty} z_{is}(t)g_{sj}(t)\Big)' = \sum_{s=1}^{\infty} z'_{is}(t)g_{sj}(t) + \sum_{s=1}^{\infty} z_{is}(t)g'_{sj}(t), \ t \in T_1.$$

This equality is indeed true because the series on its right-hand side can be majorized by the convergent numerical series

$$\sum_{s=1}^{\infty} \max_{T_1} |z'_{is}(t)| \, \|G(t)\|_{T_1} \quad \text{and} \quad \sum_{s=1}^{\infty} \max_{T_1} |z_{is}(t)| \, \|G'(t)\|_{T_1},$$

respectively, and, thus, they converge absolutely and uniformly in $t \in T_1$.

We assume that there exists a fundamental matrix $X(t)$ of equation (14.1) such that

$$X(t) = L(t)e^{Pt},$$

where $L(t)$ is a Lyapunov matrix. In equation (14.1), we perform the following change of variables:

$$x = L(t)y,$$

where $L(t) = X(t)e^{-Pt}$ and $y(t)$ is a vector function such that, on every finite interval $T_1 \subset T$, the quantities $\|y(t)\|$ and $\left\|\frac{dy(t)}{dt}\right\|$ are bounded by positive constants P_1 and P_2 dependent on T_1.

Then

$$\frac{dx}{dt} = \frac{d}{dt}(L(t))y + L(t)\frac{dy}{dt}.$$

This yields

$$\frac{d}{dt}(X(t)e^{-Pt})y + X(t)e^{-Pt}\frac{dy}{dt} = A(t)X(t)e^{-Pt}y.$$

The fundamental matrix $X(t)$ is bounded in the norm $\|\cdot\|_{T_1}$. Therefore, $\dfrac{dX(t)}{dt}$ also possesses the same property. In this case, we arrive at the equation

$$X(t)e^{-Pt}\frac{dy}{dt} + \frac{dX}{dt}e^{-Pt}y - X(t)e^{-Pt}Py = A(t)X(t)e^{-Pt}y,$$

which, in view of the fact that the matrix $X(t)$ is fundamental, yields (14.3). It is clear that the solutions of equation (14.3) are bounded in the norm $\|\cdot\|_{T_1}$ together with their derivatives.

We now prove necessity. Assume that equation (14.1) can be reduced to equation (14.3) by using the Lyapunov matrix $L(t)$. It is clear that any solution of equation (14.3) with initial values $t_0 \in T$ and $y_0 \in \mathfrak{M}$ can be represented in the form

$$y(t) = \exp\{P(t - t_0)\}y_0.$$

We denote $e^{-Pt} = C \in \mathcal{B}$. Since the matrices Pt and Pt_0 are commuting, every solution $y(t)$ of equation (14.3) can be represented in the form $y(t) = \exp\{Pt\}c$, where $c \in \mathfrak{M}$.

We denote the product $L(t)e^{Pt}$ by $X(t)$. It is clear that $X(t)$ is a fundamental matrix of equation (14.1) because all solutions of this equation are contained in the set $X(t)c,\ c \in \mathfrak{M}$, since

$$X(t)c = (L(t)e^{Pt})c = L(t)(e^{Pt} \cdot c)$$

and $\exp\{Pt\}c$ describes all solutions of equation (14.3).

Theorem 14.1 is proved.

Corollary 14.1. *For any constant matrix $P \in \mathcal{B}$, there exists a Lyapunov matrix that reduces equation (14.1) to equation (14.3).*

Proof. By virtue of Theorem 14.1, it suffices to prove that the matrizant Ω_0^t of equation (14.1) can be represented in the form $\Omega_0^t = L(t)e^{Pt}$, where $L(t)$ is a Lyapunov matrix. We set

$$L(t) = \Omega_0^t e^{-Pt}. \tag{14.4}$$

It is clear that $L(t)$ is a Lyapunov matrix.

Now assume that the matrix $A(t)$ is periodic in t with period ω: $A(t+\omega) = A(t)$, $t \in T$. In this case, Corollary 14.1 is also true for equation (14.1). The situation becomes more complicated if we require that the reducing matrix must be bounded on the entire real axis. For this purpose, it suffices to demand the periodicity of the Lyapunov matrix $L(t)$ in relation (14.4) as a function of t. In the finite-dimensional case, the existence of this matrix is guaranteed by the Floquet–Lyapunov theorem [CoL]. At the same time, it is known [DaK, MaS] that, in the infinite-dimensional case, a direct analog of this theorem is absent. In the theorem below, we formulate conditions for the existence of a periodic Lyapunov matrix of the form (14.4).

Theorem 14.2. *For the ω-periodicity of the Lyapunov matrix $L(t)$ given by equality (14.4), where Ω_0^t is the matrizant of equation (14.1) with ω-periodic matrix $A(t)$, it is necessary and sufficient that the real logarithm of the monodromy matrix Ω_0^ω exist and be such that*

$$P = \frac{1}{\omega} \ln \Omega_0^\omega .$$

This theorem is proved in exactly the same way as in [CoL].

The analysis of conditions for the existence of the logarithm of the monodromy operator in the infinite-dimensional case is in no case a trivial problem [DaK, MaS]. In the next section, we solve the problem of reducibility of equation (14.1) by using a real periodic reducing matrix without finding the logarithm of the monodromy matrix.

3.15. Periodic Systems

In the present section, we relate the solution of the problem of reducibility of equation (14.1) to equation (14.3) to the problem of reducibility of finite-dimensional systems obtained from infinite systems as a result of truncation.

Parallel with (14.1), we consider a sequence of systems

$$\frac{d\overset{(n)}{x}}{dt} = \overset{(n)}{A}(t)\overset{(n)}{x}, \quad n = 1, 2, \dots, \tag{15.1}$$

where $\overset{(n)}{x} = (\overset{(n)}{x}_1, \overset{(n)}{x}_2, \dots \overset{(n)}{x}_n) \in R^n$ and $\overset{(n)}{A}(t) = [a_{ij}(t)]_{i,\ j=1}^{n}$.

Assume that, for $n \geq N$, each equation in (15.1) can be reduced to an equation

$$\frac{d\overset{(n)}{y}}{dt} = \overset{(n)}{P}\overset{(n)}{y} \tag{15.2}$$

with the help of a reducing matrix $\overset{(n)}{\Phi}(t) = [\overset{(n)}{\varphi_{ij}}(t)]_{i,\ j=1}^{n}$ such that

$$\overset{(n)}{x}(t) = \overset{(n)}{\Phi}(t)\overset{(n)}{y}(t). \tag{15.3}$$

Let us now formulate conditions that guarantee the reducibility of the original equation (14.1) under the assumptions made above.

Definition 15.1. *A sequence of matrices $\overset{(n)}{C}(t) = [\overset{(n)}{c_{ij}}(t)]_{i,\ j=1}^{\infty}$ is called regular if, for every $t \in T$, it satisfies the following conditions:*

(i) it is jointly bounded, i.e.,

$$\|\overset{(n)}{C}(t)\| \leq P = \text{const} < \infty;$$

(ii) it weakly converges to a matrix $C(t) = [c_{ij}(t)]_{i,\ j=1}^{\infty}$;

(iii) the series $\sum_{j=1}^{\infty} |\overset{(n)}{c_{ij}}(t)| (i = 1, 2, \dots)$ converge uniformly in n.

If conditions (i)–(iii) are satisfied uniformly in $t \in T$, then the sequence of matrices $\overset{(n)}{C}$ is called uniformly regular in T.

If condition (iii) is satisfied uniformly in i, then the sequence $\overset{(n)}{C}$ is called strongly regular.

It is easy to see that the following statements are true:

1. If a sequence $\overset{(n)}{C}(t)$ is uniformly regular, then $\|C(t)\| \leq P$ for all $t \in T$.

2. Let $\overset{(n)}{C}$ be a sequence of constant infinite matrices. If there exists a matrix $S = [s_{ij}]_{i,j=1}^{\infty}$ bounded in norm and such that $|\overset{(n)}{c_{ij}}(t)| \leq s_{ij}$ $(i, j = 1, 2, \ldots)$, then the indicated sequence contains a regular subsequence.

We prove the second statement. Denote $\|S\| = Z$. Then $\|\overset{(n)}{C}(t)\| \leq Z$ for any natural n. Therefore, the sequence of matrices $\overset{(n)}{C}$ is jointly bounded. It is clear that any subsequence of this sequence possesses the same property.

The validity of condition (iii) in Definition 15.1 for any subsequence of $\{\overset{(n)}{C}\}$ is also obvious.

It remains to show that $\{\overset{(n)}{C}\}$ contains a weakly convergent subsequence.

We write the moduli of the elements of the matrix $\overset{(n)}{C}$ in the form of a column as follows:

$$\text{colon}\Big\{|\overset{(n)}{c_{11}}|,\ |\overset{(n)}{c_{21}}|,\ |\overset{(n)}{c_{12}}|,\ |\overset{(n)}{c_{31}}|,\ |\overset{(n)}{c_{22}}|,\ |\overset{(n)}{c_{13}}|, \ldots\Big\}.$$

Since every sequence $|\overset{(n)}{c_{ij}}|$ of nonnegative numbers is bounded above by a number s_{ij} $(i, j = 1, 2, \ldots)$, we can always select a convergent subsequence of this sequence.

Thus, we select a convergent subsequence $|\overset{(\alpha_i)}{c_{11}}|$ $(i = 1, 2, \ldots)$ of the sequence $|\overset{(n)}{c_{11}}|$, then a convergent subsequence $|\overset{(\beta_i)}{c_{21}}|$ of the sequence $|\overset{(\alpha_i)}{c_{21}}|$, and so on. By using the procedure of diagonalization, we form the set of sequences

$$\begin{array}{llll}
|\overset{(\alpha_1)}{c_{11}}|, & |\overset{(\beta_2)}{c_{11}}| & |\overset{(\gamma_3)}{c_{11}}|, & \ldots \\[2ex]
|\overset{(\alpha_1)}{c_{21}}|, & |\overset{(\beta_2)}{c_{21}}| & |\overset{(\gamma_3)}{c_{21}}|, & \ldots \\[2ex]
|\overset{(\alpha_1)}{c_{12}}|, & |\overset{(\beta_2)}{c_{12}}| & |\overset{(\gamma_3)}{c_{12}}|, & \ldots \\[2ex]
\multicolumn{4}{c}{\ldots\ldots\ldots\ldots\ldots\ldots,}
\end{array} \tag{15.4}$$

each of which converges as a subsequence of a convergent sequence supplemented with finitely many additional elements. We now transform the columns in table (15.4) into matrices and arrive at a weakly convergent sequence of matrices $\overset{(\alpha_1)}{C}, \overset{(\beta_2)}{C}, \overset{(\gamma_3)}{C}, \ldots$.

Theorem 15.1. *Assume that equation (14.1) is such that, in addition to the conditions formulated above, the following inequality is true:*

$$\sum_{j=n+1}^{\infty} \max_{t \in T_1} |a_{sj}(t)| \leq K_{T_1} \cdot \varepsilon(n) \quad (s = 1, 2, 3, \ldots),\ K_{T_1} = \text{const} < \infty,$$

where $\varepsilon(n) \to 0$ *as* $n \to \infty$, T_1 *is an arbitrary finite segment of* T, *the sequences* $\{\overset{(n)}{\Phi}(t)\}$ *and* $\{\overset{(n)}{\Phi}{}^{-1}(t)\}$ *are uniformly regular, and the sequence* $\{\overset{(n)}{P}\}$ *is regular in* T_1. *Then*

$$\lim_{n \to \infty} \overset{(n)}{\Phi}(t) = \Phi(t) \quad \text{and} \quad \lim_{n \to \infty} \overset{(n)}{P} = P$$

in the weak sense and equation (14.1) can be reduced to equation (14.3) with the help of the limiting matrix $\Phi(t)$.

Proof. First, we note that the sequence of matrices $\overset{(n)}{A}(t)$ in equations (15.1) is uniformly regular in T_1.

Let $x = (x_1, \ldots, x_n, x_{n+1}, x_{n+2}, \ldots)$ and $\bar{x} = (x_1, \ldots, x_n, \bar{x}_{n+1}, \bar{x}_{n+2}, \ldots)$ be two arbitrary points of the space $\mathfrak{M}$ whose coordinates coincide up to the number n inclusive. Denote $\Delta_n x = \sup\{|x_{n+1} - \bar{x}_{n+1}|, |x_{n+2} - \bar{x}_{n+2}|, \ldots\}$.

For any $s = 1, 2, 3, \ldots$, the following inequalities are true:

$$\Big|\sum_{i=1}^{\infty} a_{si}(t) x_i - \sum_{i=1}^{n} a_{si}(t) x_i - \sum_{i=n+1}^{\infty} a_{si}(t) \bar{x}_i\Big|$$

$$\leq \sum_{i=n+1}^{\infty} |a_{si}(t)|\, |x_i - \bar{x}_i| \leq \Delta_n x \sum_{i=n+1}^{\infty} \max_{T_1} |a_{si}(t)|$$

$$\leq K_{T_1} \cdot \Delta_n x \varepsilon(n),$$

where $\varepsilon(n) \to 0$ as $n \to \infty$.

This means that equation (14.1) satisfies the strong Cauchy–Lipschitz conditions with respect to x and, hence, we can use the following reasoning (see Section 1.2):

Let $x = x(t, t_0, x_0)$ and $\overset{(n)}{x} = \overset{(n)}{x}(t, t_0, \overset{(n)}{x}_0)$ be solutions of equations (14.1) and (15.1), respectively, and, in addition,

$$x(t_0, t_0, x_0) = x_0 \quad \text{and} \quad \overset{(n)}{x}(t_0, t_0, \overset{(n)}{x}_0) = \overset{(n)}{x}_0,$$

where

$$x_0 = (x_1^0, x_2^0, \dots) \in \mathfrak{M} \quad \text{and} \quad \overset{(n)}{x_0} = (x_1^0, x_2^0, \dots, x_n^0).$$

Then

$$\lim_{n\to\infty} \overset{(n)}{x}_s(t,\, t_0,\, \overset{(n)}{x_0}) = x_s(t,\, t_0,\, x_0), \quad s = 1, 2, 3, \dots,$$

uniformly in $t \in T_1$. In other words, the sequence $\overset{(n)}{x}(t,\, t_0,\, \overset{(n)}{x_0})$ weakly converges to the vector function $x(t,\, t_0,\, x_0)$ uniformly in t. According to the condition of the theorem, for each n beginning with a certain number, there exists a matrix $\overset{(n)}{\Phi}(t)$ such that $\overset{(n)}{x}(t,\, t_0,\, \overset{(n)}{x_0}) = \overset{(n)}{\Phi}(t)\, \overset{(n)}{y}(t)$ identically for all $t \in T$, where $\overset{(n)}{y}(t)$ is a solution of equation (15.2).

Let us show that the sequence $\{\overset{(n)}{y}(t)\}$ weakly converges as $n \to \infty$ (uniformly in $t \in T_1$).

For this purpose, we denote $\overset{(n)}{\Phi}{}^{-1}(t) = \overset{(n)}{\Psi}(t) = [\psi_{ij}(t)]_{i,j=1}^n$. Then the sth coordinate of the vector $\overset{(n)}{y}(t)$ has the form

$$\overset{(n)}{y_s}(t) = \sum_{i=1}^{n} \overset{(n)}{\psi_{si}}(t)\, \overset{(n)}{x_i}(t,\, t_0,\, \overset{(n)}{x_0}).$$

If we now replace the products $\overset{(n)}{\psi_{si}}(t)\, \overset{(n)}{x_i}(t,\, t_0,\, \overset{(n)}{x_0})$ with $i > n$ by zeros, then we get

$$\overset{(n)}{y_s}(t) = \sum_{i=1}^{\infty} \overset{(n)}{\psi_{si}}(t)\, \overset{(n)}{x_i}(t,\, t_0,\, \overset{(n)}{x_0}). \tag{15.5}$$

By using the Gronwall–Bellman inequality, we obtain

$$\begin{aligned} \|\overset{(n)}{x}(t,\, t_0,\, \overset{(n)}{x_0})\| &\le \|\overset{(n)}{x_0}\| \exp\Big\{\Big|\int_{t_0}^{t} \|\overset{(n)}{A}(\xi)\| d\xi\Big|\Big\} \\ &\le \|x_0\| \exp\{\Psi_{T_1}|t - t_0|\} \\ &\le M_{T_1} = \text{const} < \infty, \quad t,\, t_0, \in T_1. \end{aligned}$$

Then

$$\sum_{i=1}^{\infty} \overset{(n)}{\psi_{si}}(t)\, \overset{(n)}{x_i}(t,\, t_0,\, \overset{(n)}{x_0}) \le M_{T_1} \sum_{i=1}^{\infty} |\, \overset{(n)}{\psi_{si}}(t)|,$$

where the series on the right-hand side converges uniformly in $t \in T_1$ and n. According to the condition of the theorem, the sequence $\overset{(n)}{\Psi}(t)$ is uniformly regular and, hence, there exists a matrix $\Psi(t) = [\overset{(n)}{x_{ij}}(t)]_{i,j=1}^{\infty}$ such that $\lim_{n\to\infty} \overset{(n)}{\psi_{ij}}(t) = \psi_{ij}(t),\ i,j = 1,2,3,\ldots,$ uniformly in $t \in T_1$.

By virtue of Lemma 12.1, sequence (15.5) converges, as $n \to \infty$, to a function $y_s(t),\ s = 1,2,\ldots,$ uniformly in $t \in T_1$. Thus, it is shown that the sequence of vector functions $\{\overset{(n)}{y}(t)\}$ converges weakly to a vector function $y(t)$ uniformly in t and, in addition, $y(t) = \Psi(t)x(t,\ t_0,\ x_0)$.

Similarly, one can easily show that $x(t,\ t_0,\ x_0) = \Phi(t)y(t)$, where the matrix $\Phi(t)$ is the weak limit of the sequence $\Phi^{(n)}(t)$. It is also easy to see that, under the conditions of the theorem, the matrix $\Phi(t)$ is invertible. Furthermore, the inverse matrix $\Phi^{-1}(t)$ coincides with the matrix $\Psi(t)$ for all $t \in T$.

It remains to show that $y(t)$ is a solution of equation (14.3), where $P = \lim_{n\to\infty} \overset{(n)}{P}$ in the weak sense.

By using relation (15.2), for the sth coordinate we get

$$\frac{d\overset{(n)}{y_s}(t)}{dt} = \sum_{i=1}^{\infty} \overset{(n)}{p_{si}}\,\overset{(n)}{y_i}(t) \tag{15.6}$$

identically for all $t \in T$. We note that, for $i > n$, the products $\overset{(n)}{p_{si}}\,\overset{(n)}{y_i}(t)$ in equality (15.6) should be set equal to zero.

It is obvious that

$$\begin{aligned} \|\overset{(n)}{y}(t)\| &\leq \|\overset{(n)}{\Psi}(t)\|\,\|\overset{(n)}{x}(t,\ t_0,\ \overset{(n)}{x_0})\| \\ &\leq M_{T_1}\|\overset{(n)}{\Psi}(t)\| \leq K_1 = \text{const} < \infty. \end{aligned}$$

Then

$$\sum_{i=1}^{\infty} \overset{(n)}{p_{si}}\,\overset{(n)}{y_i}(t) \leq \sum_{i=1}^{\infty} |\overset{(n)}{p_{si}}|\,\|\overset{(n)}{y}(t)\| \leq K_1 \sum_{i=1}^{\infty} |\overset{(n)}{p_{si}}|.$$

This means that the series on the right-hand side of equality (15.6) converges uniformly in n and $t \in T_1$.

In addition,

$$\begin{aligned} |p_{si}y_i(t) - \overset{(n)}{p_{si}}\,\overset{(n)}{y_i}(t)| &\leq |p_{si}|\,|y_i(t) - \overset{(n)}{y_i}(t)| + |\overset{(n)}{y_i}(t)|\,|p_{si} - \overset{(n)}{p_{si}}| \\ &\leq \|P\|\,|y_i(t) - \overset{(n)}{y_i}(t)| + K_1|p_{si} - \overset{(n)}{p_{si}}|. \end{aligned}$$

One can always find a number N such that $|p_{si} - \overset{(n)}{p}_{si}| < \dfrac{\varepsilon}{K_1}$ and $|y_i(t) - \overset{(n)}{y}_i(t)| < \dfrac{\varepsilon}{\|P\|}$ for all $n \geq N$ and $t \in T_1$. Here, ε is an arbitrarily small positive number. This means that the sequence $\{\overset{(n)}{p}_{si}\overset{(n)}{y}_i(t)\}$ converges uniformly for all $i = 1, 2, \ldots$. In this case, we can apply Lemma 12.1 to the series on the right-hand side of relation (15.6). Therefore,

$$\lim_{n\to\infty} \frac{d\overset{(n)}{y}_s(t)}{dt} = \lim_{n\to\infty} \sum_{i=1}^{\infty} \overset{(n)}{p}_{si}\overset{(n)}{y}_i(t) = \sum_{i=1}^{\infty} p_{si} y_i(t)$$

and, moreover, the sequences in this equality converge uniformly in $t \in T_1$. Hence,

$$\frac{dy_s(t)}{dt} = \sum_{i=1}^{\infty} p_{si} y_i(t), \quad s = 1, 2, 3, \ldots,$$

which completes the proof of Theorem 15.1.

We now apply Theorem 15.1 to the case where $A(t)$ is a matrix periodic in t with period 2π.

Theorem 15.2. *Suppose that the elements of the matrix $A(t)$ are continuous, periodic in t with period 2π, and such that*

$$\sum_{j=n+1}^{\infty} \max_T |a_{sj}(t)| \leq K\varepsilon(n), \quad s = 1, 2, \ldots,$$

where $\varepsilon(n) \to 0$ as $n \to \infty$ and the sequence of matrices $\overset{(n)}{\Phi}(t)$ reducing equations (15.1) to the form (15.2) satisfies the following conditions: $\{\overset{(n)}{\Phi}(t)\}$ and $\{\overset{(n)}{\Phi}{}^{-1}(t)\}$ are uniformly regular and there exists $t^* \in T$ such that the sequences of matrices $\{\overset{(n)}{\Phi}(t^*)\}$ and $\{\frac{d}{dt}\overset{(n)}{\Phi}(t^*)\}$ are strongly regular. Then $\Phi(t) = \lim_{n\to\infty}\overset{(n)}{\Phi}(t)$ and $P = \lim_{n\to\infty}\overset{(n)}{P}$ in the weak sense and equation (14.1) can be reduced to equation (14.3) with the help of the matrix $\Phi(t)$ periodic in t with period 4π.

Proof. Clearly, it suffices to show that the sequence of matrices $\{\overset{(n)}{P}\}$ is regular and use Theorem 15.1 because the existence of reducing matrices $\overset{(n)}{\Phi}$ follows from the Floquet theorem for the finite-dimensional case.

First, we prove that if the sequences of constant matrices $\{\overset{(n)}{C}\}$ and $\{\overset{(n)}{D}\}$ are regular and strongly regular, respectively, then the sequence $\{\overset{(n)}{C}\overset{(n)}{D}\}$ is regular.

The sequence $\{\overset{(n)}{C}\overset{(n)}{D}\}$ is jointly bounded because

$$\| \overset{(n)}{C}\overset{(n)}{D} \| \leq \| \overset{(n)}{C} \| \| \overset{(n)}{D} \| \leq BB_1 = \text{const} < \infty.$$

In addition, $\lim\limits_{n\to\infty} \overset{(n)}{C}\overset{(n)}{D} = CD$ in the weak sense. Indeed, the element of the matrix $\overset{(n)}{C}\overset{(n)}{D}$ located in its ith row and pth column has the form $\sum_{j=1}^{\infty} \overset{(n)}{c_{ij}}\,\overset{(n)}{d_{jp}}$. This series converges uniformly in n because

$$\sum_{j=1}^{\infty} \overset{(n)}{c_{ij}}\,\overset{(n)}{d_{jp}} \leq \sum_{j=1}^{\infty} | \overset{(n)}{c_{ij}} | \| \overset{(n)}{D} \| = \| \overset{(n)}{D} \| \, | \sum_{j=1}^{\infty} \overset{(n)}{c_{ij}} |,$$

and the series on the right-hand side of this formula converges uniformly in n because the sequence $\{\overset{(n)}{C}\}$ is regular. Clearly, $\lim\limits_{n\to\infty} \overset{(n)}{c_{ij}}\,\overset{(n)}{d_{jp}} = c_{ij}d_{jp}$ and, hence,

$$\lim_{n\to\infty} \sum_{j=1}^{\infty} \overset{(n)}{c_{ij}}\,\overset{(n)}{d_{jp}} = \sum_{j=1}^{\infty} c_{ij}d_{jp}.$$

It remains to prove the validity of condition (iii) in Definition 15.1. Since the series $\sum\limits_{j=1}^{\infty} | \overset{(n)}{d_{ij}} |$ converges uniformly in n and i, for any (arbitrarily small) $\varepsilon > 0$, one can find a number l independent of i and n such that $| \overset{(n)}{d_{ik}} | + \ldots + | \overset{(n)}{d_{ik+m}} | \leq \dfrac{\varepsilon}{B}$ for all $k \geq l$ and $m = 1, 2, \ldots$. Then the following inequalities are true for all natural n and s:

$$\begin{aligned}
\Big| \sum_{p=k}^{k+m} \Big| \Big| \sum_{i=1}^{\infty} \overset{(n)}{c_{si}}\,\overset{(n)}{d_{ip}} \Big| &\leq \sum_{p=k}^{k+m} \sum_{j=1}^{\infty} | \overset{(n)}{c_{si}} | \, | \overset{(n)}{d_{ip}} | \\
&= \sum_{i=1}^{\infty} | \overset{(n)}{c_{si}} | \sum_{p=k}^{k+m} | \overset{(n)}{d_{ip}} | \\
&\leq \sum_{i=1}^{\infty} | \overset{(n)}{c_{si}} | \frac{\varepsilon}{B} \leq \frac{\varepsilon}{B} \| \overset{(n)}{C} \| \leq \varepsilon.
\end{aligned}$$

Hence, the series $\sum_{p=1}^{\infty} |\sum_{i=1}^{\infty} \overset{(n)}{c}_{si} \overset{(n)}{d}_{ip}|$ converges uniformly in n for $s = 1, 2, \ldots$, i.e., the sequence $\{\overset{(n)}{C}\overset{(n)}{D}\}$ is regular. Note that, for any n, we can write

$$\frac{d\overset{(n)}{y}}{dt} = \frac{d}{dt}\overset{(n)}{\Phi}(t)\cdot\overset{(n)}{x}(t) + \overset{(n)}{\Phi}(t)\frac{d\overset{(n)}{x}(t)}{dt} = \overset{(n)}{A}(t)\,\overset{(n)}{\Phi}(t)\,\overset{(n)}{x}(t), \quad t \in T.$$

This yields

$$\frac{d\overset{(n)}{x}(t)}{dt} = \overset{(n)}{\Phi}{}^{-1}(t)\Big[\overset{(n)}{A}(t)\,\overset{(n)}{\Phi}(t) - \frac{d}{dt}\overset{(n)}{\Phi}(t)\Big]\overset{(n)}{x}(t).$$

Therefore, the constant matrix $\overset{(n)}{P} = \overset{(n)}{\Phi}{}^{-1}(t)\,\overset{(n)}{A}(t)\,\overset{(n)}{\Phi}(t) - \overset{(n)}{\Phi}{}^{-1}(t)\frac{d}{dt}\overset{(n)}{\Phi}(t)$ identically for all $t \in T$ including $t = t^*$.

Since $\overset{(n)}{A}(t)$ is a strongly regular sequence for any $t \in T$ and the sum of regular matrix sequences is a regular sequence, by virtue of the statement proved above we conclude that the sequence of matrices

$$\overset{(n)}{P} = \overset{(n)}{\Phi}{}^{-1}(t^*)\,\overset{(n)}{A}(t^*)\,\overset{(n)}{\Phi}(t^*) - \overset{(n)}{\Phi}{}^{-1}(t^*)\frac{d}{dt}\overset{(n)}{\Phi}(t^*)$$

is regular.

Theorem 15.2 is proved.

Remark 15.1. The period is doubled only in the case where the period of the reducing matrices $\overset{(n)}{\Phi}(t)$ is equal to 4π $(n = 1, 2, \ldots)$.

3.16. Systems with Almost Periodic Coefficients

We consider a system of equations

$$\frac{d\varphi}{dt} = a(\varphi), \quad \frac{dx}{dt} = P(\varphi)x, \tag{16.1}$$

where $x \in R^n$, $P(\varphi) = [p_{ij}(\varphi)]_{i,j=1}^{n}$ is an $n \times n$ matrix, $\varphi = (\varphi_1, \varphi_2, \ldots)$ is a countably dimensional vector, and $a(\varphi) = \{a_1(\varphi), a_2(\varphi), \ldots\}$ is a countably dimensional function.

As a special case of this system of equations, we consider a linear system of equations with almost periodic coefficients obtained by restricting system (16.1) to the case $a(\varphi) = \omega$, where $\omega = (\omega_1, \omega_2, \ldots)$ is the frequency basis of an almost periodic function. The problem of reducibility of systems with almost periodic coefficients was studied by numerous authors, and the most interesting results in this field were obtained for systems with almost constant coefficients. A system of this sort has the form

$$\frac{d\varphi}{dt} = \omega, \quad \frac{dx}{dt} = Ax + \varepsilon P(\varphi)x, \tag{16.2}$$

where A is a constant matrix and ε is a small positive parameter.

We now study the problem of reducibility of system (16.1) to a system with constant coefficients in normal variables of the form

$$\frac{d\varphi}{dt} = a(\varphi), \quad \frac{dy}{dt} = By. \tag{16.3}$$

We relate the solution of this problem to the problem of reducibility of a sequence of truncated systems (φ is a finite number). These systems are systems with quasiperiodic coefficients and the problem of their reducibility is fairly well studied.

Let us now formulate the following useful general statement:

Consider an equation

$$\frac{dx}{dt} = A(t)x, \tag{16.4}$$

where $x = (x_1, x_2, \ldots, x_n) \in R^n$ and $t \in [a,\ b] = T_1^0 \subseteq R^1$. Assume that there exists a sequence of equations of the form

$$\frac{d\overset{(m)}{x}}{dt} = \overset{(m)}{A}(t)\overset{(m)}{x} \quad (m = 1, 2, \ldots) \tag{16.5}$$

reducible (with the help of reducing matrices $\overset{(m)}{\Phi}(t)$) in the interval T_1^0 to equations of the form

$$\frac{d\overset{(m)}{y}}{dt} = \overset{(m)}{B}\overset{(m)}{y} \quad (m = 1, 2, \ldots), \tag{16.6}$$

where $\overset{(m)}{B}$ are constant matrices, and, in addition, that, for any solution $x = x(t)$ of equation (16.4), there exists a sequence $\overset{(m)}{x} = \overset{(m)}{x}(t)$ of solutions of equations (16.5) that converges to $x(t)$ uniformly in $t \in T_1^0$.

If the sequences of matrices $\overset{(m)}{\Phi}$, $\overset{(m)}{\Phi}{}^{-1}(t)$, and $\overset{(m)}{B}$ converge as $m \to \infty$ and, in addition, the matrix $\{\overset{(m)}{\Phi}{}^{-1}(t)\}$ converges uniformly in $t \in T_1^0$, then, by the change of variables $x = \Phi(t)y$, equation (16.4) is reduced to the following equation:

$$\frac{dy}{dt} = By, \tag{16.7}$$

where $\Phi(t) = \lim_{m\to\infty} \overset{(m)}{\Phi}(t)$ and $B = \lim_{m\to\infty} \overset{(m)}{B}$.

Indeed, for every natural number m, the change $\overset{(m)}{x} = \overset{(m)}{\Phi}(t)\overset{(m)}{y}$ reduces equation (16.5) to equation (16.6). Since $\overset{(m)}{y} = \overset{(m)}{\Phi}{}^{-1}(t)\overset{(m)}{x}$, we have

$$\lim_{m\to\infty} \overset{(m)}{y} = \lim_{m\to\infty} \overset{(m)}{\Phi}{}^{-1}(t)\overset{(m)}{x} = \Phi^{-1}(t)x(t) = y(t),$$

where $\Phi^{-1}(t)$ stands for $\lim_{m\to\infty} \overset{(m)}{\Phi}{}^{-1}(t)$. In equality (16.6), we pass to the limit as $m \to \infty$ and take into account the fact that, in this case, $\overset{(m)}{B}\overset{(m)}{y} \to By(t)$ uniformly in t. This enables us to conclude that $y(t)$ is a solution of equation (16.7).

Instead of the convergence of the sequence $\{\overset{(m)}{B}\}$, it suffices to assume that the sequences $\{\overset{(m)}{A}(t)\}$ and $\{\frac{d}{dt}\overset{(m)}{\Phi}(t)\}$ converge at least at one point $t^* \in T_1^0$. This follows from the representation

$$\overset{(m)}{B} = \overset{(m)}{\Phi}{}^{-1}(t)\overset{(m)}{A}(t)\overset{(m)}{\Phi}(t) - \overset{(m)}{\Phi}{}^{-1}(t)\frac{d}{dt}\overset{(m)}{\Phi}(t).$$

Let $\varphi = \varphi_t(\varphi)$ be a solution of the equation for angular variables in system (16.1). Then we can write

$$\frac{dx}{dt} = P(\varphi_t(\varphi))x. \tag{16.8}$$

We denote the solution of this equation taking a value $x_0 \in R^n$ for $t = 0$ by $x = x_t(x_0)$.

Assume that the matrix $P(\varphi)$ and the function $a(\varphi)$ are Lipschitzian with respect to φ with coefficients ε and α, respectively, and, in addition, $\|P(0)\| = p$ and $\|a(0)\| = h$. These conditions are sufficient for the existence of a solution $\varphi = \varphi_t(\varphi)$ for all $t \in R^1$ and $\varphi \in \mathcal{F}_\infty$. In this case, the solution $x = x_t(x_0)$ exists for all $t \in R^1$ and satisfies the following inequality in any finite interval $T_1 = [a,\ b] \subset R^1$:

$$\|x_t(x_0)\| \leq \|x_0\| \exp\{(p + \varepsilon l_1)|t|\},$$

where $l_1 = (2\pi + h(b-a))\exp\{\alpha(b-a)\}$ and $0,\ t \in T_1$.

To solve the problem of reducibility of the system of equations (16.1) to system (16.3), we write, parallel with (16.1), the truncated system

$$\frac{d\varphi_i}{dt} = a_i(\varphi_1, \ldots, \varphi_m, 0, 0, \ldots), \quad \frac{dx}{dt} = P(\varphi_1, \ldots, \varphi_m, 0, 0, \ldots)x, \tag{16.9}$$

where m is an arbitrary natural number and $i = \overline{1, m}$. Let

$$\overset{(m)}{\varphi_t}(\overset{(m)}{\varphi}) = \{\overset{(m)}{\varphi_{1_t}}(\overset{(m)}{\varphi}), \ldots, \overset{(m)}{\varphi_{m_t}}(\overset{(m)}{\varphi})\}$$

be a solution of the equation for angular variables in system (16.9), where $\overset{(m)}{\varphi} = (\varphi_1, \varphi_2, \ldots, \varphi_m, 0, 0, \ldots)$. We set

$$\bar{\bar{\varphi}} = (0, \ldots, 0, \varphi_{m+1}, \varphi_{m+2}, \ldots),$$

$$\bar{\varphi}_t(\varphi) = \{\varphi_1(t, \varphi), \ldots, \varphi_m(t, \varphi), 0, 0, \ldots\},$$

$$\bar{\bar{\varphi}}_t(\varphi) = \{0, \ldots, 0, \varphi_{m+1}(t, \varphi), \varphi_{m+2}(t, \varphi), \ldots\},$$

and

$$\overset{(m)}{\varphi_0}(\overset{(m)}{\varphi}) = \overset{(m)}{\varphi}.$$

This enables us to write the equation

$$\frac{dx}{dt} = P(\overset{(m)}{\varphi_t}(\overset{(m)}{\varphi}))x.$$

Its solution with initial values $t = 0$ and $x = x^0 \in R^n$ is denoted by $\overset{(m)}{x} = \overset{(m)}{x_t}(\overset{(m)}{\varphi}, x^0)$.

We now prove several auxiliary statements.

Lemma 16.1 *Suppose that a function $a(\varphi) \in \widehat{C}^0_{\text{Lip}}(\mathcal{F}_\infty)$ with coefficient $\alpha(m)$. Then, for all $t \in T_1 \subset R^1$,*

$$V(t) = \|\bar{\varphi}_t(\varphi) - \bar{\varphi}_t(\overset{(m)}{\varphi})\| \leq \alpha(m) K_1 \exp\{2\alpha(0)|t|\},$$

$$V_1(t) = \|\bar{\varphi}_t(\varphi) - \overset{(m)}{\varphi_t}(\overset{(m)}{\varphi})\| \leq \alpha(m) K_2 \exp\{\alpha(0)|t|\},$$

where K_1 and K_2 are constants independent of the order of truncation m.

Proof. For $t > 0$, we have

$$V(t) \leq \int_0^t \{\|a(\varphi_s(\varphi)) - a(\bar{\varphi}_s(\varphi) + \bar{\bar{\varphi}}_s(\overset{(m)}{\varphi}))\|$$

$$+\|a(\bar{\varphi}_s(\varphi) + \bar{\bar{\varphi}}_s(\overset{(m)}{\varphi})) - a(\varphi_s(\overset{(m)}{\varphi}))\|\}ds$$

$$\leq \int_0^t \alpha(m)\|\varphi_s(\varphi) - \varphi_s(\overset{(m)}{\varphi})\|ds + \int_0^t \alpha(0)\|\bar{\varphi}_s(\varphi) - \bar{\varphi}_s(\overset{(m)}{\varphi})\|ds.$$

Since

$$\|\varphi_s(\varphi) - \varphi_s(\overset{(m)}{\varphi})\| \leq \exp\{\alpha(0)|s|\}\|\bar{\bar{\varphi}}\|,$$

we obtain

$$V(t) \leq \int_0^t \alpha(m)\exp\{\alpha(0)s\}\|\bar{\bar{\varphi}}\|ds + \int_0^t \alpha(0)V(s)ds$$

for $t \geq 0$. By using the Gronwall–Bellman lemma and the inequality

$$\int_0^t \alpha(m)\exp\{\alpha(0)s\}\|\bar{\bar{\varphi}}\|ds \leq 2\pi\frac{\alpha(m)}{\alpha(0)}\exp\{\alpha(0)t\}$$

we arrive at the estimate

$$V(t) \leq 2\pi\frac{\alpha(m)}{\alpha(0)}\left(\exp\{\alpha(0)t\}\right)^2.$$

The case $t < 0$ is considered in exactly the same way. This proves the first inequality of Lemma 16.1 with $K_1 = \dfrac{2\pi}{\alpha(0)}$.

We also have

$$V_1(t) \leq \int_0^{|t|} \|a(\varphi_s(\varphi)) - a(\overset{(m)}{\varphi_s}(\overset{(m)}{\varphi}))\|ds$$

$$\leq \int_0^{|t|} \|a(\varphi_s(\varphi)) - a(\bar{\varphi}_s(\varphi))\|ds + \int_0^{|t|} \|a(\varphi_s(\varphi)) - a(\overset{(m)}{\varphi_s}(\overset{(m)}{\varphi}))\|ds$$

$$\leq \int_0^{|t|} \alpha(m)\bar{\bar{\varphi}}_s(\varphi)ds + \int_0^{|t|} \alpha(0)\|\bar{\varphi}_s(\varphi) - \overset{(m)}{\varphi_s}(\overset{(m)}{\varphi})\|ds$$

$$\leq \alpha(m)l_1(b-a) + \int_0^{|t|} \alpha(0)V_1(s)ds, \quad t \in T_1,$$

which yields the second inequality of Lemma 16.1 with $K_2 = l_1(b-a)$.

Lemma 16.2. *Assume that the above-mentioned conditions are satisfied and* $P(\varphi) \in \widehat{C}^0_{\mathrm{Lip}}(\mathcal{F}_\infty)$ *with coefficient* $\varepsilon(m)$*. Then, for all* $t \in T_1$,

$$\|P(\varphi_t(\varphi)) - P(\overset{(m)}{\varphi}(\overset{(m)}{\varphi}))\| \leq \varepsilon(m)M_1 + \alpha(m)M_2,$$

where M_1 *and* M_2 *are constants independent of* m.

Proof. The following chain of inequalities is true:

$$\|P(\varphi_t(\varphi)) - P(\overset{(m)}{\varphi_t}(\overset{(m)}{\varphi}))\|$$

$$\leq \|P(\varphi_t(\varphi)) - P(\varphi_t(\overset{(m)}{\varphi}))\| + \|P(\varphi_t(\overset{(m)}{\varphi})) - P(\bar{\varphi}_t(\overset{(m)}{\varphi}))\|$$

$$+ \|P(\bar{\varphi}_t(\overset{(m)}{\varphi})) - P(\overset{(m)}{\varphi_t}(\overset{(m)}{\varphi}))\|$$

$$\leq \|P(\varphi_t(\varphi)) - P(\varphi_t(\overset{(m)}{\varphi}))\| + \varepsilon(m)\|\bar{\bar{\varphi}}(\overset{(m)}{\varphi})\|$$

$$+ \|P(\bar{\varphi}_t(\overset{(m)}{\varphi})) - P(\bar{\varphi}_t(\varphi))\| + \|P(\bar{\varphi}_t(\varphi)) - P(\overset{(m)}{\varphi_t}(\overset{(m)}{\varphi}))\|$$

$$\leq \|P(\varphi_t(\varphi)) - P(\varphi_t(\overset{(m)}{\varphi}))\| + \varepsilon(m)\|\varphi_t(\overset{(m)}{\varphi})\| + \varepsilon(0)V(t) + \varepsilon(0)V_1(t).$$

We estimate the quantity $L(t) = \|P(\varphi_t(\varphi)) - P(\varphi_t(\overset{(m)}{\varphi}))\|$. Thus, we can write

$$\begin{aligned} L(t) \;&\leq\; \|P(\varphi_t(\varphi)) - P(\bar{\varphi}_t(\varphi) + \bar{\bar{\varphi}}_t(\overset{(m)}{\varphi}))\| \\ &\qquad + \|P(\bar{\varphi}_t(\varphi) + \bar{\bar{\varphi}}_t(\overset{(m)}{\varphi})) - P(\varphi_t(\overset{(m)}{\varphi}))\| \\ &\leq\; \varepsilon(m)\|\bar{\bar{\varphi}}_t(\varphi) - \bar{\bar{\varphi}}_t(\overset{(m)}{\varphi}))\| + \varepsilon(0)\|\bar{\varphi}_t(\varphi) - \bar{\varphi}_t(\overset{(m)}{\varphi})\| \\ &\leq\; \varepsilon(m)\|\varphi_t(\varphi) - \varphi_t(\overset{(m)}{\varphi})\| + \varepsilon(0)\alpha(m)K_1\exp\{2\alpha(0)|t|\} \\ &\leq\; \varepsilon(m)2\pi\exp\{\alpha(0)|t|\} + \varepsilon(0)\alpha(m)K_1\exp\{2\alpha(0)|t|\} \end{aligned}$$

and, hence,

$$\begin{aligned}&\|P(\varphi_t(\varphi)) - P(\overset{(m)}{\varphi_t}(\overset{(m)}{\varphi}))\|\\ &\quad\leq \varepsilon(m)2\pi \exp\{\alpha(0)|t|\} + \varepsilon(0)\alpha(m)K_1 \exp\{2\alpha(0)|t|\}\\ &\qquad + \varepsilon(m)l_1 + \varepsilon(0)(K_1 \exp\{\alpha(0)|t|\} + K_2)\alpha(m) \exp\{\alpha(0)|t|\},\end{aligned}$$

which completes the proof of Lemma 16.2.

Let us now formulate the principal statement of the present section.

Theorem 16.1. *Assume that the system of equations (16.1) is such that* $a(\varphi)$, $P(\varphi) \in \widehat{C}^0_{\text{Lip}}(\mathcal{F}_\infty)$ *and the following conditions are satisfied:*

(i) for any natural m*, there exists a matrix* $\overset{(m)}{\Phi}(\overset{(m)}{\varphi})$ *that reduces the truncated system (16.9) to the form*

$$\frac{d\overset{(m)}{\varphi}}{dt} = \bar{a}(\overset{(m)}{\varphi}), \qquad \frac{d\overset{(m)}{y}}{dt} = \overset{(m)}{B}(\overset{(m)}{y});$$

(ii) $\overset{(m)}{\Phi}(\overset{(m)}{\varphi})$, $\overset{(m)}{\Phi}{}^{-1}(\overset{(m)}{\varphi}) \in \widehat{C}^0_{\text{Lip}}(\mathcal{F}_\infty)$ *are matrices with constant coefficients independent of* m*;*

(iii) the sequences $\{\overset{(m)}{\Phi}(\overset{(m)}{\varphi})\}$, $\{\overset{(m)}{B}\}$, *and* $\{\overset{(m)}{\Phi}{}^{-1}(\overset{(m)}{\varphi})\}$ *converge in norm as* $m \to \infty$*; moreover, the last sequence converges uniformly in* $\overset{(m)}{\varphi}$.

Then the system of equations (16.1) can be reduced to system (16.3) on the entire real axis.

Proof. It follows from the equations for $x_t(x^0)$ and $\overset{(m)}{x_t}(\overset{(m)}{\varphi}, x^0)$ that

$$x_t(x^0) = x^0 + \int_0^t P(\varphi_s(\varphi))x_s(x^0)ds,$$

$$\overset{(m)}{x_t}(\overset{(m)}{\varphi}, x^0) = x^0 + \int_0^t P(\overset{(m)}{\varphi_s}(\overset{(m)}{\varphi}))\, \overset{(m)}{x_s}(\overset{(m)}{\varphi}, x^0)ds.$$

This implies that, for any $t \in R^1$,

$$\|x_t - \overset{(m)}{x_t}\| \le \int_0^{|t|} \|P(\varphi_s(\varphi))\| \|x_s - \overset{(m)}{x_s}\| ds + \int_0^{|t|} \|P(\varphi_s(\varphi)) - P(\overset{(m)}{\varphi_s}(\overset{(m)}{\varphi}))\| \|\overset{(m)}{x_s}\| ds.$$

At the same time, $\|P(\varphi_t(\varphi))\| \le p + \varepsilon(0)l_1$ and $\|\overset{(m)}{x_t}\| \le d_1$ for all $t \in T_1$, $m = 1, 2, \ldots$. Thus, by using Lemma 16.2, we obtain

$$\|x_t - \overset{(m)}{x_t}\| \le \int_0^{|t|} (p + \varepsilon(0)l_1)\|x_s - \overset{(m)}{x_s}\| ds + d_1(\varepsilon(m)M_1 + \alpha(m)M_2)(b-a),$$

and, hence,

$$\|x_t - \overset{(m)}{x_t}\| \le (\varepsilon(m)d_1M_1 + \alpha(m)d_1M_2)(b-a)\exp\{(p + \varepsilon(0)l_1)(b-a)\} = \xi(m).$$

Since $\xi(m) \to 0$ as $m \to \infty$, we have

$$\|x_t(x^0) - \overset{(m)}{x_t}(\overset{(m)}{\varphi}, x^0)\| \to 0 \quad \text{as} \quad m \to \infty \tag{16.10}$$

uniformly in t for any finite interval T_1.

We now set $\lim_{m\to\infty} \overset{(m)}{\Phi}(\overset{(m)}{\varphi}) = \Phi(\varphi)$ and $\lim_{m\to\infty} \overset{(m)}{\Phi}{}^{-1}(\overset{(m)}{\varphi}) = \Phi^{-1}(\varphi)$ and denote a constant for which $\overset{(m)}{\Phi}(\overset{(m)}{\varphi}) \in \widehat{C}^0_{\text{Lip}}(\mathcal{F}_\infty)$ by $L(k)$.

In view of (16.10), to prove the theorem, it suffices to show that

$$\overset{(m)}{\Phi}(\overset{(m)}{\varphi_t}(\overset{(m)}{\varphi})) \to \Phi(\varphi_t(\varphi)), \quad \overset{(m)}{\Phi}{}^{-1}(\overset{(m)}{\varphi_t}(\overset{(m)}{\varphi})) \to \Phi^{-1}(\varphi_t(\varphi)) \tag{16.11}$$

as $m \to \infty$ uniformly in $t \in T_1$.

Thus, we have

$$
\begin{aligned}
&\|\Phi(\varphi_t(\varphi)) - \overset{(m)}{\Phi}(\overset{(m)}{\varphi_t}(\overset{(m)}{\varphi}))\| \\
&\qquad \le \|\Phi(\varphi_t(\varphi)) - \overset{(m)}{\Phi}(\bar{\varphi}_t(\varphi))\| + \|\overset{(m)}{\Phi}(\bar{\varphi}_t(\varphi)) - \overset{(m)}{\Phi}(\overset{(m)}{\varphi_t}(\overset{(m)}{\varphi}))\| \\
&\qquad \le \|\Phi(\varphi_t(\varphi)) - \Phi(\bar{\varphi}_t(\varphi))\| + \|\Phi(\bar{\varphi}_t(\varphi)) - \overset{(m)}{\Phi}(\bar{\varphi}_t(\varphi))\| \\
&\qquad\qquad + \|\overset{(m)}{\Phi}(\bar{\varphi}_t(\varphi)) - \overset{(m)}{\Phi}(\overset{(m)}{\varphi_t}(\overset{(m)}{\varphi}))\|. \qquad (16.12)
\end{aligned}
$$

According to Lemma 16.1, for $t \in T_1$, we can write

$$
\begin{aligned}
\|\overset{(m)}{\Phi}(\bar{\varphi}_t(\varphi)) - \overset{(m)}{\Phi}(\overset{(m)}{\varphi_t}(\overset{(m)}{\varphi}))\| &\le L(0)\|\bar{\varphi}_t(\varphi) - \overset{(m)}{\varphi_t}(\overset{(m)}{\varphi})\| \\
&\le L(0)\alpha(m)K_2 \exp\{\alpha(0)|t|\} \\
&\le L(0)\alpha(m)K_3 \to 0 \quad \text{as} \quad m \to \infty.
\end{aligned}
$$

Further, by the condition of the theorem,

$$
\|\Phi(\bar{\varphi}_t(\varphi)) - \overset{(m)}{\Phi}(\bar{\varphi}_t(\varphi))\| \to 0 \quad \text{as} \quad m \to \infty.
$$

Let us now show that if $\{\overset{(m)}{\Phi}(\overset{(m)}{\varphi})\} \in \widehat{C}^0_{\text{Lip}}(\mathcal{F}_\infty)$, then $\Phi(\varphi) \in \widehat{C}^0_{\text{Lip}}(\mathcal{F}_\infty)$ with the same coefficient $L(k) \to 0$ as $k \to \infty$ independent of m. Denote $\varphi'_k = (\varphi_1, \ldots, \varphi_k, \varphi'_{k+1}, \varphi'_{k+2}, \ldots)$. This yields

$$
\begin{aligned}
&\left\|\overset{(k+s)}{\Phi}\left(\overset{(k+s)}{\varphi}\right) - \overset{(k+s)}{\Phi}\left(\overset{(k+s)}{\varphi'_k}\right)\right\| \\
&\quad = \left\|\overset{(k+s)}{\Phi}(\varphi_1, \ldots, \varphi_k, \ldots, \varphi_{k+s}) - \overset{(k+s)}{\Phi}(\varphi_1, \ldots, \varphi_k, \varphi'_{k+1}, \ldots, \varphi'_{k+s})\right\| \\
&\quad \le L(k) \sup\{|\varphi_{k+1} - \varphi'_{k+1}|, \ldots, |\varphi_{k+s} - \varphi'_{k+s}|\} \\
&\quad \le L(k)\|\varphi - \varphi'_k\|. \qquad (16.13)
\end{aligned}
$$

This inequality is true for all $m = k + s$, where s is an arbitrary natural number. Clearly, for all $m \le k$,

$$
\left\|\overset{(m)}{\Phi}\left(\overset{(m)}{\varphi}\right) - \overset{(m)}{\Phi}\left(\overset{(m)}{\varphi'_k}\right)\right\| = 0.
$$

Since

$$\|\Phi(\varphi) - \Phi(\varphi'_k)\| = \lim_{m\to\infty}\left\| \overset{(m)}{\Phi}\left(\overset{(m)}{\varphi}\right) - \overset{(m)}{\Phi}\left(\overset{(m)}{\varphi'_k}\right)\right\|$$

$$= \lim_{s\to\infty}\left\| \overset{(k+s)}{\Phi}\left(\overset{(k+s)}{\varphi}\right) - \overset{(k+s)}{\Phi}\left(\overset{(k+s)}{\varphi'_k}\right)\right\|,$$

by using inequality (16.13) we obtain

$$\|\Phi(\varphi) - \Phi(\varphi'_k)\| \le L(k)\|\varphi - \varphi'_k\|,$$

i.e., $\Phi(\varphi) \in \widehat{C}^0_{\mathrm{Lip}}(\mathcal{F}_\infty)$. In this case,

$$\|\Phi(\varphi_t(\varphi)) - \Phi(\bar{\varphi}_t(\varphi))\| \le L(m)\|\,\bar{\bar{\varphi}}_t(\varphi)\| \le L(m)\|\varphi_t(\varphi)\| \le L(m)l_1 \to 0$$

as $m \to \infty$.

Thus, in view of inequality (16.2), we have

$$\|\Phi(\varphi_t(\varphi)) - \overset{(m)}{\Phi}(\overset{(m)}{\varphi_t}(\overset{(m)}{\varphi}))\| \to 0 \quad \text{as} \quad m \to \infty$$

uniformly in $t \in T_1$, which completes the proof of the first part of condition (16.11).

The second part of this condition is proved in exactly the same way.

Remark 16.1. Theorem 16.1 remains true if the number m appearing in its conditions takes the values $m_1 < m_2 < \ldots < m_\nu < \ldots$, where $m_\nu \to \infty$ as $\nu \to \infty$.

To illustrate this assertion, we consider, as an example, a system of the form (16.1) containing a scalar normal variable:

$$\frac{d\varphi_\nu}{dt} = \omega_\nu, \quad \frac{dx}{dt} = \Big[a + \sum_{\nu=1}^{\infty} \alpha_\nu \sin\varphi_\nu\Big]x \quad (\nu = 1, 2, \ldots). \tag{16.14}$$

The truncated system corresponding to this system has the form

$$\frac{d\varphi_\nu}{dt} = \omega_\nu, \quad \frac{dx}{dt} = \Big[a + \sum_{\nu=1}^{m} \alpha_\nu \sin\varphi_\nu\Big]x \quad (\nu = \overline{1, m}). \tag{16.15}$$

Its solution $\overset{(m)}{x_t}(\overset{(m)}{\varphi}, x_0)$ can be represented in the form

$$\overset{(m)}{x_t}(\overset{(m)}{\varphi}, x_0) = x_0 \exp\Big\{at + \sum_{\nu=1}^{m} \frac{\alpha_\nu}{\omega_\nu}[\cos\varphi^0_\nu - \cos(\omega_\nu t + \varphi^0_\nu)]\Big\}.$$

The change $\overset{(m)}{x} = \exp\left\{-\sum\limits_{\nu=1}^{m} \frac{\alpha_\nu}{\omega_\nu} \cos\varphi_\nu\right\} \overset{(m)}{y}$ reduces system (16.15) to the system

$$\frac{d\varphi_\nu}{dt} = \omega_\nu, \quad \frac{d\overset{(m)}{y}}{dt} = a\overset{(m)}{y} \quad (\nu = \overline{1, m}).$$

We assume that the series $\sum\limits_{\nu=1}^{\infty} \left|\frac{\alpha_\nu}{\omega_\nu}\right|$ converges and set

$$\overset{(m)}{\Phi} = \exp\left\{-\sum_{\nu=1}^{m} \frac{\alpha_\nu}{\omega_\nu} \cos\varphi_\nu\right\}.$$

In this case, it is clear that the sequences $\{\overset{(m)}{\Phi}\}$ and $\{\overset{(m)}{\Phi}{}^{-1}\}$ converge as $m \to \infty$ and, moreover,

$$\Phi = \lim_{m\to\infty} \overset{(m)}{\Phi} = \exp\left\{-\sum_{\nu=1}^{\infty} \frac{\alpha_\nu}{\omega_\nu} \cos\varphi_\nu\right\}.$$

The change of variables $x = \Phi y$ reduces system (16.14) to the system

$$\frac{d\varphi_\nu}{dt} = \omega_\nu, \quad \frac{dy}{dt} = ay \quad (\nu = 1, 2, \ldots).$$

In the system of equations (16.1), we now assume that $x \in \mathfrak{M}$ and $\varphi \in \mathfrak{M}$ and denote the system obtained as a result by (16.16). In this case, one can use different procedures of truncation of the variables x and φ. Thus, in particular, if we truncate the system only with respect to x, then we get a system of the form (16.1) and, by virtue of Theorem 16.1, the following assertion is true:

Theorem 16.2. *Suppose that the system of equations (16.16) is such that $a(\varphi) \in C^0_{\mathrm{Lip}}(\mathcal{F}_\infty)$, $P(\varphi) \in C^0(\mathcal{F}_\infty)$, and, for any natural n, the system*

$$\frac{d\varphi}{dt} = a(\varphi), \quad \frac{d\overset{(n)}{y}}{dt} = \overset{(n)}{P}(\varphi)\overset{(n)}{x}$$

truncated with the help of a reducing matrix $\overset{(n)}{\Phi}(\varphi)$ can be reduced to a system of the form

$$\frac{d\varphi}{dt} = a(\varphi), \quad \frac{d\overset{(n)}{x}}{dt} = \overset{(n)}{B}\overset{(n)}{y}.$$

If, in addition, $\{\overset{(n)}{\Phi}(\varphi)\}$ *and* $\overset{(n)}{\Phi}{}^{-1}(\varphi)\}$ *are uniformly regular sequences and* $\{\overset{(n)}{B}\}$ *is a regular sequence, then system (16.16) is reducible and its reducing matrix* $\Phi(\varphi) = \lim_{n\to\infty} \overset{(n)}{\Phi}(\varphi)$.

Combining the data on the reducibility of systems with quasiperiodic coefficients of the form (16.2) with the theorem on reducibility (Theorem 16.1), one can obtain new results concerning the reducibility of systems with almost periodic coefficients. In what follows, we formulate one of these results.

Assume that the matrix A in system (16.2) with $x \in R^n$ and $\varphi \in \mathfrak{M}$ is a constant matrix with different eigenvalues and $P(\varphi) \in \widehat{C}^0_{\text{Lip}}(\mathcal{F}_\infty)$ is a matrix analytic as a function of $\overset{(m)}{\varphi}$ in the domain $|\text{Im}\varphi| \le \rho$ and real for real φ. In this case, the frequency basis of the almost periodic function ω is such that, for any natural m, there exist positive numbers $\delta(m)$ and $d(m)$ such that $|(\overset{(m)}{k}, \overset{(m)}{\omega})| \ge \delta(m)|\overset{(m)}{k}|^{-d(m)}$ for all nonzero integral vectors $\overset{(m)}{k}$.

It is known that, under these conditions, there exist $\varepsilon(m) > 0$ such that any system truncated with respect to φ to the order m and corresponding to system (16.2) can be reduced, for all $\varepsilon < \varepsilon(m)$, to a system of the form

$$\frac{d\overset{(m)}{\varphi}}{dt} = \overset{(m)}{\omega}, \quad \frac{dy}{dt} = \overset{(m)}{A_0}\, y$$

by using the matrix $\overset{(m)}{\Phi}(\overset{(m)}{\varphi})$ analytic as a function of $\overset{(m)}{\varphi}$ for $|\text{Im}\,\overset{(m)}{\varphi}| \le \dfrac{\rho}{2}$ and real for real $\overset{(m)}{\varphi}$.

Theorem 16.3. *Assume that system (16.2) satisfies the conditions presented above, the matrices* $\overset{(m)}{\Phi}(\overset{(m)}{\varphi})$, $\overset{(m)}{\Phi}{}^{-1}(\overset{(m)}{\varphi}) \in \widehat{C}^0_{\text{Lip}}(\mathcal{F}_\infty)$ *with coefficients independent of* m, *the sequences* $\{\overset{(m)}{\Phi}(\overset{(m)}{\varphi})\}$, $\{\overset{(m)}{A_0}\}$, *and* $\overset{(m)}{\Phi}{}^{-1}(\overset{(m)}{\varphi})$ *converge as* $m \to \infty$, *and, moreover, the last of these sequences converges uniformly in* $\overset{(m)}{\varphi}$.

Then, for all $\varepsilon \le \inf_{m=1,2\ldots} \varepsilon(m)$, *the system of equations (16.2) is reducible with the help of the reducing matrix* $\Phi(\varphi) = \lim_{m\to\infty} \overset{(m)}{\Phi}(\overset{(m)}{\varphi})$.

Note that if the matrix A in system (16.2) is infinite and unbounded in norm, then it is impossible to use the proposed method for the solution of the problem of reducibility. The next section is devoted to the analysis of this case.

3.17. Quasiperiodic Systems with Unbounded Right-Hand Side

In the present section, using the method of accelerated convergence of iterations, we construct reducing matrices for systems of differential equations of the form

$$\frac{dx}{dt} = Ax + P(\varphi)x, \quad \frac{d\varphi}{dt} = \omega \tag{17.1}$$

under the assumption that the norm of the matrix $P(\varphi)$ is sufficiently small.

We assume that A is an infinite constant real diagonal matrix, $x \in \mathfrak{M}$, $\varphi \in R^m$, $\omega \in R^m$, and $P(\varphi)$ is an infinite matrix 2π-periodic in φ_i $(i = \overline{1, m})$, analytic in the domain $|\text{Im}\varphi| = \max\{|\text{Im}\varphi_1|, \ldots, |\text{Im}\varphi_m|\} \leq \rho_0$, real for real φ, and such that

$$\|P(\varphi)\|_\varphi = \sup_i \sum_{j=1}^{\infty} \sup_\varphi |p_{ij(\varphi)}| \leq M_0,$$

where M_0 is a sufficiently small constant.

Our aim is to find an invertible matrix $\Phi(\varphi)$ analytic in the domain $|\text{Im}\varphi| \leq \frac{\rho_0}{2}$, real for real φ, 2π-periodic in φ_i $(i = \overline{1, m})$, bounded in the norm $\|\cdot\|_\varphi$, and such that the change of variables

$$x = \Phi(\varphi)y, \quad \varphi = \omega t + \varphi_0$$

transforms the system of equations (17.1) into a system

$$\frac{dy}{dt} = A_0 y, \quad \frac{d\varphi}{dt} = \omega, \tag{17.2}$$

where A_0 is a constant matrix.

An important role in the construction of the matrix $\Phi(\varphi)$ by the method with accelerated convergence of iterations is played by a matrix equation of the form

$$\sum_{\alpha=1}^{m} \frac{\partial U(\varphi)}{\partial \varphi_\alpha} \cdot \omega_\alpha + UA = AU + P(\varphi), \tag{17.3}$$

where $P(\varphi)$ is given by the series

$$P(\varphi) = \sum_{(k)} P_k e^{i(k,\, \varphi)},$$

the diagonal elements of the matrix

$$P_0 = \frac{1}{(2\pi)^m} \int_0^{2\pi} \dots \int_0^{2\pi} P(\varphi) d\varphi_1 \dots d\varphi_m$$

are equal to zero, $k = (k_1, k_2, \dots, k_m)$ is an integral vector, and $(k) = \sum_{i=1}^{m} k_i$.

We set $|k| = \sum_{i=1}^{m} |k_i|$ and $(k, \omega) = \sum_{i=1}^{m} k_i \omega_i$. The following statement concerning the existence of periodic solutions of equation (17.3) is true:

Lemma 17.1. *Assume that the matrix* $A = \operatorname{diag}\{a_1, a_2, \dots\}$ *is such that* $\inf_{i \neq j} |a_i - a_j| = r > 0$ *and, moreover, one can find numbers* $\varepsilon > 0$ *and* $d > 1$ *such that the inequality*

$$|(k,\, \omega)| \geq \varepsilon |k|^{-d}$$

holds for all $|k| \neq 0$.

Then equation (17.3) possesses a solution $U(\varphi)$ 2π*-periodic in* φ_i $(i = \overline{1,\, m})$ *in the form of an analytic (real for real* φ*) matrix.*

Proof. The required solution of equation (17.3) is sought in the form of a series

$$U(\varphi) = \sum_{(k)} U_k e^{i(k,\, \varphi)}$$

whose coefficients satisfy the matrix equation

$$U_k(A + E \cdot i(\omega,\, k)) - AU_k = P_k,$$

where E is the identity matrix. Note that this equation has the form $XB - AX = C$, where A and B are diagonal matrices. This enables us to write the following relation for the coefficients U_k :

$$U_k = \begin{pmatrix} \dfrac{p_{11}^{(k)}}{i(k,\, \omega)} & \dfrac{p_{12}^{(k)}}{a_2 - a_1 + i(k,\, \omega)} & \dots \\ \dfrac{p_{21}^{(k)}}{a_1 - a_2 + i(k,\, \omega)} & \dfrac{p_{22}^{(k)}}{i(k,\, \omega)} & \dots \\ \dots & \dots & \dots \end{pmatrix},$$

where $P_k = [p_{ij}^{(k)}]_{i,j=1}^{\infty}$.

Let us now study the problem of the existence of the matrix U_k for different k.

It is clear that if $|k| \neq 0$, then the solution U_k exists and is unique. For $|k| = 0$, the solution U_k either does not exist or cannot be defined [for $p_{ii}^{(k)} = 0$ $(i = 1, 2, \ldots)$]. As U_0, we take the matrix

$$U_0 = \begin{pmatrix} 0 & \dfrac{p_{12}^0}{a_2 - a_1} & \cdots \\ \dfrac{p_{21}^0}{a_1 - a_2} & 0 & \cdots \\ \cdots & \cdots & \cdots \end{pmatrix}.$$

Since $|a_i - a_j + i(k,\, \omega)| \geq \varepsilon |k|^{-d}$, we have

$$\begin{aligned} \|U_k\| &= \sup_i \sum_{j=1}^{\infty} \frac{|p_{ij}^{(k)}|}{|a_j - a_i + i(k,\, \omega)|} \leq \sup_i \sum_{j=1}^{\infty} \frac{|p_{ij}^{(k)}|}{\varepsilon |k|^{-d}} \\ &\leq \frac{1}{\varepsilon} |k|^d \sup_i \sum_{j=1}^{\infty} |p_{ij}^{(k)}| = \frac{1}{\varepsilon} |k|^d \|P_k\| \quad (|k| \neq 0), \end{aligned}$$

$$\|U_0\| \leq \frac{1}{r} \|P_0\|.$$

The following assertions are true (they are proved in [BMS]):

1°. Consider an analytic function $f(\varphi)$ of an angular variable $\varphi = (\varphi_1, \varphi_2, \ldots, \varphi_m)$ defined in the domain $|\mathrm{Im}| \leq \rho$ and bounded by a constant M_0. This function can be expanded in the Fourier series

$$\sum_{(k)} f_k e^{i(k,\, \varphi)}$$

and its Fourier coefficients satisfy inequality $|f_k| \leq M_0 \exp\{-\rho |k|\}$.

2°. The inequality

$$\sum_{(k)} |k|^{\nu} e^{-2\delta |k|} < \left(\frac{\nu}{e}\right)^{\nu} \frac{1}{\delta^{\nu+m}} (1 + e)^m$$

holds for $0 < \delta < 1,\ \nu > 1$.

We set $|\mathrm{Im}\varphi|_0 = \sup_\varphi |\mathrm{Im}\varphi|$.
By using assertions 1° and 2°, we conclude that the inequalities

$$\|U(\varphi) - U_0(\varphi)\|_\varphi \le \sum_{|k|\neq 0} \|U_k\| e^{|\mathrm{Im}\varphi|_0|k|} \le \sum_{|k|\neq 0} \frac{1}{\varepsilon}|k|^d \|P_k\| e^{|\mathrm{Im}\varphi|_0|k|}$$

$$\le \frac{1}{\varepsilon} M_0 \sum_{|k|\neq 0} |k|^{-d} e^{-2\delta_0|k|} \le \frac{M_0}{\varepsilon}\left(\frac{d}{e}\right)^d \frac{(1+e)^m}{\delta_0^{d+m}},$$

$$\left\|\frac{\partial U(\varphi)}{\partial \varphi_s}\right\|_\varphi \le \sum_{\alpha=1}^{m} \left\|\frac{\partial U(\varphi)}{\partial \varphi_\alpha}\right\|_\varphi \le \sum_{\alpha=1}^{m} \sum_{|k|\neq 0} |k_\alpha| \|U_k\| e^{|\mathrm{Im}\varphi|_0|k|}$$

$$\le m \sum_{|k|\neq 0} |k| \frac{1}{\varepsilon} |k|^d M_0 e^{-\rho_0|k| + |\mathrm{Im}\varphi|_0|k|}$$

$$\le \frac{m}{\varepsilon} M_0 \sum_{|k|\neq 0} |k|^{d+1} e^{2\delta_0|k|} \le \frac{mM_0}{\varepsilon}\left(\frac{d+1}{e}\right)^{d+1} \frac{(1+e)^m}{\delta_0^{d+m+1}}$$

hold for $|\mathrm{Im}\varphi| \le \rho_0 - 2\delta_0 = \rho_1 > 0$ and $0 < \delta_0 < 1$.

To show that the matrix $U(\varphi)$ is real for real φ, it suffices to use the fact that its coefficients U_k and U_{-k} are complex conjugate. This follows from the same property of the Fourier coefficients P_k and P_{-k} of the matrix $P(\varphi)$, which is real for $|\mathrm{Im}\varphi| = 0$.

The lemma proved above is used for the construction of the iterative process studied in what follows.

The space of functions $z(t) = \{z_1(t), z_2(t), \dots\}$ such that $\sup\limits_{i,t\in T}\left\{|z_i(t)|\right\} < \infty$ and $\sup\limits_{i,t\in T}\left\{\left|\frac{dz_i(t)}{dt}\right|\right\} < \infty$, where T is a (finite or infinite) interval of time, is denoted by M.

A function $x = x(t)$ is called a solution of the system of equations (17.1) on T if it turns equations (17.1) into identities for all $t \in T$ and belongs to the space M for any finite interval $T_1 \subset T$.

The diagonal matrix whose elements coincide with the diagonal elements of the matrix $P(\varphi)$ is denoted by $D(\varphi)$. We set

$$\overline{D} = \frac{1}{(2\pi)^m} \int_0^{2\pi} \dots \int_0^{2\pi} D(\varphi) d\varphi_1 \dots d\varphi_m.$$

Further, we perform a change of variables according to the formula

$$x = (E + U^{(1)}(\varphi))y^{(1)},$$

where $U^{(1)}(\varphi)$ is a solution of the equation

$$\Big(\frac{\partial U^{(1)}(\varphi)}{\partial\varphi}, \omega\Big) + U^{(1)}A = AU^{(1)} + P(\varphi) - \overline{D}. \tag{17.4}$$

By virtue of Lemma 17.1, equation (17.4) is indeed solvable.

For $\|U^{(1)}(\varphi)\| < 1$, there exists the inverse matrix

$$(E + U^{(1)}(\varphi))^{-1} = E + \sum_{k_0=1}^{\infty}(-1)^{k_0}(U^{(1)}(\varphi))^{k_0}.$$

Moreover, the series on the right-hand side of this equality is uniformly convergent.

Then $y^{(1)} = (E+U^{(1)}(\varphi))^{-1}x$. Note that the analytic matrix $(E+U^{(1)}(\varphi))^{-1}$ satisfies the Cauchy estimate for $|\mathrm{Im}\varphi| < \rho_1 - \sigma_1 > 0$, i.e., the derivative of this matrix with respect to φ_i is bounded in norm.

Indeed, we now denote $(E + U^{(1)}(\varphi))^{-1} = W = [\omega_{ij}(\varphi)]_{i,j=1}^{\infty}$ and show that $\Big\|\dfrac{\partial W(\varphi)}{\partial\varphi_s}\Big\|_{\varphi} \le \dfrac{1}{\sigma_1}\cdot P$ for $\|W\|_{\varphi} \le P = \text{const}$ $(s = \overline{1,m})$. Consider an arbitrary element $\omega_{ij}(\varphi)$ of the matrix W. It is clear that $\sup\limits_{\varphi}|\omega_{ij}(\varphi)|$ is bounded in the domain $|\mathrm{Im}\varphi| < \rho_1$. Otherwise, the constant P does not exist. Any point φ_s of the strip $|\mathrm{Im}\varphi_s| < \rho_1 - \sigma_1$ can be regarded as the center of a disk completely contained in the strip $|\mathrm{Im}\varphi_s| \le \rho_1$, where $\omega_{ij}(\varphi)$ is an analytic function of φ_s. This yields the inequality $\Big|\dfrac{\partial\varphi_{ij}(\varphi)}{\partial\varphi_s}\Big| \le \dfrac{1}{\sigma_1}P_{\sigma_1}$, where P_{σ_1} is the maximum value of the modulus of $\omega_{ij}(\varphi)$ on the boundary of this disk. Thus, we have

$$\Big|\frac{\partial\omega_{ij}(\varphi)}{\partial\varphi_s}\Big| \le \frac{1}{\sigma_1}\sup_{\varphi}|\omega_{ij}(\varphi)|$$

for all $i,\ j = 1, 2, 3, \ldots$ and $s = \overline{1,\ m}$. This gives

$$\Big\|\frac{\partial W(\varphi)}{\partial\varphi_s}\Big\|_{\varphi} = \sup_i\sum_{j=1}^{\infty}\sup_{\varphi}\Big|\frac{\partial\omega_{ij}(\varphi)}{\partial\varphi_s}\Big|$$

$$\le \frac{1}{\sigma_1}\sup_i\sum_{j=1}^{\infty}\sup_{\varphi}|\omega_{ij}(\varphi)| = \frac{1}{\sigma_1}\|W\|_{\varphi}.$$

Let $x = x(t)$ be a solution of the system of equations (17.1) in T. Then, for all $t \in T$, we have

$$\frac{d}{dt}\Big(\sum_{j=1}^{\infty}\omega_{ij}(\varphi)x_j(t)\Big) = \sum_{j=1}^{\infty}\frac{d\omega_{ij}(\varphi)}{dt}x_j(t) + \sum_{j=1}^{\infty}\omega_{ij}(\varphi)\frac{dx_j}{dt} \quad (i = 1, 2, \ldots),$$

since the series on the right-hand side of this equality converge absolutely and uniformly in a sufficiently small neighborhood T_1 of the point t. This is true because the second series is majorized by the convergent numerical series

$$\sum_{j=1}^{\infty}\sup_{\varphi}|\omega_{ij}(\varphi)|N_2,$$

where

$$\Big\|\frac{dx(t)}{dt}\Big\| \leq N_2,$$

and the first series can be represented in the form

$$\sum_{k_0=1}^{m}\sum_{j=1}^{\infty}\frac{\partial\omega_{ij}(\varphi)}{\partial\varphi_{k_0}}\omega_{k_0}x_j(t),$$

where the inner series is also absolutely and uniformly convergent because it can be majorized by the convergent numerical series

$$\sum_{j=1}^{\infty}\|W\|_{\varphi}\max_{\varphi}\Big|\frac{\partial\omega_{ij}(\varphi)}{\partial\varphi_{k_0}}\Big|N_1,$$

where $\|x(t)\| \leq N_1$. This means that, for $t \in T$ and $y^{(1)} \in M$, we have

$$\frac{d}{dt}\Big((E + U^{(1)}(\varphi))y^{(1)}\Big) = \frac{d}{dt}\Big(U^{(1)}(\varphi)\Big)y^{(1)} + \Big(E + U^{(1)}(\varphi)\Big)\frac{dy^{(1)}}{dt},$$

and the system of equations (17.1) reduces to the form

$$\frac{dy^{(1)}}{dt} = (A + \overline{D})y^{(1)} + P^{(1)}(\varphi)y^{(1)}, \quad \frac{d\varphi}{dt} = \omega, \tag{17.5}$$

where $P^{(1)}(\varphi) = (E + U^{(1)}(\varphi))^{-1}(P(\varphi)U^{(1)}(\varphi) - U^{(1)}(\varphi)\overline{D})$. Further, we have

$$\begin{aligned}\|U^{(1)}(\varphi)\|_{\varphi} &\leq \frac{2}{r}M_0 + \frac{2M_0}{\varepsilon}\left(\frac{d}{e}\right)^{d}\frac{(1+e)^m}{\delta_0^{d+m}} \\ &= 2M_0\left(\frac{1}{r} + \frac{1}{\varepsilon}\left(\frac{d}{e}\right)^{d}\frac{(1+e)^m}{\delta_0^{d+m}}\right).\end{aligned}$$

We choose sufficiently small M_0 such that

$$16M_0^{2-\kappa}\left(\frac{1}{r}+\frac{1}{\varepsilon}\left(\frac{d}{e}\right)^d\frac{(1+e)^m}{\delta_0^{d+m}}\right)\leq 1,$$

where $1<\kappa<2$ is a fixed number. Then

$$\|U^{(1)}(\varphi)\|_\varphi\leq\frac{M_0^{\kappa-1}}{8}\quad\text{and}\quad\|U^{(1)(\varphi)}\|_\varphi<1$$

for small M_0. Therefore, the matrix $(E+U^{(1)}(\varphi))^{-1}$ indeed exists.

We estimate the norm of the matrix $P^{(1)}(\varphi)$:

$$\begin{aligned}
&\|P^{(1)}(\varphi)\|_\varphi\\
&\leq\|(E+U^{(1)}(\varphi))^{-1}\|_\varphi(\|P(\varphi)\|_\varphi\|U^{(1)}(\varphi)\|_\varphi+\|U^{(1)}(\varphi)\|_\varphi\|\overline{D}\|_\varphi)\\
&\leq\Big(1+\sum_{k=1}^{\infty}\|U^{(1)}(\varphi)\|_\varphi^{k_0}\Big)\Big(M_0\frac{M_0^{\kappa-1}}{8}+M_0^{\kappa-1}\frac{M_0}{8}\Big)\\
&\leq\left(1+\frac{1}{1-M_0^{\kappa-1}/8}\right)\frac{M_0^\kappa}{8}\leq M_0^\kappa=M_1.
\end{aligned}$$

Despite the fact that, in order to get the last inequality, it is sufficient that $|\mathrm{Im}\varphi|\leq\rho_0-2\delta_0$ (because, in this domain, the matrix $P^{(1)}(\varphi)$ is analytic and bounded by the constant M_1), we must formally say that the system of equations (17.1) is reduced to a system of the form (17.5) in a somewhat narrower domain $|\mathrm{Im}\varphi|<\rho_1-\sigma_1$. Since we consider the real case ($|\mathrm{Im}\varphi|=0$), we have $y^{(1)}\in M$ for all $t\in T$.

Denote $A^{(1)}=A+\overline{D}$. It is easy to see that, for small M_0, the diagonal elements of the matrix $A^{(1)}$ have the same property as the elements of the matrix A. Indeed,

$$\inf_{i\neq j}|a_i^{(1)}-a_j^{(1)}|\geq\inf_{i\neq j}|a_i-a_j|-\sup_{i\neq j}|\bar{d}_i-\bar{d}j|\geq r_0-2M_0=r_1>0.$$

Thus, it is shown that the system of equations

$$\frac{dy^{(1)}}{dt}=A^{(1)}y^{(1)}+P^{(1)}(\varphi)y^{(1)},\quad\frac{d\varphi}{dt}=\omega$$

obtained in the first step of the process admits the second step of the iterative process.

We define a sequence σ_s $(s = 1, 2, 3, \ldots)$ by the inequality $\sigma_s < \rho_s$. Assume that, after s steps of the process, we arrive at the following system of equations:

$$\frac{dy^{(s)}}{dt} = A^{(s)}y^{(s)} + P^{(s)}(\varphi)y^{(s)}, \quad \frac{d\varphi}{dt} = \omega, \tag{17.6}$$

where the matrices $A^{(s)}$ and $P^{(s)}(\varphi)$ are defined in the domain $|\text{Im}\varphi| \leq \rho_s$ and the vector function $y^{(s)} \in M$ is defined for $t \in T$.

We now realize the $(s+1)$th step of the process for system (17.6). To this end, in the last system, we perform the change of variables according to the formula

$$y^{(s)} = (E + U^{(s+1)}(\varphi))y^{(s+1)},$$

where $U^{(s+1)}(\varphi)$ is a solution of the matrix equation

$$\Big(\frac{\partial U^{(s+1)}(\varphi)}{\partial \varphi}, \omega\Big) + U^{(s+1)}(\varphi)A^{(s)} = A^{(s)}U^{(s+1)}(\varphi) + P^{(s)}(\varphi) - \overline{D}^{(s)}$$

defined in the domain $|\text{Im}\varphi| \leq \rho_{s+1}$.

As shown above,

$$\|U^{(s+1)}(\varphi)\|_\varphi \leq 2M_s \left(\frac{1}{r_s} + \frac{1}{\varepsilon}\left(\frac{d}{e}\right)^d \frac{(1+e)^m}{\delta_0^{(d+m)s}}\right).$$

Setting $M_{s+1} = M_s^\kappa$, $\rho_{s+1} = \rho_s - 2\delta_0^{s+1}$, $\delta_0 \leq \dfrac{\rho_0}{4+\rho_0}$, $1 < \kappa < 2$, and $r_{s+1} = r_s - M_s^{\kappa-1} > 0$ and taking sufficiently small M_0 such that

$$16M_0^{(2-\kappa)\kappa^s} \left(\frac{1}{r_0 - \sum_{k_0=0}^{s-1} M_0^{(\kappa-1)\kappa^{k_0}}} + \frac{1}{\varepsilon}\left(\frac{d}{e}\right)^d \frac{(1+e)^m}{\delta_0^{(d+m)s}}\right) \leq 1,$$

$$r_0 - \sum_{k_0=0}^{\infty} M_0^{(\kappa-1)\kappa^{k_0}} > 0,$$

we easily obtain the inequality

$$\|U^{(s+1)}(\varphi)\|_\varphi \leq \frac{M_s^{\kappa-1}}{8}$$

and, hence,

$$\begin{aligned}\|P^{(s+1)}(\varphi)\|_\varphi &\leq \left(1 + \frac{1}{1 - M_s^{\kappa-1}/8}\right)\left(M_s\frac{M_s^{\kappa-1}}{8} + M_s^{\kappa-1}\frac{M_s}{8}\right) \\ &= \left(\frac{1}{8} + \frac{1}{8 - M_s^{\kappa-1}}\right) M_s^\kappa \leq M_{s+1}.\end{aligned}$$

If, in this case, $\|\mathrm{Im}\varphi\| < \rho_{s+1} - \sigma_{s+1}$, then $y^{(s+1)} \in M$ for $t \in T$. The function $y^{(s+1)}$ is indeed a solution of the system of equations

$$\frac{dy^{(s+1)}}{dt} = A^{(s+1)} y^{(s+1)} + P^{(s+1)}(\varphi) y^{(s+1)}, \quad \frac{d\varphi}{dt} = \omega \tag{17.7}$$

and the matrices $A^{(s+1)}$ and $P^{(s+1)}$ are defined for $\|\mathrm{Im}\varphi\| \le \rho_s - 2\delta_0^{s+1}$.

The reasoning and estimates presented above prove the following auxiliary statement:

Lemma 17.2. *Assume that the system of equations (17.6) satisfies the following conditions:*

(i) the matrix $P^{(s)}(\varphi)$ is 2π-periodic in φ_i $(i = \overline{1, m})$, analytic as a function of φ in the domain $\|\mathrm{Im}\varphi\| \le \rho_s$, real for real φ, and bounded in norm by a constant M_s;

(ii) the matrix $A^{(s)}$ is diagonal and real and, in addition,

$$\inf_{i \ne j} |a_i^{(s)} - a_j^{(s)}| \ge r_s > 0;$$

(iii) there exist $\varepsilon > 0$ and $d > 1$ such that $|(k, \omega)| \ge \varepsilon |k|^{-d}$ for all integral vectors k such that $|k| \ne 0$.

Then there exists a change of variables $y^{(s)} = (E + U^{(s+1)}(\varphi))\, y^{(s+1)}$ that reduces the system of equations (17.6) to a system of the form (17.7) in the domain $|\mathrm{Im}\varphi| < \rho_{s+1} - \sigma_{s+1}$. Moreover, the matrices $U^{(s+1)}(\varphi)$ and $P^{(s+1)}(\varphi)$ are periodic in φ_i $(i = \overline{1, m})$ with period 2π, analytic in the domain $|\mathrm{Im}\varphi| < \rho_{s+1}$, and such that

$$\|P^{(s+1)}(\varphi)\|_\varphi \le M_{s+1} \quad \text{and} \quad \|U^{(s+1)}(\varphi)\|_\varphi \le \frac{M_s^{\kappa-1}}{8},$$

the matrix $A^{(s+1)}$ is diagonal, real, and such that

$$\inf_{i \ne j} |\, a_i^{(s+1)} - a_j^{(s+1)}\,| \ge r_{s+1} > 0,$$

and, in addition, the following relations are true:

$$M_{s+1} = M_s^\kappa, \quad \rho_{s+1} = \rho_s - 2\delta_0^s,$$

$$r_{s+1} = r_s - M_s^{\kappa-1} > 0, \quad 1 < \kappa < 2, \quad \delta_0 \le \frac{\rho_0}{4 + \rho_0}.$$

Lemma 17.2 enables us to continue the iterative process infinitely. The superposition of s changes of the indicated type, i.e., the change

$$x = (E + U^{(1)}(\varphi))(E + U^{(2)}(\varphi)) \dots (E + U^{(s)}(\varphi))y^{(s)}, \tag{17.8}$$

turns the system of equations (17.1) into a system of the form (17.6). We set

$$\Phi^{(s)} = \prod_{\alpha=1}^{s} (E + U^{(\alpha)}(\varphi)).$$

By virtue of the definition of the constants ρ_s, and σ_s, all matrices $\Phi^{(s)}(\varphi)$ are defined and analytic in the domain $|\mathrm{Im}\varphi| \le \dfrac{\rho_0}{2}$ and all $y^{(s)}$ belong to the set M for $t \in T$.

Let us now prove that the sequence $\{\Phi^{(s)}\}$ of matrices converges in the domain $|\mathrm{Im}\varphi| \le \dfrac{\rho_0}{2}$ to a certain matrix $\Phi(\varphi)$ periodic in φ with period 2π, real for $|\mathrm{Im}\varphi| = 0$, and analytic in the indicated domain.

Since, for all $s = 1, 2, \dots$, the matrices $\Phi^{(s)}$ possess the indicated properties, it suffices to show that the convergence of $\{\Phi^{(s)}(\varphi)\}$ to the matrix Φ is uniform in the matrix norm $\|\cdot\|_\varphi$. Thus, we have

$$\begin{aligned}
\|\Phi^{(s+1)}(\varphi) - \Phi^{(s)}(\varphi)\|_\varphi &\le \|\Phi^{(s)}(\varphi)\|_\varphi \|U^{(s+1)}(\varphi)\|_\varphi \\
&\le \Big[\prod_{\alpha=1}^{s}\Big(1 + \frac{M_{\alpha-1}^{\kappa-1}}{8}\Big)\Big] \frac{M_s^{\kappa-1}}{8} \\
&= \frac{1}{8}\prod_{\alpha=1}^{s}\Big(1 + \frac{M_{\alpha-1}^{\kappa-1}}{8}\Big) M_s^{\kappa-1} \\
&\le \frac{1}{8}\prod_{\alpha=1}^{\infty}\Big(1 + \frac{M_{\alpha-1}^{\kappa-1}}{8}\Big) M_s^{\kappa-1} \\
&= cM_s^{\kappa-1}, \quad c = \text{const} > 0.
\end{aligned}$$

This yields the validity of the following criterion of uniform convergence of the sequence $\{\Phi^{(s)}\}$:

$$\|\Phi^{(s+k_0)}(\varphi) - \Phi^{(s)}(\varphi)\|_\varphi < c \sum_{\alpha=s}^{s+k_0-1} M_\alpha^{\kappa-1} < c \sum_{\alpha=s}^{\infty} M_\alpha^{\kappa-1}.$$

Lemma 17.3. *The sequence* $y^{(s)}(t)$ $(s = 1, 2, \ldots)$ *is uniformly bounded on* T_1 *and*

$$\lim_{s\to\infty} y^{(s)}(t) = z(t) \in M$$

uniformly in $t \in T_1$, *where* T_1 *is an arbitrary finite segment from* T.

Proof. In the domain $|\mathrm{Im}\varphi| \leq \dfrac{\rho_0}{2}$, we can write

$$\begin{aligned}\|\Phi^{(s)}(\varphi) - E\|_\varphi &\leq \prod_{\alpha=1}^{s}\Big(1 + \frac{M_{\alpha-1}^{\kappa-1}}{8}\Big) - 1\\ &\leq \prod_{\alpha=1}^{s}\Big(1 + \frac{M_{\alpha-1}}{8}\Big) - 1\\ &\leq \prod_{\alpha=1}^{\infty}\Big(1 + \frac{M_{\alpha-1}}{8}\Big) - 1.\end{aligned}$$

Since the series $\sum\limits_{\alpha=1}^{\infty} \dfrac{M_{\alpha-1}}{8}$ converges, the infinite product $\prod\limits_{\alpha=1}^{\infty}\Big(1 + \dfrac{M_{\alpha-1}}{8}\Big)$ is also convergent and, moreover, $\prod\limits_{\alpha=1}^{\infty}\Big(1 + \dfrac{M_{\alpha-1}}{8}\Big) - 1 = l < 1$ for sufficiently small M_0. Then the series $\sum\limits_{\alpha=1}^{\infty}(E - \Phi^{(s)}(\varphi))^\alpha$ is uniformly convergent and defines the inverse matrix $\overset{(s)}{\Phi}{}^{-1}(\varphi)$. Since

$$\begin{aligned}\|\overset{(s)}{\Phi}{}^{-1}(\varphi)\|_\varphi &\leq \sum_{\gamma=0}^{\infty}\Big\{\prod_{\alpha=1}^{s}\Big(1 + \frac{M_{\alpha-1}^{\kappa-1}}{8}\Big) - 1\Big\}^\gamma\\ &\leq \sum_{\gamma=0}^{\infty}\Big\{\prod_{\alpha=1}^{\infty}\Big(1 + \frac{M_{\alpha-1}}{8}\Big) - 1\Big\}^\gamma \leq \sum_{\gamma=0}^{\infty} l^\gamma = K_1,\end{aligned}$$

where $K_1 = \text{const} < \infty$, the sequence $\{\overset{(s)}{\Phi}{}^{-1}(\varphi)\}$ is uniformly bounded.

Let $T_1 \subset T$ be an arbitrary finite segment. In this segment, we have $\|x(t)\| \leq N_1^0$ and $\|\dfrac{dx(t)}{dt}\| \leq N_2^0$, where N_1^0 and N_2^0 are positive constants. Hence, for $t \in T_1$,

$$\|\overset{(s)}{y}(t)\| \leq \|\overset{(s)}{\Phi}{}^{-1}(\varphi_t)\|\|x(t)\| \leq K_1 N_1^0 = \text{const} < \infty.$$

We chose $\sigma = \text{const}$ such that $\dfrac{\rho_0}{2} - \sigma > 0$. By using the Cauchy estimate, we obtain

$$\begin{aligned}\Big\|\frac{\partial \overset{(s)}{\Phi}{}^{-1}(\varphi)}{\partial\varphi_i}\Big\|_{\varphi} &= \sup_g \sum_{j=1}^{\infty} \sup_{\varphi}\Big|\frac{\partial \overset{(s)}{\varphi}{}^{-1}_{gj}(\varphi)}{\partial\varphi_i}\Big| \\ &\le \sup_g \sum_{j=1}^{\infty} \frac{1}{\sigma} \sup_{\varphi} |\overset{(s)}{\varphi}{}^{-1}_{gj}(\varphi)| \le \frac{K_1}{\sigma},\end{aligned}$$

where $[\overset{(s)}{\varphi}{}^{-1}_{gj}(\varphi)]_{g,j=1}^{\infty} = \overset{(s)}{\Phi}{}^{-1}(\varphi)$ and $|\text{Im}\varphi| \le \dfrac{\rho_0}{2} - \sigma$. Since $x(t) \in M$, we get

$$\frac{d\overset{(s)}{y}(t)}{dt} = \frac{d\overset{(s)}{\Phi}{}^{-1}(\varphi(t))}{dt}x(t) + \overset{(s)}{\Phi}{}^{-1}(\varphi(t))\frac{dx(t)}{dt}$$

for $t \in T$, whence

$$\Big\|\frac{d\overset{(s)}{y}(t)}{dt}\Big\| \le \frac{mK_1N_1^0}{\sigma}\|\omega\| + K_1N_2^0 = K_2 = \text{const} < \infty, \quad t \in T_1.$$

Relation (17.8) implies that

$$\begin{aligned}\|\overset{(s+1)}{y}(t) - \overset{(s)}{y}(t)\| &= \|\overset{(s+1)}{y}(t) - (E + \overset{(s+1)}{U}(\varphi))\overset{(s+1)}{y}(t)\| \\ &\le \|\overset{(s+1)}{U}(\varphi)\|\|\overset{(s+1)}{y}(t)\| \\ &\le K_1N_1^0\frac{1}{8}M_s^{\kappa-1} = c_1M_s^{\kappa-1}, \quad c_1 = \text{const} > 0.\end{aligned}$$

This yields the criterion of uniform convergence of the sequence $\{\overset{(s)}{y}(t)\}$ on T_1 as $s \to \infty$. We denote

$$\lim_{s\to\infty} \overset{(s)}{y}(t) = z(t) \quad \text{and} \quad \lim_{s\to\infty} \overset{(s)}{\Phi}(\varphi) = \Phi(\varphi).$$

Then we get $x(t) = \Phi(\varphi(t))z(t)$.

Since $\|\overset{(s)}{\Phi}(\varphi) - E\| \le l < 1$, by passing to the limit we obtain

$$\|\Phi(\varphi) - E\| < 1.$$

This means that the series $\sum_{\alpha=0}^{\infty}(E-\Phi(\varphi))^{\alpha}$ is uniformly convergent and defines the matrix $\Phi^{-1}(\varphi)$ bounded and analytic in the domain $|\mathrm{Im}\varphi| \leq \frac{\rho_0}{2}$ and inverse to the matrix $\Phi(\varphi)$. Therefore,

$$z(t) = \Phi^{-1}(\varphi(t))x(t) \in M.$$

Let us now show that the function $z(t)$ is a solution of the system of equations (17.2), where $A_0 = \lim_{s\to\infty}(A + \overline{D} + \overline{D}^{(1)} + \ldots + \overline{D}^{(s)})$. To this end, it suffices to prove the validity of the equality

$$\lim_{s\to\infty}\frac{d\overset{(s)}{y}(t)}{dt} = \frac{d}{dt}(\lim_{s\to\infty}\overset{(s)}{y}(t)).$$

This follows from the uniform coordinatewise convergence of the sequence $\{\overset{(s)}{A}\overset{(s)}{y} + \overset{(s)}{P}(\varphi(t))\overset{(s)}{y}(t)\}$ to the function $A_0 z(t)$ as $s \to \infty$ for any finite segment $T_1 \subset T$.

We summarize the results obtained in the present section in the form of the following statement:

Theorem 17.1. *Assume that the conditions of Lemma 17.1 are satisfied. Then there exists a sufficiently small constant $M^0 > 0$ such that, for all $M_0 \leq M^0$, any solution $x(t)$ of the system of equations (17.1) can be represented in the form $x(t) = \Phi(\varphi_t)z(t)$, where $z(t) \in M$ is a solution of the system of equations (17.2). Moreover, the invertible matrix $\Phi(\varphi)$ is 2π-periodic in $\varphi_i (i = \overline{1,m})$, analytic in the domain $|\mathrm{Im}\varphi| \leq \frac{\rho_0}{2}$, and real for real φ.*

Note that, for sufficiently small M_0, the fact that all elements a_i of the matrix A are positive implies that all elements A_i^0 of the matrix A_0 are also positive.

The solution of the system of equations (17.2) $z = z(t)$ with initial values z_0 and t_0 has the form

$$z_i(t) = z_i^0 \exp\{a_i^0(t - t_0)\} \quad (i = 1, 2, \ldots).$$

We choose a sequence $\{z_i^*\} \in \mathfrak{M}$ so that $\{|z_i^*|\,|a_i^0|\} \in \mathfrak{M}$. Then $z(t_0, z^*, t) \in M$ for all $t \in (-\infty,\ t_0]$. Furthermore, for all initial values t_0 and z^0 such that $|z_i^0| \leq |z_i^*|$, the corresponding solution belongs to M for all $t \in (-\infty,\ t_0]$.

A similar situation is encountered in the case where all elements of the matrix A are negative. The case where the matrix A contains finitely many positive or negative elements can also be studied without difficulties.

Corollary 17.1. *Assume that the system of equations (17.2) has a solution* $z = z(t) \in M$ *on the segment* T *with initial values* $t_0 \in T$ *and* $\Phi^{-1}(\varphi(t_0))x^0$, $x^0 \in \mathfrak{M}$. *Then, under the conditions of Theorem 17.1, the system of equations (17.1) possesses a solution* $x = x(t) \in M$ *on* T *with the initial values* t_0 *and* x^0.

To prove this assertion, it suffices to show that, for all $t \in T$,

$$\frac{d\Phi(\varphi(t))}{dt} + \Phi(\varphi(t))A_0 - (A + P(\varphi(t)))\Phi(\varphi(t)) = 0. \tag{17.9}$$

We set

$$L_s(t) = \frac{d\Phi^{(s)}}{dt} + \Phi^{(s)}A^{(s)} + \Phi^{(s)}P^{(s)} - A\Phi^{(s)} - P\Phi^{(s)}$$

and prove that, for all natural s and $t \in T$,

$$L_s(t) = 0. \tag{17.10}$$

Thus, for $s = 1$,

$$\begin{aligned} L_1(t) &= \frac{d\Phi^{(1)}}{dt}(E + U^{(2)}) + \Phi^{(1)}\frac{dU^{(2)}}{dt} \\ &\quad + \Phi^{(1)}\Big(E + U^{(2)}\Big)(A^{(1)} + \overline{D}_1) + \Phi^{(1)}(P^{(1)}U^{(2)} - U^{(2)}\overline{D}_1) \\ &\quad - A\Phi^{(1)}(E + U^{(2)}) - P\Phi^{(1)}(E + U^{(2)}) = 0. \end{aligned}$$

We assume that equalities (17.10) are true for all $s \leq k$. Then

$$\begin{aligned} L_{k+1}(t) &= L_k(t) + L_k U^{(k+1)} \\ &\quad + \Phi^{(k)}\Big(\frac{dU^{(k+1)}}{dt} + U^{(k+1)}A^{(k)} - A^{(k)}U^{(k+1)} - P^{(k)} + \overline{D}_k\Big) \\ &= 0. \end{aligned}$$

This means that equality (17.10) is proved for all natural s by induction.

In view of the uniform convergence $\Phi^{(s)} \to \Phi$, $A^{(s)} \to A_0$, and $P^{(s)} \to 0$ as $s \to \infty$, it follows from relation (17.10) that $\dfrac{d\Phi^{(s)}}{dt} \to \dfrac{d\Phi}{dt}$.

Finally, in relation (17.10), we pass to the limit as $s \to \infty$ and obtain the required equality (17.9).

3.18. Decomposition of Countable Systems

We consider a system of equations

$$\frac{d\varphi}{dt} = a(\varphi), \quad \frac{dx}{dt} = P_1(\varphi)x + P_{12}(\varphi)y, \tag{18.1}$$

$$\frac{dy}{dt} = P_{21}(\varphi)x + P_2(\varphi)y,$$

where $x,\ y \in \mathfrak{M},\ \varphi \in R^m$, and $P_1,\ P_2,\ P_{12},\ P_{21} \in C^0(\mathcal{F}_m)$ are infinite matrices bounded in the norm

$$\|P(\varphi)\|_{\mathcal{F}_m} = \sup_i \sum_{j=1}^{\infty} \max_{\varphi \in \mathcal{F}_m} |p_{ij}(\varphi)|.$$

The case where x or y belongs to R^n is not excluded.

The set of matrices from $C^0(\mathcal{F}_m)$ bounded in the norm $\|\cdot\|_{\mathcal{F}_m}$ together with their derivatives with respect to φ_i $(i = \overline{1,\ m})$ is denoted by $C^1_{\mathcal{F}_m}(\mathcal{F}_m)$. Our aim is to find a change of variables

$$x = x_1 + U_1(\varphi)y_1, \quad y = y_1 + U_2(\varphi)x_1, \tag{18.2}$$

where $U_1(\varphi)$ and $U_2(\varphi)$ are infinite matrices, that reduces the system of equations (18.1) to a system of the form

$$\frac{d\varphi}{dt} = a(\varphi), \quad \frac{dx_1}{dt} = \mathcal{P}_1(\varphi)x_1, \quad \frac{dy_1}{dt} = \mathcal{P}_2(\varphi)y_1. \tag{18.3}$$

Let $\varphi_t = \varphi_t(\varphi)$ be a solution of the equation for angular variables in system (18.1). We denote the matrices $P(\varphi)$ and $P(\varphi_t(\varphi))$ by P and $P(t)$, respectively.

If $U_1,\ U_2 \in C^1_{\mathcal{F}_m}(\mathcal{F}_m)$ are solutions of the matrix Riccati equations

$$\sum_{i=1}^{m} \frac{\partial U_1}{\partial \varphi_i} a_i(\varphi) = P_1U_1 - U_1P_2 - U_1P_{21}U_1 + P_{12}, \tag{18.4}$$

$$\sum_{i=1}^{m} \frac{\partial U_2}{\partial \varphi_i} a_i(\varphi) = P_2U_2 - U_2P_1 - U_2P_{12}U_2 + P_{21}, \tag{18.5}$$

then

$$\frac{dU_1(t)}{dt} = P_1(t)U_1(t) - U_1(t)P_2(t) - U_1(t)P_{21}(t)U_1(t) + P_{12}(t),$$

$$\frac{dU_2(t)}{dt} = P_2(t)U_2(t) - U_2(t)P_1(t) - U_2(t)P_{12}(t)U_2(t) + P_{21}(t), \quad (18.5^*)$$

and the change of variables (18.2) reduces system (18.1) to a system of the form

$$\frac{d\varphi}{dt} = a(\varphi),$$

$$(E - U_1U_2)\frac{dx_1}{dt} = (P_1 + P_{12}U_2 - U_1U_2P_1 - U_1U_2P_{12}U_2)x_1,$$

$$(E - U_2U_1)\frac{dy_1}{dt} = (P_2 + P_{21}U_1 - U_2U_1P_2 - U_2U_1P_{21}U_1)y_1.$$

If the matrices $(E - U_1U_2)$ and $(E - U_2U_1)$ are invertible, then the system of equations (18.1) is reduced to a system of the form (18.3), where

$$\mathcal{P}_1(\varphi) = (E - U_1U_2)^{-1}(P_1 + P_{12}U_2 - U_1U_2P_1 - U_1U_2P_{12}U_2),$$

$$\mathcal{P}_2(\varphi) = (E - U_2U_1)^{-1}(P_2 + P_{21}U_1 - U_2U_1P_2 - U_2U_1P_{21}U_1). \quad (18.6)$$

The reasoning presented above can be summarized in the form of the following statement:

Lemma 18.1. *If the solutions $U_1(\varphi)$, $U_2(\varphi) \in C^1_{\mathcal{F}_m}(\mathcal{F}_m)$ of equations (18.4) and (18.5) exist and are such that*

$$\|U_1(\varphi)\|_{\mathcal{F}_m} \cdot \|U_2(\varphi)\|_{\mathcal{F}_m} < 1,$$

then, by the change of variables (18.2), the system of equations (18.1) is reduced to a system of the form (18.3), where the matrices $\mathcal{P}_1(\varphi)$ and $\mathcal{P}_2(\varphi)$ are given by relations (18.6).

The problem of the existence of solutions of the Riccati equations can be reduced to the finite-dimensional case. For this purpose, we consider the following system of equations truncated with respect to x and y and corresponding to system (18.1):

$$\frac{d\varphi}{dt} = a(\varphi), \qquad \frac{d\overset{(n)}{x}}{dt} = \overset{(n)}{P_1}(\varphi)\overset{(n)}{x} + \overset{(n)}{P_{12}}(\varphi)\overset{(n)}{y},$$

$$\frac{d\overset{(n)}{y}}{dt} = \overset{(n)}{P_{21}}(\varphi)\overset{(n)}{x} + \overset{(n)}{P_2}(\varphi)\overset{(n)}{y}. \quad (18.7)$$

System (18.7) is finite-dimensional and if the norm of the matrices $\overset{(n)}{P}_{12}(\varphi)$ and $\overset{(n)}{P}_{21}(\varphi)$ is sufficiently small, then the corresponding Riccati equations

$$\sum_{i=1}^{m} \frac{\partial \overset{(n)}{U}_1}{\partial \varphi_i} a_i(\varphi) = \overset{(n)}{P}_1 \overset{(n)}{U}_1 - \overset{(n)}{U}_1 \overset{(n)}{P}_2 - \overset{(n)}{U}_1 \overset{(n)}{P}_{21} \overset{(n)}{U}_1 + \overset{(n)}{P}_{12},$$

$$\sum_{i=1}^{m} \frac{\partial \overset{(n)}{U}_2}{\partial \varphi_i} a_i(\varphi) = \overset{(n)}{P}_2 \overset{(n)}{U}_2 - \overset{(n)}{U}_2 \overset{(n)}{P}_1 - \overset{(n)}{U}_2 \overset{(n)}{P}_{12} \overset{(n)}{U}_2 + \overset{(n)}{P}_{21} \tag{18.8}$$

are solvable with respect to the matrices $\overset{(n)}{U}_1(\varphi)$ and $\overset{(n)}{U}_2(\varphi)$, provided that

$$\begin{aligned} &\|\Omega_0^t(\overset{(n)}{P}_1)\| \le K_1(n) \exp\{-\gamma(n)|t|\}, \; t \in R^1, \\ &\|\Omega_0^t(\overset{(n)}{P}_2)\| \le K_2(n) \exp\{-\gamma(n)|t|\}, \; t \in R^1. \end{aligned} \tag{18.9}$$

This enables us to formulate the following theorem:

Theorem 18.1. *Assume that there exist* $\varepsilon_0 = \text{const} > 0$ *and a sequence of natural numbers* $m_1 < m_2 < \ldots < m_\nu < \ldots$ *such that inequalities (18.9) are true for* $n = m_\nu$ $(\nu = 1, 2, 3, \ldots)$ *and the solutions* $\overset{(m_\nu)}{U_1}$ *and* $\overset{(m_\nu)}{U_2}$ *of equations (18.8) for* $\max\{\|P_{12}(\varphi)\|_{\mathcal{F}_m}, \|P_{21}(\varphi)\|_{\mathcal{F}_m}\} \le \varepsilon_0$ *are such that* $\|\overset{(m_\nu)}{U_i}\|_{\mathcal{F}_m} \le l < 1$ *and* $\lim_{\nu \to \infty} \overset{(m_\nu)}{U_i}(\varphi) = U_i(\varphi)$ $(i = 1, 2)$ *in the norm* $\|\cdot\|_{\mathcal{F}_m}$.

Then system (18.1) is reduced to a system of the form (18.3) by the change of variables (18.2), provided that $U_1(\varphi), U_2(\varphi) \in C^1_{\mathcal{F}_m}(\mathcal{F}_m)$.

The proof of the theorem reduces to the verification of equalities (18.5*). We now prove the first equality. For this purpose, we set $m_\nu = s$ and write the identity

$$\frac{d\overset{(s)}{U}_1(t)}{dt} = \overset{(s)}{P}_1(t) \overset{(s)}{U}_1(t) - \overset{(s)}{U}_1(t) \overset{(s)}{P}_2(t) - \overset{(s)}{U}_1(t) \overset{(s)}{P}_{21}(t) \overset{(s)}{U}_1(t) + \overset{(s)}{P}_{12}(t). \tag{18.10}$$

It suffices to show that relation (18.10) admits the componentwise limit transition as $s \to \infty$ and that the operations of limit transition "$\lim\limits_{s \to \infty}$" and differentiation "$\frac{d}{dt}$" on the left-hand side of this relation can be interchanged.

We set

$$\overset{(s)}{U}_1(t) = [\overset{(s)}{u}_{ij}]_{i,j=1}^{s}, \qquad U_1(t) = [u_{ij}]_{i,j=1}^{\infty},$$

$$P_l(t) = [\overset{(l)}{p}_{ij}]_{i,j=1}^{\infty}, \qquad P_{lk}(t) = [\overset{(lk)}{p}_{ij}]_{i,j=1}^{\infty},$$

where l and k take values 1 and 2. For the sake of definiteness, we write equation (18.10) for the element $\overset{(s)}{u}_{11}$:

$$\frac{d\overset{(s)}{u}_{11}}{dt} = \sum_{i=1}^{s} \overset{(1)}{p}_{1i}\,\overset{(s)}{u}_{i1} - \sum_{i=1}^{s} \overset{(1)}{u}_{1i}\,\overset{(2)}{p}_{i1} - \sum_{r=1}^{s}\sum_{i=1}^{s} \overset{(1)}{u}_{r1}\,\overset{(s)}{u}_{1i}\,\overset{(21)}{p}_{ir} + \overset{(12)}{p}_{11}\,. \tag{18.11}$$

Since

$$\lim_{s\to\infty} \sum_{i=1}^{s} \overset{(1)}{p}_{1i}\,u_{i1} = \sum_{i=1}^{\infty} \overset{(1)}{p}_{1i}\,u_{i1}$$

uniformly in t and

$$\Big|\sum_{i=1}^{s} \overset{(1)}{p}_{1i}\,\overset{(s)}{u}_{i1} - \sum_{i=1}^{s} \overset{(1)}{p}_{1i}\,u_{i1}\Big| \le \sum_{i=1}^{\infty} |\,\overset{(1)}{p}_{1i}\,||u_{i1} - \overset{(s)}{u}_{i1}\,| \le \|P_1\|\|U_1 - \overset{(s)}{U}_1\|,$$

we conclude that

$$\lim_{s\to\infty} \sum_{i=1}^{s} \overset{(1)}{p}_{1i}\,\overset{(s)}{u}_{i1} = \sum_{i=1}^{\infty} \overset{(1)}{p}_{1i}\,u_{i1} \tag{18.12}$$

uniformly in t.

Similarly,

$$\lim_{s\to\infty} \sum_{i=1}^{s} \overset{(s)}{u}_{1i}\,\overset{(2)}{p}_{i1} = \sum_{i=1}^{\infty} u_{1i}\,\overset{(2)}{p}_{i1} \tag{18.13}$$

uniformly in t.

We set

$$\sum_{r=1}^{s}\sum_{i=1}^{s} \overset{(s)}{u}_{r1}\,\overset{(s)}{u}_{1i}\,\overset{(21)}{p}_{ir} = J(t,\ s), \qquad \sum_{r=1}^{s}\sum_{i=1}^{s} u_{r1}u_{1i}\,\overset{(21)}{p}_{ir} = J^*(t,\ s),$$

$$\mathcal{L}(t,\ s) = J^*(t,\ s) - J(t,\ s).$$

The following inequalities are true:

$$|\mathcal{L}(t,\ s)| \le \sum_{r=1}^{s}\sum_{i=1}^{s} |u_{r1}u_{1i} + u_{r1}\overset{(s)}{u_{1i}} - u_{r1}\overset{(s)}{u_{1i}} - \overset{(s)}{u_{r1}}\,\overset{(s)}{u_{1i}}\,||\,\overset{(21)}{p_{ir}}\,|$$

$$\le \sum_{i=1}^{s} |u_{1i} - \overset{(s)}{u_{1i}}\,| \sum_{r=1}^{s} |u_{r1}||\,\overset{(21)}{p_{ir}}\,| + \sum_{i=1}^{s} |\,\overset{(s)}{u_{1i}}\,| \sum_{r=1}^{s} |u_{r1} - \overset{(s)}{u_{r1}}\,||\,\overset{(21)}{p_{1r}}\,|$$

$$\le \|P_{21}\|\Big\{\|U_1\|\sum_{i=1}^{\infty} |u_{1i} - \overset{(s)}{u_{1i}}\,| + \|U_1 - \overset{(s)}{U_1}\,\|\sum_{i=1}^{\infty} |\,\overset{(s)}{u_{1i}}\,|\Big\}$$

$$\le\ 2\varepsilon_0\|U_1 - \overset{(s)}{U_1}\,\|_{\mathcal{F}_m}.$$

It follows from these inequalities that

$$\lim_{s\to\infty} (J(t,\ s) - J^*(t,\ s)) = 0$$

uniformly in t.

At the same time,

$$\lim_{s\to\infty} J^*(t,\ s) = \sum_{r=1}^{\infty}\sum_{i=1}^{\infty} u_{r1}u_{1i}\,\overset{(21)}{p_{ir}}$$

and, hence,

$$\lim_{s\to\infty} J(t,\ s) = \sum_{r=1}^{\infty}\sum_{i=1}^{\infty} u_{r1}u_{1i}\,\overset{(21)}{p_{ir}} \tag{18.14}$$

uniformly in t.

Further, by using relations (18.12), (18.13), and (18.14), we pass to the limit as $s \to \infty$ in relation (18.11) and obtain the identity

$$\frac{du_{11}}{dt} = \sum_{i=1}^{\infty} \overset{(1)}{p_{1i}}\,u_{i1} - \sum_{i=1}^{\infty} u_{1i}\,\overset{(2)}{p_{i1}} - \sum_{r=1}^{\infty}\sum_{i=1}^{\infty} u_{r1}u_{1i}\,\overset{(21)}{p_{ir}} + \overset{(12)}{p_{11}}\,.$$

This completes the proof of the validity of the first equality in (18.5*).

The validity of the second equality in (18.5*) is proved by analogy.

We now consider a system of equations

$$\frac{dx}{dt} = P_1(t)x + P_{12}(t)y, \quad \frac{dy}{dt} = P_{21}(t)x + P_2(t)y, \tag{18.15}$$

where $x, y \in \mathfrak{M}$ and $P_1(t), P_2(t), P_{12}(t)$, and $P_{21}(t)$ are infinite matrices continuous as functions of t.

Assume that $P_1(t)$ and $P_2(t)$ are bounded in the norm $\|\cdot\|_{T_1}$, where T_1 is an arbitrary finite segment from the semiaxis $T^- = (-\infty, 0]$ and the matrizants $\Omega_\tau^t(P_1)$ and $\Omega_\tau^t(P_2)$ are differentiable with respect to $\tau \in T^-$. We also assume that, for all $t, \tau \in T^-$,

$$\|\Omega_\tau^t(P_i)\| \leq K \exp\{-\gamma|t-\tau|\} \quad (i = 1, 2)$$

with constant K and γ, and the matrices $P_{12}(t)$ and $P_{21}(t)$ are such that

$$\max\{\|P_{12}(t)\|, \quad \|P_{21}(t)\|\} \leq \frac{\alpha\gamma}{K^2},$$

where $0 < \alpha = \text{const} < \min\left\{1, \dfrac{1}{K^2}\right\}$.

Let us now formulate conditions under which system (18.15) is decomposable without truncation. Let G be a set of matrices $P(t)$ such that

$$\|P(t)\|_0 = \sup_{t \in T^-} \|P(t)\| \leq \alpha$$

for $t \in T^-$.

We now seek solutions of equations (18.5*) corresponding to system (18.15) and restrict ourselves, e.g., to the second equation. We set

$$U_2(t) = \Omega_0^t(P_2)V_2(t)\Omega_t^0(P_1), \tag{18.16}$$

where $V_2(t) \in G$, and, in addition,

$$\|V_2(t)\|_{T_1} < \infty, \quad \left\|\frac{dV_2}{dt}\right\|_{T_1} < \infty. \tag{18.17}$$

Substituting relation (18.16) in (18.5*), we arrive at the equation

$$\frac{dV_2}{dt} = \Omega_t^0(P_2)P_{21}\Omega_0^t(P_1) - V_2\Omega_t^0(P_1)P_{12}\Omega_0^t(P_2)V_2. \tag{18.18}$$

Since $\lim\limits_{t \to -\infty} V_2(t) = 0$, we can pass from equation (18.18) to the following integral equation:

$$V_2(t) = \int_{-\infty}^{t} \{\Omega_s^0(P_2)P_{21}\Omega_0^s(P_1) - V_2\Omega_s^0(P_1)P_{12}\Omega_0^s(P_2)V_2\}ds.$$

Consider the operator L acting in the set G as follows:

$$LS = \int_{-\infty}^{t} \{\Omega_s^0(P_2)P_{21}\Omega_0^s(P_1) - S\Omega_s^0(P_1)P_{12}\Omega_0^s(P_2)S\}ds.$$

Since, for any $S \in G$, we have $\|LS\|_0 \leq \alpha$, the operator L maps the set G into itself.

For any $S_1,\ S_2 \in G$, we obtain

$$\|LS_1 - LS_2\|_0 \leq \sup_{t\in T^-}\Big\{\int_{-\infty}^{t} \|S_1\|K^2e^{2\gamma s}\|P_{12}\|\|S_1 - S_2\|ds$$

$$+ \int_{-\infty}^{t} \|P_{12}\|K^2e^{2\gamma s}\|S_2\|\|S_1 - S_2\|ds\Big\}$$

$$\leq \sup_{t\in T^-} \alpha^2 e^{2\gamma t}\|S_1 - S_2\|_0$$

$$\leq \alpha\|S_1 - S_2\|_0.$$

Thus, L is a contraction operator. Let us show that the space G is complete in the metric $\rho(S_1, S_2) = \|S_1 - S_2\|_0$. Let $\{\overset{(n)}{P}(t)\}$ be a fundamental sequence in G. Then, for any $\varepsilon > 0$, there exists a number N such that

$$\sup_{t\in T^-}\sup_i \sum_{j=1}^{\infty} \left|\overset{(n)}{p}_{ij}(t) - \overset{(m)}{p}_{ij}(t)\right| < \varepsilon$$

for $n > N$ and $m > n$. This immediately implies that the sequence $\{\overset{(n)}{P}(t)\}$ converges coordinatewise to a matrix $P(t)$.

Let us show that this sequence converges to $P(t)$ in the norm $\|\cdot\|_0$. In the inequality $\sum_{j=1}^{k} |\overset{(n)}{p}_{ij} - \overset{(m)}{p}_{ij}| < \varepsilon$, we pass to the limit as $m \to \infty$ and conclude that $\sum_{j=1}^{k} |\overset{(n)}{p}_{ij} - p_{ij}| < \varepsilon$ for any natural k. If we now pass in this inequality to the limit as $k \to \infty$, then we get

$$\sum_{j=1}^{\infty} |\overset{(n)}{p}_{ij}(t) - p_{ij}(t)| \leq \varepsilon$$

for any natural i, which yields the required convergence. Since $\| \overset{(n)}{P}(t) \|_0 \leq \alpha$ $(n = 1, 2, \ldots)$, we have $\|P(t)\|_0 \leq \alpha$, i.e., $P(t) \in G$.

Since the set G is complete, the equation $LS = S$ possesses a unique solution $V_2 \in G$. It can be represented in the form $V_2(t) = \lim_{n\to\infty} \overset{(n)}{V_2}(t)$, where $\overset{(n)}{V_2}(t)$ are given by the recurrence relation $\overset{(n+1)}{V_2}(t) = L\overset{(0)}{V_2}(t)$ and the role of $\overset{(0)}{V_2}(t)$ can be played by an arbitrary matrix from G.

The matrix $U_2(t)$ given by relation (18.16) is a solution of the second equation in (18.5*) if $V_2(t)$ satisfies inequalities (18.17) and, in addition, $\|U_2(t)\| < 1$ (since $K^2\alpha < 1$).

The first equation in (18.5*) is solved by analogy if, instead of relation (18.16), we set

$$U_1(t) = \Omega_0^t(P_1)V_1(t)\Omega_t^0(P_2).$$

The results established above are summarized in the following theorem:

Theorem 18.2. *Under the assumptions made above, system (18.15) is decomposable on the semiaxis T^- if the solution of equation (18.18) and the solution V_1 of a similar equation corresponding to the first equation in (18.5*) satisfy inequalities (18.17).*

This result can easily be generalized to the case of the right semiaxis.

4. IMPULSIVE SYSTEMS

Countable systems of differential equations with pulse action are introduced by analogy with finite-dimensional impulsive systems. The terminology and notation accepted in the theory of finite-dimensional systems with pulse action are naturally generalized to the case of countable systems of differential equations. In the present chapter, we consider only some problems of the theory of countable systems with pulse action naturally connected with problems close to the problems considered in the previous sections. We present some results from the theory of reducibility of linear pulse systems, prove theorems on the existence of invariant sets and invariant tori for quasilinear systems with pulse action, and substantiate an algorithm used for the construction of periodic solutions of pulse systems with small parameter. For finite-dimensional systems, these and related problems were studied in [Per, SaP1–SaP3].

4.19. Some Results of the Theory of Linear Systems

We represent a countable system of differential equations with pulse action in the form

$$\frac{dx}{dt} = A(t)x, \quad t \neq \tau_j, \qquad \Delta x|_{t=\tau_j} = B_j x(\tau_j - 0), \tag{19.1}$$

where $x \in \mathfrak{M}$, $A(t) = [a_{ik}(t)]_{i,k=1}^{\infty}$ and $B_j = [b_{ik}^{(j)}]_{i,k=1}^{\infty}$ are infinite matrices, $\ldots, \tau_{-2}, \tau_{-1}, \tau_0, \tau_1, \ldots, \tau_j, \ldots$ is an increasing sequence of real numbers from the set $R^1 = (-\infty, \infty)$, and $j = \ldots, -1, 0, 1, \ldots$.

Assume that equation (19.1) is such that the following conditions are satisfied:

1°. The functions $a_{is}(t)$ $(i, s = 1, 2, \ldots)$ are continuous in t for $t \in T = R^1$. Moreover, for any finite segment $T_1 = [a, b] \subset T$,

$$\|A(t)\|_{T_1} = \sup_i \sum_{s=1}^{\infty} \max_{t \in T_1} |\, a_{is}(t)\,| \leq \Psi_{[a,b]} = \text{const} < \infty.$$

2°. The matrices B_j $(j = \ldots, -1, 0, 1, \ldots)$ are constant and bounded in the norm $\|\cdot\|$. Moreover, the matrices $B_j + E$, where E is the identity matrix, are invertible and the inverse matrices $(B_j + E)^{-1}$ are bounded.

3°. The pulses are separated in time, i.e., $\tau_{j+1} - \tau_j \geq c = \text{const} > 0$ $(j = \ldots, -1, 0, 1, \ldots)$.

Then for any point (x_0, t_0) in the domain $D = \mathfrak{M} \times T$, there exits a unique solution $x = x(x_0, t)$, $x_0 = x(x_0, t_0)$ of equation (19.1) extendable to the entire segment T. Furthermore, for $t \geq t_0$, equation (19.1) is equivalent to the equation

$$x(t, x_0) = x_0 + \int_{t_0}^{t} A(\sigma)x(\sigma, x_0)d\sigma + \sum_{t_0 < \tau_j < t} B_j x(\tau_j - 0, x_0). \tag{19.2}$$

The existence of the fundamental matrix of solutions of equation (19.1) readily follows from conditions 1°–3° if we prove that the matrizant $X(t, t_0)$ of this equation is bounded in the norm $\|\cdot\|_{T_1}$. For this purpose, we assume that the times of pulses are arranged in the segment T_1 as follows:

$$a < \tau_{j-l} < \ldots < \tau_{j-2} < \tau_{j-1}$$
$$< \tau_j < \ldots < \tau_{j+k} < \tau_{j+k+1} < \ldots < \tau_{j+s} < b$$

and split the segment $[a, b]$ into $s + l + 2$ intervals. For the sake of definiteness, we assume that $t_0 \in (\tau_{j-1}, \tau_j]$ and $t \in (\tau_{j+k}, \tau_{j+k+1})$. Then

$$X(t, t_0) = U(t, \tau_{j+k})(E + B_{j+k})U(\tau_{j+k}, \tau_{j+k-1}) \ldots (E + B_j)U(\tau_j, t_0),$$

where $U(t, \tau)$ is the matrizant of equation (19.1) without pulses, i.e., for $\Delta x = 0$.

For t lying to the left of t_0, e.g., for $t \in (\tau_{j-l}, \tau_{j-l+1}]$, we can write

$$\begin{aligned} X(t, t_0) &= U(t, \tau_{j-l+1})(E + B_{j-l+1})^{-1}U(\tau_{j-l+1}, \tau_{j-l+2}) \\ &\quad \times \ldots \times (E + B_{j-1})^{-1}U(\tau_{j-1}, t_0). \end{aligned}$$

This yields

$$\|X(t, t_0)\|_{T_1} \leq \exp\{\Psi_{[a,b]}(b - a)(k + 2)\} \max_{j \leq i \leq j+k} \|E + B_i\|^{k+1},$$

$$t \in (\tau_{j+k}, \tau_{j+k+1}],$$

$$\|X(t,t_0)\|_{T_1} \leq \exp\{\Psi_{[a,b]}(b-a)l\} \max_{j-l+1\leq i\leq j-1} \|(E+B_i)^{-1}\|^{l-1},$$

$$t \in (\tau_{j-l}, \tau_{j-l+1}].$$

Note that the constants on the right-hand sides of these inequalities are positive. Since the number of analyzed intervals from T_1 is finite, there exists a constant $0 < K^0 < \infty$ such that

$$\|X(t,t_0)\|_{T_1} \leq K^0 \quad \forall t \in T_1.$$

Definition 19.1. *A matrix $L(t)$ is called a Lyapunov matrix if it satisfies the following conditions:*

(i) it is piecewise continuous on the segment T and has discontinuities of the first kind at the points τ_j ($j = \ldots, -1, 0, 1, \ldots$);

(ii) it is invertible on T, continuously differentiable on the set $T \setminus \{\tau_j\}$, and such that

$$\max\{\|L(t)\|_{T_1};\ \|L^{-1}(t)\|_{T_1};\ \|\frac{d}{dt}L(t)\|_{T_1}\} < \infty$$

for any finite segment $T_1 \subset T$.

In the case of a finite segment T, it suffices to set $T = T_1$ and indicate finitely many discontinuity points τ_j.

We say that equation (19.1) is reducible on the segment T to an equation of the form

$$\frac{dy}{dt} = Py, \quad y \in \mathfrak{M}, \tag{19.3}$$

where P is a constant matrix from the space $\mathcal{B}$, in the Lyapunov sense if there exists a Lyapunov matrix $L(t)$ such that any solution $x = x(t)$ of equation (19.1) can be represented in the form $x(t) = L(t)y(t)$, $t \in T$, where $y(t)$ is a solution of equation (19.3).

Let us now prove an analog of the Erugin theorem for impulsive systems of equations.

Theorem 19.1. *Equation (19.1) is reducible to an equation of the form (19.3) in the Lyapunov sense if and only if its fundamental matrix $X(t)$ can be represented in the form $X(t) = L(t)\exp\{Pt\}$, where $L(t)$ is a Lyapunov matrix.*

Proof. Assume that there exists a fundamental matrix $X(t)$ of equation (19.1) such that

$$X(t) = L(t)\exp\{Pt\},$$

where $L(t)$ is a Lyapunov matrix. In equation (19.1), we perform the change of variables

$$x = L(t)y,$$

where $L(t) = X(t)e^{-Pt}$ and the function $y(t)$ is such that, on every finite segment $T_1 \subset T$, the quantities $\|y(t)\|$ and $\left\|\dfrac{dy(t)}{dt}\right\|$ are bounded by constants P_1 and P_2 dependent on T_1. Then

$$\frac{dx}{dt} = \frac{d}{dt}(L(t))y + L(t)\frac{dy}{dt}, \quad t \neq \tau_j.$$

This yields

$$\frac{d}{dt}\Big(X(t)\exp\{-Pt\}\Big)y + X(t)\exp\{-Pt\}\frac{dy}{dt} = A(t)X(t)\exp\{-Pt\}y.$$

The fundamental matrix $X(t)$ is bounded in the norm $\|\cdot\|_{T_1}$ for any finite segment T_1. Therefore, this is also true for $\dfrac{dX(t)}{dt}$. In this case, we arrive at the equations

$$\begin{aligned} X(t)\exp\{-Pt\}\frac{dy}{dt} + \frac{dX}{dt}\exp\{-Pt\}y - X(t)\exp\{-Pt\}Py& \\ = A(t)X(t)\exp\{-Pt\}y, \quad t \neq \tau_j,& \end{aligned} \tag{19.4}$$

$$\Delta X(t)|_{t=\tau_j}\exp\{-P\tau_j\}y(\tau_j) = B_jX(\tau_j)\exp\{-P\tau_j\}y(\tau_j).$$

Since $X(t)$ is a fundamental matrix of the system of equations (19.1), we have

$$\frac{dX}{dt} = A(t)X \quad \text{for} \quad t \neq \tau_j, \quad \Delta X|_{t=\tau_j} = B_jX.$$

By using these equalities, we represent system (19.4) in the form (19.3) because the second equation in system (19.4) is identically satisfied for all $y(t) \in \mathfrak{M}$. It is clear that the solutions of the last equation from the space $\mathfrak{M}$ are bounded together with their derivatives in the norm $\|\cdot\|_{T_1}$ for any finite segment T_1.

The proof of necessity repeats the corresponding part of the proof of Theorem 14.1.

Remark 19.1. In the case where a matrix $\Phi(t)$ that reduces equation (19.1) to an equation of the form (19.3) is not a Lyapunov matrix, one can also find a fundamental matrix $X(t)$ of equation (19.1) such that $X(t) = \Phi(t)\exp\{Pt\}$.

Indeed, let $y(t) = y(t, t_0, y^0)$ be a solution of equation (19.3). We also consider the solution of equation (19.1) $x = x(t, t_0, \Phi(t_0)y^0)$. There exists a solution $\tilde{y}(t)$ of equation (19.3) such that $x(t, t_0, \Phi(t_0)y^0) = \Phi(t)\tilde{y}(t)$. At the same time, for $t = t_0$, we have $\Phi(t_0)y^0 = \Phi(t_0)\tilde{y}(t_0)$ and, in view of the invertibility of the matrix $\Phi(t)$, we have $y^0 = \tilde{y}(t_0)$. This means that $\tilde{y}(t) = y(t)$ identically on T and, hence, for any solution $y(t)$ of equation (19.3), the quantity $\Phi(t)y(t)$ is a solution of equation (19.1). At the same time, for any solution $y(t)$, we have $y(t) = e^{Pt}c$, $c \in \mathfrak{M}$. Since

$$(\Phi(t)e^{Pt})c = \Phi(t)(e^{Pt}c),$$

we conclude that any solution

$$x(t) = (\Phi(t)e^{Pt})c = X(t)c,$$

i.e., $X(t)$ is a fundamental matrix of equation (19.1).

Let us now present an analog of the theorem on reducibility (Theorem 15.1) for differential equations with pulses of the form

$$\frac{dx}{dt} = A(t)x,\ t \neq \tau_j, \quad \Delta x|_{t=\tau_j} = B_j x(\tau_j - 0), \tag{19.5}$$

where $A(t)$ is an ω-periodic matrix and one can always find a natural number p such that

$$B_{j+p} = B_j \quad \text{and} \quad \tau_{j+p} = \tau_j + \omega$$

for all integer j; the times of pulses are enumerated as follows: $t_0 < \tau_1 < \ldots < \tau_p < \omega + t_0$. In addition, equation (19.5) satisfies the same conditions as equation (19.1).

The truncated equation corresponding to equation (19.5) has the form

$$\frac{d\overset{(n)}{x}}{dt} = \overset{(n)}{A}(t)\overset{(n)}{x},\ t \neq \tau_j, \quad \Delta\overset{(n)}{x}|_{t=\tau_j} = \overset{(n)}{B_j}\overset{(n)}{x}(\tau_j - 0). \tag{19.6}$$

It is known [SaP3] that, for any natural n, the finite-dimensional equation (19.6) is reducible. At the same time, no direct analogs of the Floquet–Lyapunov theorem can be formulated for equation (19.6) in the infinite-dimensional space.

Without loss of generality, we set $T = [t_0, +\infty)$ and enumerate the times of pulses τ_j by natural numbers 1, 2, 3, ... The solution $\overset{(n)}{x}(t, \overset{(n)}{x^0}, t_0)$ of equation (19.6) satisfies the relation

$$\overset{(n)}{x}\Big(t, \overset{(n)}{x^0}, t_0\Big) = \overset{(n)}{x^0} + \int_{t_0}^{t} \overset{(n)}{A}(\sigma)\overset{(n)}{x}\Big(\sigma, \overset{(n)}{x^0}, t_0\Big)d\sigma$$

$$+ \sum_{t_0<\tau_j<t} \overset{(n)}{B_j}\overset{(n)}{x}\Big(\tau_j - 0, \overset{(n)}{x^0}, t_0\Big). \qquad (19.7)$$

Below, we present the statement of Lemma 2.1 from [SaP3].

Assume that, for $t \geq t_0$, a nonnegative piecewise-continuous function $u(t)$ satisfies the inequality

$$u(t) \leq c + \int_{t_0}^{t} v(\tau)u(\tau)d\tau + \sum_{t_0<\tau_j<t} \beta_j u(\tau_j), \qquad (19.\text{V})$$

where $c \geq 0$, $\beta_j \geq 0$, $v(\tau) > 0$, and τ_j are the discontinuity points (of the first kind) of the function $u(t)$. Then the function $u(t)$ satisfies the inequality

$$u(t) \leq c \prod_{t_0<\tau_j<t} (1 + \beta_j) \exp\Big\{\int_{t_0}^{t} v(\tau)d\tau\Big\}.$$

In what follows, this assertion is called Statement 19.V.
First, we prove the following auxiliary fact:

Lemma 19.1. *The sequence $\{\overset{(n)}{x}\}$ contains a subsequence that converges to the solution $x(t, x^0, t_0)$ of equation (19.5) uniformly in the coordinates on each segment $[t_0, \tau_1]$, $(\tau_{j-1}, \tau_j]$ $(j = 2, 3, 4, \ldots)$.*

Proof. By using (19.7), we obtain the following estimate for the sequence $\{\overset{(n)}{x}\}$ on the segment $[t_0, \tau_1]$:

$$\|\overset{(n)}{x}\| \leq \Big\|\overset{(n)}{x^0}\Big\| \exp\{\Psi_{[t_0,\tau_1]}(\tau_1 - t_0)\} = K_0.$$

Further, in the segment $(\tau_j, \tau_{j+1}]$, we get

$$\|\overset{(n)}{x}\| \le \left\|\overset{(n)}{x^0}\right\| + \int_{t_0}^{t} \Psi_{[t_0,\tau_{j+1}]}\|\overset{(n)}{x}(\sigma)\|d\sigma$$

$$+ \sum_{1\le k\le j} \|B_k\|\,\|\overset{(n)}{x}(\tau_k - 0)\|,$$

whence, in view of Statement 19.V, we conclude that

$$\|\overset{(n)}{x}\| \le \left\|\overset{(n)}{x^0}\right\| \exp\{\Psi_{[t_0,\tau_{j+1}]}(\tau_{j+1} - t_0)\} \prod_{1\le k\le j} (1 + \|B_k\|)$$

$$= K_j = \text{const} < \infty.$$

This means that the sequence $\{\overset{(n)}{x}\}$ is uniformly bounded in each segment $[t_0, \tau_1]$, $(\tau_j, \tau_{j+1}]$ $(j = 1, 2, 3, 4, \ldots)$. Moreover, as follows from the inequality

$$\|\overset{(n)}{x}(t_1) - \overset{(n)}{x}(t_2)\| \le \left|\int_{t_1}^{t_2} \|\overset{(n)}{A}(\sigma)\|\,\|\overset{(n)}{x}(\sigma)\|d\sigma\right| \le \Psi_{[t_0,\tau_{j+1}]}K_j\,|\,t_1 - t_2\,|\,,$$

$$t_1,\ t_2 \in (\tau_j, \tau_{j+1}] \quad (j = 0, 1, 2, \ldots),$$

where $\tau_0 = t_0$, this sequence is equicontinuous on each of these segments.

In this case, on the segment $[t_0, \tau_1]$, we select a convergent subsequence $\overset{(\alpha_1)}{x_1}, \overset{(\alpha_2)}{x_1}, \ldots, \overset{(\alpha_s)}{x_1}, \ldots$ of the sequence $\{\overset{(n)}{x}\}$ (of the first coordinates of $\overset{(n)}{x}$). In the sequence of the second coordinates $\overset{(\alpha_1)}{x_2}, \overset{(\alpha_2)}{x_2}, \ldots, \overset{(\alpha_s)}{x_2}, \ldots$, we select a convergent subsequence $\overset{(\beta_1)}{x_2}, \overset{(\beta_2)}{x_2}, \ldots, \overset{(\beta_s)}{x_2}, \ldots$. This process is repeated infinitely many times. By using the procedure of diagonalization, we select a subsequence $\overset{(\alpha_1)}{x}, \overset{(\beta_2)}{x}, \overset{(\gamma_3)}{x}, \ldots$ of the sequence $\{\overset{(n)}{x}\}$ that converges uniformly in the coordinates on the segment $[t_0, \tau_1]$. This subsequence is denoted by $\{\overset{(n_1)}{x}\}$.

Further, we consider the segment $(\tau_1, \tau_2]$ and the sequence $\{\overset{(n_1)}{x}\}$ defined on this segment. Acting in exactly the same way as above, we select a subsequence $\{\overset{(n_2)}{x}\}$ of this sequence uniformly convergent on this segment with respect to the coordinates. We repeat this procedure infinitely many times. As a result of the procedure of diagonalization, we arrive at a subsequence $\{\overset{(s)}{x}\}$ of the sequence

$\{\overset{(n)}{x}\}$ that converges uniformly in the coordinates on each segment indicated in the conditions of Lemma 19.1 to a function $z(t)$. Moreover, it is clear that $z(t) \in \mathfrak{M}$ for any $t \in T$.

It remains to prove that $z(t) \equiv x(t, x^0, t_0)$, $t \in T$. We fix an arbitrary value of $t \in T$ and denote a segment containing t_0 and t by $[t_0, b] = T_2$.

Clearly,

$$\frac{d\overset{(s)}{x}_i(t)}{dt} = \sum_{k=1}^{s} \overset{(s)}{a}_{ik}(t)\overset{(s)}{x}_k(t), \quad t \neq \tau_j, \qquad \Delta\overset{(s)}{x}_i\big|_{t=\tau_j} = \sum_{k=1}^{s} \overset{(s)}{b}_{ik}\overset{(s)}{x}(\tau_j - 0).$$

If we now set $\overset{(s)}{a}_{ik}(t)\overset{(s)}{x}_k(t) = 0$ for $k > s$, then we get

$$\begin{aligned}\overset{(s)}{a}_{ik}(t)\overset{(s)}{x}_k(t) &\leq \sum_{k=1}^{\infty} |\overset{(s)}{a}_{ik}(t)|\, \|\overset{(s)}{x}(t)\| \leq K^0 \sum_{k=1}^{\infty} |a_{ik}(t)| \\ &\leq K^0 \sum_{k=1}^{\infty} \max_{T_2} |a_{ik}(t)| \leq K^0 \|A(t)\|_{T_2} \leq K^0 \Psi_{[t_0,b]}.\end{aligned}$$

This enables us to use Lemma 12.1 because the sequence $\overset{(s)}{a}_{ik}(t)\overset{(s)}{x}_k(t)$ converges uniformly in $t \in T_2$. Indeed,

$$\begin{aligned}\left| a_{ik}(t)z_k(t) - \overset{(s)}{a}_{ik}(t)\overset{(s)}{x}_k(t) \right| &\leq |a_{ik}(t)| \left| z_k(t) - \overset{(s)}{x}_k(t) \right| \\ &\leq \|A(t)\|_{T_2} \left| z_k(t) - \overset{(s)}{x}_k(t) \right|,\end{aligned}$$

and $\overset{(s)}{x}_k(t) \to z_k(t)$ uniformly in $t \in T_2$. Thus, we get

$$\lim_{s\to\infty} \frac{d\overset{(s)}{x}_i(t)}{dt} = \frac{d}{dt}\Big(\lim_{s\to\infty} \overset{(s)}{x}_i(t)\Big) = \frac{dz_i(t)}{dt} = \sum_{k=1}^{\infty} a_{ik}(t)z_k(t)$$

for $t \neq \tau_j$, $t \in T_2$. Finally, we obtain

$$\lim_{s\to\infty} \Delta\overset{(s)}{x}_i = \lim_{s\to\infty} \sum_{k=1}^{s} \overset{(s)}{b}_{ik}\overset{(s)}{x}_k(\tau_j - 0) = \sum_{k=1}^{\infty} b_{ik}z_k(\tau_j - 0),$$

which completes the proof of Lemma 19.1 because $z(t)$ is a solution of equation (19.5) and $z(t_0) = x^0$, whence $z(t) \equiv x(t, x^0, t_0)$ for all $t \in T_2$, including the fixed value of t.

Let $L^{(n)}$ be a matrix reducing equation (19.6) to an equation of the form

$$\frac{d\overset{(n)}{y}}{dt} = \overset{(n)}{P}\,\overset{(n)}{y}. \tag{19.8}$$

The following theorem is true:

Theorem 19.2. *Assume that the sequence of matrices* $\{\overset{(n)}{L}(t)\}$ *that reduces the truncated equations (19.6) to equations of the form (19.8) and the sequence of inverse matrices* $\{\overset{(n)}{L^{-1}}(t)\}$ *are uniformly regular on each segment* $[t_0, \tau_1]$, $(\tau_1, \tau_2], \ldots$ *and, furthermore, the sequence* $\{\overset{(n)}{P}\}$ *is regular. Then* $\lim\limits_{n\to\infty} \overset{(n)}{P} = P$ *and* $\lim\limits_{n\to\infty} \overset{(n)}{L}(t) = L(t)$ *in the weak sense and equation (19.5) is reduced to an equation of the form (19.3) with the help of the limiting matrix* $L(t)$.

Proof. Let t_1 be an arbitrarily large fixed value of t. We denote $T_0 = \{[t_0, \tau_1], (\tau_1, \tau_2], \ldots, (\tau_{l-1}, \tau_l], (\tau_l, t_1]\}$, where $\tau_i\ (i = \overline{1, l})$ are the times of pulses lying in the segment $[t_0, t_1]$.

Let $x = x(t, x_0)$, $x(t_0, x_0) = x_0 \in \mathfrak{M}$, be a solution of equation (19.5). By virtue of Lemma 19.1, the sequence $\overset{(n)}{x} = \overset{(n)}{x}(t, \overset{(n)}{x_0})$, $\overset{(n)}{x}(t_0, \overset{(n)}{x_0}) = \overset{(n)}{x_0}$, of solutions of truncated equations of the form (19.6) always contains a subsequence $\{\overset{(r)}{x}(t, \overset{(r)}{x_0})\}$ that converges to $x(t, x_0)$ (uniformly in the coordinates) on every segment from the set T_0.

For each number r, there exists a reducing matrix $\overset{(r)}{L}(t)$ such that the change of variables $\overset{(r)}{x}(t) = \overset{(r)}{L}(t)\overset{(r)}{y}(t)$ reduces the truncated equation (19.6) of order r to an equation of the form (19.8) with $n = r$. This means that there exists a solution $\overset{(r)}{y}(t)$ of equation (19.8) (with $n = r$) such that, on every segment from the set T_0, we have

$$\overset{(r)}{x}(t, x_0) = \overset{(r)}{L}(t)\overset{(r)}{y}(t).$$

Let us show that, on every segment indicated, the sequence $\{\overset{(r)}{y}(t)\}$ converges uniformly in t in the weak sense.

Denote $\overset{(n)}{L^{-1}}(t) = [l_{ij}^{-1}(t)]_{i,j=1}^{\infty}$. Since $\overset{(r)}{y}(t) = \overset{(r)}{L^{-1}}(t) \times \overset{(r)}{x}(t, x_0)$, the sth coordinate of the vector $\overset{(r)}{y}(t)$ has the form

$$\overset{(r)}{y}_s(t) = \sum_{i=1}^{\infty} \overset{(r)}{l_{si}^{-1}}(t) \overset{(r)}{x_i}(t, \overset{(r)}{x_0}), \tag{19.9}$$

where $\overset{(r)}{l_{si}^{-1}}(t)\overset{(r)}{x_i}(t, \overset{(r)}{x_0})$ should be set equal to zero for $i > r$. At the same time,

$$\overset{(r)}{x}(t, \overset{(r)}{x_0}) = \overset{(r)}{x_0} + \int_{t_0}^{t} \overset{(r)}{A}(\sigma) \overset{(r)}{x}(\sigma, \overset{(r)}{x_0}) d\sigma + \sum_{t_0 < \tau_j < t} \overset{(r)}{B_j} \overset{(r)}{x}(\tau_j - 0, \overset{(r)}{x_0}),$$

whence it follows that

$$\begin{aligned} \|\overset{(r)}{x}(t, \overset{(r)}{x_0})\| \;&\leq\; \|x_0\| + \int_{t_0}^{t} \|A(\sigma)\| \, \|\overset{(r)}{x}(\sigma, \overset{(r)}{x_0})\| d\sigma \\ &\quad + \sum_{t_0 < \tau_j < t} \|B_j\| \, \|\overset{(r)}{x}(\tau_j - 0, \overset{(r)}{x_0})\|. \end{aligned}$$

By using Statement 19.V, we conclude that

$$\begin{aligned} \|\overset{(r)}{x}(t, \overset{(r)}{x_0})\| \;&\leq\; \|x_0\| \prod_{1 \leq j \leq l} (1 + \|B_j\|) \exp\{\Psi_{[t_0, t_1]}(t_1 - t_0)\} \\ &= C_{T_0} = \text{const} < \infty \end{aligned}$$

for any $t \in [t_0, t_1]$. Then

$$\sum_{i=1}^{\infty} \left| \overset{(r)}{l_{si}^{-1}}(t) \overset{(r)}{x_i}(t, \overset{(r)}{x_0}) \right| \leq C_{T_0} \sum_{i=1}^{\infty} \left| \overset{(r)}{l_{si}^{-1}}(t) \right|,$$

where the series on the right-hand side converges uniformly in t and r for any segment from the set T_0. According to the condition of the theorem, the sequence $\{\overset{(n)}{L^{-1}}(t)\}$ is uniformly regular on each segment indicated and, hence, it uniformly converges on each of these segments in the weak sense to a matrix denoted by $L^{-1}(t) = [l_{ij}^{-1}(t)]_{i,j=1}^{\infty}$.

By virtue of Lemma 12.1, sequence (19.9) converges to a function $y_s(t)$ uniformly in t on every segment from the set T_0 that does not contain the times of pulses. The required statement is proved. We also have $\{\overset{(r)}{y}\} \to y(t)$ weakly for $t \in T_0$ and, in addition, $y(t) = L^{-1}(t)x(t, x_0)$.

According to the condition of the theorem, the sequence of matrices $\{\overset{(n)}{L}(t)\}$ converges to a matrix $L(t)$ and, hence, $x(t, x_0) = L(t)y(t)$.

It remains to show that $y(t)$ is a solution of equation (19.3), where $P = \lim\limits_{r\to\infty} \overset{(r)}{P}$ in the weak sense. This fact is proved in exactly the same way as in Theorem 14.1.

Note that, for every solution $x(t, x_0)$ of equation (19.5), one can indicate, generally speaking, its own sequence $\{\overset{(s)}{x}(t)\}$ of solutions of truncated equations of the form (19.6) that converges to this solution in the weak sense. Therefore, for every solution $x(t)$ of equation (19.5), one can indicate its own sequences $\{\overset{(s)}{L}(t)\}$ and $\{\overset{(s)}{P}\}$ corresponding to the subsequence $\{\overset{(s)}{x}(t)\}$ that converges to $x(t)$. However, any subsequence of a regular (uniformly regular) matrix sequence is regular (uniformly regular) itself and weakly converges to the same limit. This means that, for all solutions $x(t)$ of equation (19.5), the reducing matrix $L(t)$ and the matrix P in equation (19.3) are identical and coincide with the matrices specified in the conditions of the theorem.

4.20. Integral Sets and Invariant Tori

Consider a system of equations

$$\frac{d\varphi}{dt} = a(\varphi), \quad \frac{dy}{dt} = A(\varphi)y + c(\varphi) \quad \text{for} \quad t \neq t_j, \qquad (20.1)$$
$$\Delta y|_{t=t_j} = J_j(\varphi_{t_j}(\varphi)) \quad (j = \ldots, -1, 0, 1, \ldots),$$

where $\varphi = (\varphi_1, \varphi_2, \ldots) \in \mathfrak{M}$, $y = (y_1, y_2, \ldots) \in \mathfrak{M}$, the matrix $A(\varphi)$ and the functions $c(\varphi)$ and $J_j(\varphi)$ belong to the space $C^0(\mathcal{F}_\infty)$ for all integer j, $a(\varphi) \in C^0_{\text{Lip}}(\mathcal{F}_\infty)$, and $\varphi = \varphi_t(\varphi)$ is a solution of the equation $\dfrac{d\varphi}{dt} = a(\varphi)$ such that $\varphi_0(\varphi) = \varphi \in \mathcal{F}_\infty$.

Assume that the matrix $A = [a_{is}(\varphi)]_{i,s=1}^{\infty}$ satisfies the inequality

$$\sum_{s=1}^{\infty} \sup_{\varphi\in\mathcal{F}_\infty} |a_{is}(\varphi)| \leq a^0 = \text{const} < \infty \quad (i = 1, 2, \ldots) \qquad (20.2)$$

and the times of pulses are separated, i.e., for all integer j,

$$t_{j+1} - t_j \geq \theta, \qquad (20.3)$$

where θ is a real positive number.

Definition 20.1. *A set* $\Gamma_0(t)\colon y = u(\varphi, t)$ *is called an integral set of the system of equations (20.1) if* $u(\varphi, t)$ *is a function bounded in the norm* $\|\cdot\|$ *and such that* $y = u(\varphi_t(\varphi), t)$ *is a bounded solution of this system, i.e.,*

$$\frac{du(\varphi_t(\varphi), t)}{dt} = A(\varphi_t(\varphi))u(\varphi_t(\varphi), t) + c(\varphi_t(\varphi)), \quad t \neq t_j, \ t \in R^1,$$

$$\Delta u(\varphi_t(\varphi), t)|_{t=t_j} = J_j(\varphi_{t_j}(\varphi)) \quad (j = \ldots, -1, 0, 1, \ldots).$$

The sum of the series $\theta_1 = \sum\limits_{-\infty<i<\infty} f_i(t)$ is determined as the sum of the following two series:

$$\theta_2 = \sum_{i=1}^{\infty} f_i(t) \quad \text{and} \quad \theta_3 = \sum_{-i=0}^{\infty} f_i(t), \quad i \leq 0,$$

provided that they are convergent for the corresponding values of t. It is clear that the sum θ_1 does not depend on the point at which the series θ_1 is split into two parts. We also say that the series θ_1 converges uniformly in t if this is true for the series θ_2 and θ_3.

We now prove the following assertion:

Theorem 20.1. *Assume that conditions (20.2) and (20.3) are satisfied for the system of equations (20.1) and* $\|J_j(\varphi)\| \leq J = \text{const} < \infty$ *for all integer* j. *If the Green function* $G_0(\tau, \varphi)$ *of the problem of invariant tori for the equation* $\dfrac{dy}{dt} = A(\varphi_t)y$ *exists and satisfies inequality (10.5), then system (20.1) possesses an integral set* Γ_0 *specified by the formula*

$$u(\varphi, t) = \int_{-\infty}^{\infty} G_0(\tau, \varphi)c(\varphi_\tau(\varphi))\, d\tau + \sum_{-\infty<t_j<\infty} G_0(t_j - t, \varphi)J_j(\varphi_{t_j - t}(\varphi)),$$

where $u(\varphi, t)$ *is a function* 2π*-periodic in* φ_i $(i = 1, 2, \ldots)$.

Proof. Under the conditions of the theorem, the system of equations (20.1) possesses a unique solution defined on the entire real axis for all initial values $\varphi, y_0 \in \mathfrak{M}$ and $t_0 \in R^1$. Let us show that the function

$$y(t, \varphi) = \int_{-\infty}^{\infty} G_t(\tau, \varphi)c(\varphi_\tau(\varphi))d\tau + \sum_{-\infty<t_j<\infty} G_t(t_j, \varphi)J_j(\varphi_{t_j}(\varphi)) \tag{20.4}$$

specifies a family of bounded solutions of the system of equations (20.1) dependent on φ (regarded as a parameter).

We consider a series

$$\sum_{-\infty<t_j<\infty} G_t(t_j,\varphi)J_j(\varphi_{t_j}(\varphi)) \tag{20.5}$$

and show that, for all $t \neq t_j$, it converges uniformly in t on each interval (t_j, t_{j+1}), where j is an arbitrary integer number. For the sake of definiteness, we assume that $t \in (t_p, t_{p+1})$. Series (20.5) can be represented as the sum of the two series

$$Z_1 = \sum_{-\infty<t_j\le t_p} G_t(t_j,\varphi)J_j(\varphi_{t_j}(\varphi)),$$

$$Z_2 = \sum_{t_p<t_j<\infty} G_t(t_j,\varphi)J_j(\varphi_{t_j}(\varphi)),$$

provided that both these series converge. In the case where these series are uniformly convergent on the interval (t_p, t_{p+1}), series (20.5) is also uniformly convergent on this interval.

We have

$$\begin{aligned}
\|Z_1\| &\le \sum_{-\infty<t_j\le t_p} JK\exp\{-\gamma(t-t_j)\} \\
&\le JK\exp\{-\gamma t_p\}\sum_{j=p}^{-\infty}\exp\{\gamma t_j\} \\
&\le JK\exp\{-\gamma t_p\} \\
&\quad\times\Big(\exp\{\gamma t_p\} + \exp\{\gamma(t_p-\theta)\} + \exp\{\gamma(t_p-2\theta)\} + \ldots\Big) \\
&= \frac{JK}{1-\exp\{-\gamma\theta\}},
\end{aligned}$$

$$\begin{aligned}
\|Z_2\| &\le JK\exp\{\gamma t_{p+1}\}\sum_{j=p+1}^{\infty}\exp\{-\gamma t_j\} \\
&\le JK\exp\{\gamma t_{p+1}\}\Big(\exp\{-\gamma t_{p+1}\} + \exp\{-\gamma(t_{p+1}+\theta)\} \\
&\quad+\exp\{-\gamma(t_{p+1}+2\theta)\} + \ldots\Big) \\
&= \frac{JK}{1-\exp\{-\gamma\theta\}}.
\end{aligned}$$

These inequalities prove that the series Z_1 and Z_2 are uniformly convergent on the interval (t_p, t_{p+1}). Thus, one can easily show that the series

$$Z_3 = \sum_{-\infty<t_j<\infty} \frac{d}{dt}\Big(G_t(t_j,\varphi)\Big) J_j(\varphi_{t_j}(\varphi))$$

is also uniformly convergent on the interval (t_p, t_{p+1}).

Indeed,

$$\begin{aligned} \|Z_3\| &= \left\| \sum_{-\infty<t_j<\infty} A(\varphi_t(\varphi)) G_t(t_j,\varphi) J_j(\varphi_{t_j}(\varphi)) \right\| \\ &\leq JKa^0 \sum_{-\infty<t_j<\infty} \exp\{-\gamma\,|\,t-t_j\,|\,\}, \end{aligned}$$

and we arrive at the case analyzed above.

By using Theorem 7.1, we can write the following formula for $\dfrac{dy}{dt}$ (valid for $t \neq t_j$):

$$\begin{aligned} \frac{dy}{dt} &= \frac{d}{dt}\int_{-\infty}^{\infty} G_t(\tau,\varphi) c(\varphi_\tau(\varphi)) d\tau + \frac{d}{dt} \sum_{-\infty<t_j<\infty} G_t(t_j,\varphi) J_j(\varphi_{t_j}(\varphi)) \\ &= A(\varphi_t(\varphi)) \int_{-\infty}^{\infty} G_t(\tau,\varphi) c(\varphi_\tau(\varphi)) d\tau + c(\varphi_t(\varphi)) \\ &\quad + \sum_{-\infty<t_j<\infty} \frac{dG_t(t_j,\varphi)}{dt} J_j(\varphi_{t_j}(\varphi)) \\ &= A(\varphi_t(\varphi)) y(t,\varphi) + c(\varphi_t(\varphi)). \end{aligned}$$

Hence, the function specified by relation (20.4) satisfies the first equation in (20.1).

Let us show that this function satisfies the second equation of this system for any integer j.

For definiteness, we fix the value of j. We have

$$\Delta y\Big|_{t=t_j}$$

$$= \int_{-\infty}^{\infty} G_t(\tau,\varphi)c(\varphi_\tau(\varphi))d\tau\Big|_{t=t_j} + \sum_{-\infty<t_\nu\le t_{j-1}} G_t(t_\nu,\varphi)J_\nu(\varphi_{t_\nu}(\varphi))\Big|_{t=t_j}$$

$$+ \sum_{t_{j+1}\le t_\nu<\infty} G_t(t_\nu,\varphi)J_\nu(\varphi_{t_\nu}(\varphi))\Big|_{t=t_j} + G_t(t_j,\varphi)J_j(\varphi_{t_j}(\varphi))\Big|_{t=t_j}$$

$$= G_t(t_j,\varphi)|_{t=t_j} J_j(\varphi_{t_j}(\varphi))$$

$$= EJ_j(\varphi_{\varphi_{t_j}}(\varphi)) = J_j(\varphi_{t_j}(\varphi)).$$

These relations show that the function given by relation (20.4) satisfies the second equation of system (20.1). We also have

$$\|y(t,\varphi)\| \le \frac{2K}{\gamma}\Big(C^0 + \frac{\gamma J}{1-\exp\{-\gamma\theta\}}\Big),$$

where $\|c(\varphi)\| \le C^0 = \text{const} < \infty$.

Consider the formula

$$\begin{aligned} u(\varphi_t(\varphi),t) &= \int_{-\infty}^{\infty} G_0(\tau,\varphi_t(\varphi))c(\varphi_\tau(\varphi_t(\varphi)))d\tau \\ &\qquad + \sum_{-\infty<t_j<\infty} G_0(t_j-t,\varphi_t(\varphi))J_j(\varphi_{t_j-t}(\varphi_t(\varphi))) \\ &= \int_{-\infty}^{\infty} G_t(\tau,\varphi))c(\varphi_\tau(\varphi))d\tau + \sum_{-\infty<t_j<\infty} G_t(t_j,\varphi)J_j(\varphi_{t_j}(\varphi)). \end{aligned}$$

Its right-hand side coincides with the right-hand side of (20.4) and, therefore, $u(\varphi_t(\varphi),t) = y(t,\varphi)$. Moreover, the function $u(\varphi,t)$ is bounded in norm.

We now consider a system of equations of the form

$$\frac{d\varphi}{dt} = a(\varphi_1,\psi), \quad \frac{dy}{dt} = A(\varphi_1,\psi)y + c(\varphi_1,\psi), \quad t \ne t_j(\varphi),$$

$$\Delta y|_{t=t_j(\varphi)} = J_j(\varphi_{t_j(\varphi)}(\varphi_1,\psi)) \quad (j=\ldots,-1,0,1,\ldots), \tag{20.6}$$

where φ_1 is a scalar, $\varphi = (\varphi_1,\psi) = (\varphi_1,\varphi_2,\varphi_3,\ldots) \in \mathfrak{M}$, $t_j(\varphi)$ is a solution of the equation

$$\varphi 1_{t_j(\varphi)}(\varphi) = \varphi^{(j)}, \tag{20.7}$$

and $\varphi^{(j)}$ are real constants.

In what follows, we study integral sets of the system of equations (20.6) generated by functions independent of t.

Definition 20.2. *A function $h = u(\varphi)$ is called an invariant torus of the system of equations (20.6) if it specifies an integral set of this system and is 2π-periodic in φ_i $(i = 1, 2, \ldots)$.*

First, we prove the following auxiliary assertion:

Lemma 20.1. *Assume that all conditions of Theorem 20.1 except inequality (20.3) are satisfied for the system of equations (20.6), the function $a(\varphi_1, \psi) = \{a_1(\varphi_1, \psi), a_2(\varphi_1, \psi), \ldots\}$ satisfies the inequality*

$$a_1(\varphi_1, \psi) \geq \alpha_0 = \text{const} > 0, \tag{20.8}$$

the constants $\varphi^{(j)}$ are such that $0 < \varphi^{(1)} < \varphi^{(2)} < \ldots < \varphi^{(p)} < 2\pi$, and

$$\varphi^{(j+p)} = \varphi^{(j)} + 2\pi \tag{20.9}$$

for $j = 0, \pm 1, \pm 2, \ldots$. Then the function $u(\varphi, t)$ specifying an integral set of system (20.6) exists.

Proof. It follows from inequality (20.8) that the function $\varphi_1 = \varphi 1_t(\varphi)$ monotonically increases in R^1 and, hence, for any integer j, equation (20.7) has at most one solution $t_j = t_j(\varphi)$. In addition, in view of condition (20.9), the times of pulses $t_j(\varphi)$ are separated, i.e., there exists a real number $\theta > 0$ such that inequality (20.3) holds for all integer j. This enables us to write

$$y(t, \varphi)$$

$$= \int_{-\infty}^{\infty} G_t(\tau, \varphi) c(\varphi_\tau(\varphi)) d\tau + \sum_{-\infty < t_j(\varphi) < \infty} G_t(t_j(\varphi), \varphi) J_j(\varphi_{t_j(\varphi)}(\varphi)). \tag{20.10}$$

The fact that this relation is a bounded solution of the system of equations (20.6) is proved by analogy with the proof of Theorem 20.1.

Substituting $\varphi_{-t}(\varphi)$ for φ in equality (20.7), we obtain

$$\varphi 1_{t_j(\varphi_{-t}(\varphi))}(\varphi_{-t}(\varphi)) = \varphi^{(j)} \quad (j = \ldots, -1, 0, 1, \ldots)$$

for all $t \in R^1$.

Consider the function

$$u(\varphi,t) = \int_{-\infty}^{\infty} G_t(\tau,\varphi_{-t}(\varphi))c(\varphi_{\tau-t}(\varphi))d\tau$$

$$+ \sum_{-\infty<t_j(\varphi_{-t}(\varphi))<\infty} G_t(t_j(\varphi_{-t}(\varphi)),\varphi_{-t}(\varphi))J_j(\varphi_{t_j(\varphi_{-t}(\varphi))}(\varphi_{-t}(\varphi))).$$

It is easy to see that this function specifies an integral set of the system of equations (20.6) because $u(\varphi_t(\varphi),t)$ takes the form (20.10).

We now consider a special case of the system of equations (20.6) obtained by setting $a_1(\varphi_1,\psi) = a_1(\varphi_1)$. The system of equations obtained as a result is denoted by (20.6^0).

We fix an arbitrary value of the parameter φ_1 and denote the set of all vectors $\varphi = (\varphi_1,\psi) \in \mathfrak{M}$ whose first coordinate is equal to φ_1 by $G(\varphi_1)$. It is clear that the times of pulses $t_j(\varphi_1)$ are identical for all values of the parameter $\varphi \in G(\varphi_1)$. In this case, the function

$$u(\varphi_1,\psi,t) = \int_{-\infty}^{\infty} G_0(\tau,\varphi)c(\varphi_\tau(\varphi))d\tau$$

$$+ \sum_{-\infty<t_j(\varphi_1)<\infty} G_0(t_j(\varphi_1)-t,\varphi)[J_j(\varphi)]_{\varphi_1=\varphi^{(j)}}$$

specifies an integral set of the system of equations (20.6) for all $\varphi \in G(\varphi_1)$. Moreover, for any fixed φ_1, this function specifies its own integral set periodic in ψ with period 2π.

In particular, if $a_1(\varphi_1) = \omega = \text{const} > 0$, then $\varphi_{1_{-t}} = \varphi_1 - \omega t$ and $t_j(\varphi_{1_{-t}}) = t_j(\varphi_1) + t$. The following theorem solves the problem of the existence of an invariant torus for the system of equations (20.6):

Theorem 20.2. *Assume that the system of equations (20.6) is such that the conditions of Lemma 20.1 are satisfied and, for all integer j, the jump functions satisfy the identity*

$$J_{j+p}(\varphi) = J_j(\varphi). \qquad (20.11)$$

Then the system of equation (20.6) possesses the invariant torus $\mathcal{F}: h = u(\varphi)$, $\varphi \in \mathcal{F}_\infty$, where

$$u(\varphi)$$
$$= \int_{-\infty}^{\infty} G_0(\tau,\varphi)c(\varphi_\tau(\varphi))d\tau + \sum_{-\infty<t_j(\varphi)<\infty} G_0(t_j(\varphi),\varphi)[J_j(\varphi)]_{\varphi_1=\varphi^{(j)}}. \quad (20.12)$$

Proof. Let $\varphi = \varphi_t(\xi)$ be an arbitrary solution of the equation $\frac{d\varphi}{dt} = a(\varphi_1,\psi)$ with initial values 0, $\xi \in \mathfrak{M}$. Substituting this solution in equality (20.12), we obtain

$$u(\varphi_t(\xi)) = \int_{-\infty}^{\infty} G_0(\tau,\varphi_t(\xi))c(\varphi_\tau(\varphi_t(\xi)))d\tau$$
$$+ \sum_{-\infty<t_j(\varphi_t(\xi))<\infty} G_0(t_j(\varphi_t(\xi)),\varphi_t(\xi))[J_j(\varphi_t(\xi))]_{\varphi_1=\varphi^{(j)}}.$$

Since

$$\varphi_{t_j(\varphi_t(\xi))}(\varphi_t(\xi)) = \varphi_{t_j(\varphi_t(\xi))+t}(\xi) = \varphi^{(j)},$$

we conclude that $t_j(\varphi_t(\xi)) + t = t_j(\xi)$. Hence,

$$u(\varphi_t(\xi)) = \int_{-\infty}^{\infty} G_t(\tau,\xi)c(\varphi_\tau(\xi))d\tau$$
$$+ \sum_{-\infty<t_j(\xi)<\infty} G_t(t_j(\xi),\xi)J_j(\varphi_{t_j(\xi)}(\xi)).$$

This expression is a bounded solution of the system of equations (20.6) and has the same form as (20.10).

Let us show that $u(\varphi)$ is a function 2π-periodic in φ_i $(i = 1,2,3,\ldots)$. Since

$$\varphi 1_{t_j(\varphi+2\pi)}(\varphi + 2\pi) = \varphi 1_{t_j(\varphi+2\pi)}(\varphi) + 2\pi = \varphi^{(j)},$$

we obtain

$$\varphi 1_{t_j(\varphi+2\pi)}(\varphi) = \varphi^{(j)} - 2\pi = \varphi^{(j-p)} = \varphi 1_{t_{j-p}(\varphi)}(\varphi)$$

and, therefore,

$$t_j(\varphi + 2\pi) = t_{j-p}(\varphi).$$

Further, by using condition (20.11), we obtain

$$
\begin{aligned}
&u(\varphi + 2\pi) \\
&\quad = \int_{-\infty}^{\infty} G_0(\tau, \varphi + 2\pi) c(\varphi_\tau(\varphi) + 2\pi) d\tau \\
&\quad + \sum_{-\infty < t_j(\varphi + 2\pi) < \infty} G_0(t_j(\varphi + 2\pi), \varphi + 2\pi)[J_j(\varphi + 2\pi)]_{\varphi_1 + 2\pi = \varphi^{(j)}} \\
&\quad = \int_{-\infty}^{\infty} G_0(\tau, \varphi) c(\varphi_\tau(\varphi)) d\tau \\
&\quad + \sum_{-\infty < t_{j-p}(\varphi) < \infty} G_0(t_{j-p}(\varphi), \varphi)[J_{j-p}(\varphi)]_{\varphi_1 = \varphi^{(j-p)}} = u(\varphi).
\end{aligned}
$$

This competes the proof of Theorem 20.2.

Finally, we consider a system of equations

$$
\frac{d\varphi}{dt} = \omega + \mu a(\varphi), \quad \frac{dy}{dt} = A(\varphi) y + c(\varphi) \tag{20.13}
$$

subjected to a pulse action obeying the law

$$
\Delta y|_{\langle r, \varphi \rangle = 2\pi k} = H_k(\varphi), \tag{20.14}
$$

where $y \in \mathfrak{M}$, $\varphi \in R^m$, $\omega = (\omega_1, \ldots, \omega_m) \in R^m$, $r = (r_1, \ldots, r_m) \in R^m$, $a(\varphi) \in C^0_{\text{Lip}}(\mathcal{F}_m)$, $c(\varphi)$, $A(\varphi)$, $H_k(\varphi) \in C^0(\mathcal{F}_m)$, $\langle r, \varphi \rangle = \sum_{\nu=1}^{m} r_\nu \varphi_\nu$, μ is a positive parameter, and $k = \ldots, -1, 0, 1, \ldots$.

Assume that the matrix $A(\varphi)$ satisfies condition (20.2) and $\|H_k(\varphi)\| \leq H = \text{const} < \infty$ for all integer k.

The surface $\mathcal{F}$: $H = u(\varphi)$, $\varphi \in \mathcal{F}_m$, is an invariant torus of the system of equations (20.13), (20.14) if the function $u(\varphi)$ is 2π-periodic in φ_i $(i = \overline{1, m})$, bounded in the norm $\| \cdot \|$, and such that the following equalities are true for all $t \in R^1$:

$$
\begin{aligned}
&\frac{du(\varphi_t(\varphi))}{dt} = A(\varphi_t(\varphi)) u(\varphi_t(\varphi)) + c(\varphi_t(\varphi)), \quad t \neq t_k(\varphi), \\
&\Delta u(\varphi_{t_k(\varphi)}(\varphi)) = H_k(\varphi_{t_k(\varphi)}(\varphi)) \quad (k = \ldots, -1, 0, 1, \ldots),
\end{aligned}
$$

where $t_k(\varphi)$ are times corresponding to the points where the trajectory $\varphi = \varphi_t(\varphi)$ crosses the surfaces $\langle r, \varphi\rangle = 2\pi k$.

Let us now formulate conditions under which the system of equations (20.13), (20.14) possesses an invariant torus. To this end, we prove the following theorem:

Theorem 20.3. *Assume that equation (20.13) possesses the Green function $G_0(\tau, \varphi)$ of the problem of invariant tori satisfying inequality (10.5), $a_i(\varphi) > 0$ and $r_i > 0$ for all $i = \overline{1, m}$, $\varphi \in \mathfrak{M}_m$, $\sum_{i=1}^{m} r_i = p$, where p is an integer number, and $\langle \omega, r\rangle > 0$. If, in addition, $H_{k+p}(\varphi) = H_k(\varphi)$ for all integer k, then the system of equations (20.13), (20.14) possesses an invariant torus $\mathcal{F}$ for all $0 < \mu < \mu^0$, where μ^0 is a sufficiently small positive constant.*

Proof. We choose an arbitrary solution $\varphi = \varphi_t(\varphi)$ of the first equation in system (20.13). Let $t = t_k(\varphi)$ be a solution of the equation

$$\langle r, \varphi_{t_k}(\varphi)\rangle = 2\pi k. \tag{20.15}$$

For each integer k, this equation has at most one solution. Indeed, assume that $\langle r, \varphi_{t_{k_0}}(\varphi)\rangle = 2\pi k$, where $t_{k_0} \neq t_k$. Then

$$\langle r, \varphi_{t_k}(\varphi) - \varphi_{t_{k_0}}(\varphi)\rangle = 0.$$

This yields

$$\langle r, \omega\rangle (t_k - t_{k_0}) + \left\langle r, \int_{t_{k_0}}^{t_k} a(\varphi_s(\varphi))\, ds\right\rangle = 0.$$

Since $\langle r, \omega\rangle > 0$, one can easily see that the last equality is possible only for $t_k = t_{k_0}$.

Clearly, $t_k(\varphi)$ is a solution of the equation

$$t_k(\varphi) = \frac{2k\pi - \langle r, \varphi\rangle}{\langle \omega, r\rangle} - \frac{\mu}{\langle \omega, r\rangle}\left\langle r, \int_0^{t_k(\varphi)} a(\varphi_s(\varphi)) ds\right\rangle. \tag{20.16}$$

For this solution, we construct the following iterative process:

$$t_k^{(0)} = t_k^{(0)}(\varphi) = \frac{2k\pi - \langle r, \varphi\rangle}{\langle \omega, r\rangle},$$

$$t_k^{(m)} = t_k^{(m)}(\varphi) = \frac{2k\pi - \langle r, \varphi\rangle}{\langle \omega, r\rangle} - \frac{\mu}{\langle \omega, r\rangle}\left\langle r, \int_0^{t_k^{(m-1)}} a(\varphi_s(\varphi)) ds\right\rangle.$$

We set $\|a(\varphi)\| \leq M = \text{const} < \infty$ and $\mu^0 = \dfrac{\langle \omega, r\rangle}{Mm\|r\|}$. By induction, for all $\mu < \mu^0$, we prove that

$$\begin{aligned} \left| t_k^{(n)} \right| &\leq \left| t_k^0 \right| + \frac{\mu m\|r\|M}{\langle \omega, r\rangle} \, | \, t_k^{(n-1)} \, | \\ &\leq \frac{| \, t_k^{(0)} \, |}{1 - \mu m\|r\|M/\langle \omega, r\rangle} \\ &= \frac{| \, 2k\pi - \langle r, \varphi\rangle \, |}{\langle \omega, r\rangle - \mu m\|r\|M}. \end{aligned}$$

Further, we get

$$\begin{aligned} | \, t_k^{(n)} - t_k^{(n-1)} \, | &\leq \frac{\mu}{\langle \omega, r\rangle} \left| \left\langle r, \int\limits_{t_k^{(n-2)}}^{t_k^{(n-1)}} a(\varphi_s(\varphi)) ds \right\rangle \right| \\ &\leq \frac{\mu m\|r\|M}{\langle \omega, r\rangle} \left| t_k^{(n-1)} - t_k^{(n-2)} \right| \\ &< \ldots < \left(\frac{\mu m\|r\|M}{\langle \omega, r\rangle} \right)^{(n-1)} \left| t_k^{(1)} - t_k^{(0)} \right| \\ &\leq \left(\frac{\mu m\|r\|M}{\langle \omega, r\rangle} \right)^{n} \left| \, t_k^{(0)} \, \right| \\ &= \left(\frac{\mu m\|r\|M}{\langle \omega, r\rangle} \right)^{n} \left| \frac{2k\pi - \langle r, \varphi\rangle}{\langle \omega, r\rangle} \right|. \end{aligned}$$

Since $\dfrac{\mu m\|r\|M}{\langle \omega, r\rangle} < 1$, the sequence $\{t_k^{(n)}(\varphi)\}$ is uniformly convergent as $n \to \infty$. Denote $\lim\limits_{n\to\infty} t_k^{(n)} = t_k(\varphi)$. It is clear that $t_k(\varphi)$ is a solution of equation (20.16).

It is easy to see that

$$\begin{aligned} t_{k+1}(\varphi) - t_k(\varphi) &= \frac{2\pi}{\langle \omega, r\rangle} - \frac{\mu}{\langle \omega, r\rangle} \left\langle r, \int\limits_{t_k}^{t_{k+1}} a(\varphi_s(\varphi)) ds \right\rangle \\ &\geq \frac{2\pi}{\langle \omega, r\rangle} - \frac{\mu m\|r\|M}{\langle \omega, r\rangle} \, | \, t_{k+1}(\varphi) - t_k(\varphi) \, | \, . \end{aligned}$$

For $t_{k+1} - t_k > 0$, this yields

$$t_{k+1}(\varphi) - t_k(\varphi) \geq \frac{2\pi}{\langle \omega, r\rangle + \mu m\|r\|M} = \theta > 0,$$

which proves that the times $t_k(\varphi)$ are separated. For $t_{k+1}(\varphi) < t_k(\varphi)$, we obtain

$$t_{k+1} - t_k \geq \frac{2\pi}{\langle \omega, r\rangle} : \left(1 - \frac{\mu m\|r\|M}{\langle \omega, r\rangle}\right) > 0,$$

i.e., we arrive at a contradiction.

Thus, $t_{k+1}(\varphi) > t_k(\varphi)$ for all integer k. This enables us to write the following relations for the functions $u(\varphi)$ and $y(t,\varphi)$:

$$\begin{aligned} u(\varphi) &= \int_{-\infty}^{\infty} G_0(\tau,\varphi)c(\varphi_\tau(\varphi))d\tau \\ &\quad + \sum_{-\infty<t_k(\varphi)<\infty} G_0(t_k(\varphi),\varphi)[H_k(\varphi)]_{\langle r,\varphi\rangle=2\pi k}, \end{aligned}$$

$$\begin{aligned} y(t,\varphi) &= \int_{-\infty}^{\infty} G_t(\tau,\varphi)c(\varphi_\tau(\varphi))d\tau \\ &\quad + \sum_{-\infty<t_k(\varphi)<\infty} G_t(t_k(\varphi),\varphi)H_k(\varphi_{t_k(\varphi)}(\varphi)). \end{aligned}$$

The fact that the functions $u(\varphi)$ and $y(t,\varphi)$ are bounded in the norm $\|\cdot\|$ is proved in exactly the same way as in Theorem 20.1. By analogy with Theorem 20.1, we can easily prove that

$$\frac{dy(t,\varphi)}{dt} = A(\varphi_t(\varphi))y(t,\varphi) + c(\varphi_t(\varphi)), \quad t \neq t_k(\varphi),$$

$$\Delta y(t,\varphi)|_{t=t_k(\varphi)} = H_k(\varphi_{t_k(\varphi)}(\varphi)) \quad (k = \ldots,-1,0,1,\ldots).$$

It remains to show that $u(\varphi_t(\varphi)) = y(t,\varphi)$ and $u(\varphi)$ is a function 2π-periodic in φ_i $(i = \overline{1,m})$.

Substituting $\varphi_t(\varphi)$ for φ in (20.15), we obtain

$$\left\langle \varphi_{t_k(\varphi_t(\varphi))}(\varphi_t(\varphi)), r\right\rangle = \left\langle \varphi_{t_k(\varphi_t(\varphi))+t}(\varphi), r\right\rangle = 2\pi k.$$

This implies that $t_k(\varphi_t(\varphi)) + t = t_k(\varphi)$ and, hence,

$$u(\varphi_t(\varphi))$$
$$= \int_{-\infty}^{\infty} G_0(\tau, \varphi_t(\varphi))c(\varphi_\tau(\varphi_t(\varphi)))d\tau$$
$$+ \sum_{-\infty<t_k(\varphi_t(\varphi))<\infty} G_0(t_k(\varphi_t(\varphi)), \varphi_t(\varphi))[H_k(\varphi_t(\varphi))]_{\langle r,\varphi_t(\varphi)\rangle=2\pi k}$$
$$= \int_{-\infty}^{\infty} G_t(\tau, \varphi)c(\varphi_\tau(\varphi))d\tau$$
$$+ \sum_{-\infty<t_k(\varphi_t(\varphi))<\infty} G_t(t_k(\varphi), \varphi)H_k(\varphi_{t_k(\varphi)}(\varphi)) = y(t, \varphi).$$

Further, we have

$$\begin{aligned}\Big\langle \varphi_{t_k(\varphi+2\pi)}(\varphi + 2\pi), r\Big\rangle &= \Big\langle \varphi_{t_k(\varphi+2\pi)}(\varphi) + 2\pi, r\Big\rangle \\ &= \Big\langle \varphi_{t_k(\varphi+2\pi)}(\varphi), r\Big\rangle + \langle 2\pi, r\rangle = 2\pi k,\end{aligned}$$

whence it follows that

$$\Big\langle \varphi_{t_k(\varphi+2\pi)}(\varphi), r\Big\rangle = 2\pi k - 2\pi\sum_{i=1}^{m} r_i = 2\pi(k-p).$$

Therefore, $t_k(\varphi + 2\pi) = t_{k-p}(\varphi)$. However, in this case,

$$u(\varphi + 2\pi)$$
$$= \int_{-\infty}^{\infty} G_0(\tau, \varphi + 2\pi)c(\varphi_\tau(\varphi) + 2\pi)d\tau$$
$$+ \sum_{-\infty<t_k(\varphi+2\pi)<\infty} G_0(t_k(\varphi + 2\pi), \varphi + 2\pi)[H_k(\varphi + 2\pi)]_{\langle r,\varphi+2\pi\rangle=2\pi k}$$

$$= \int_{-\infty}^{\infty} G_0(\tau,\varphi)c(\varphi_\tau(\varphi))d\tau$$

$$+ \sum_{-\infty<t_{k-p}(\varphi)<\infty} G_0(t_{k-p}(\varphi),\varphi)[H_{k-p}(\varphi)]_{\langle\tau,\varphi\rangle=2\pi(k-p)} = u(\varphi),$$

which completes the proof Theorem 20.3.

We preserve the same notation as in the case of the system of equations (20.6) and proceed to the analysis of a quasilinear system

$$\frac{d\varphi}{dt} = a(\varphi_1,\psi), \quad \frac{dy}{dt} = A(\varphi_1,\psi)y + c(\varphi_1,\psi,y,\mu), \quad t \neq t_j(\varphi),$$

$$\Delta y|_{t=t_j(\varphi)} = J_j(\varphi_t(\varphi),y,\mu)|_{t=t_j(\varphi)} (j = \ldots,-1,0,1,\ldots), \tag{20.17}$$

where, for $\|y\| \leq d$ and $\mu \in [0,\mu_0]$ (μ is a positive parameter), we have $A(\varphi)$, $c(\varphi,y,\mu)$, $J_j(\varphi,y,\mu) \in C^0(\mathcal{F}_\infty)$, $a(\varphi) \in C^0_{\text{Lip}}(\mathcal{F}_\infty)$, $\|c(\varphi,y,\mu)\| \leq c^0 =$ const, and $\|J_j(\varphi,y,\mu)\| \leq J = \text{const} < \infty$ for all integer j.

Assume that, for all $\varphi \in \mathcal{F}_\infty$, $c(\varphi,0,0) = J_j(\varphi,0,0) = 0 \in \mathfrak{M}$, inequality (20.2) is true, $t_j(\varphi)$ is a solution of equation (20.7), and $G_0(\tau,\varphi)$ is the Green function of the problem of invariant tori for the equation $\dfrac{dy}{dt} = A(\varphi_t(\varphi))y$ satisfying inequality (10.5).

We want to establish conditions under which the system of equations (20.17) possesses an invariant torus. The following assertion is true:

Theorem 20.4. *Assume that, for all $\|y\| \leq d$, $\mu \in [0,\mu_0]$, $\varphi \in \mathcal{F}_\infty$ and $j = \ldots,-1,0,1,\ldots$, one has $J_{j+p}(\varphi,y,\mu) = J_j(\varphi,y,\mu)$, conditions (20.8) and (20.9) are satisfied, and the following inequality is true:*

$$\max\{\|c(\varphi,y,\mu) - c(\varphi,\bar{y},\bar{\mu})\|, \|J_j(\varphi,y,\mu) - J_j(\varphi,\bar{y},\bar{\mu})\|\}$$

$$\leq N(d,\mu_0)\|y-\bar{y}\| + \mathcal{L}\|\mu-\bar{\mu}\|, \tag{20.18}$$

where $\|\bar{y}\| \leq d$, $\bar{\mu} \in [0,\mu_0]$, $\mathcal{L} = \text{const} > 0$, and $N(d,\mu_0) \to 0$ as $d \to 0$ and $\mu_0 \to 0$. Moreover, for any fixed $0 \leq d_0 \leq d$, the function $N(d_0,\mu)$ increases on the segment $[0,\mu_0]$.

Then there exists $\mu^0 \in (0,\mu_0]$ such that, for all $0 \leq \mu \leq \mu^0$, the system of equations (20.17) possesses an invariant torus $\mathcal{F}\colon y = u(\varphi,\mu)$, $\varphi \in \mathcal{F}_\infty$, such that $\|u(\varphi,\mu)\| \to 0$ as $\mu \to 0$.

Proof. Consider a sequence of functions $\{u^{(i+1)}(\varphi,\mu)\}$ $(i = 0, 1, 2, \ldots)$ specifying the invariant tori $\mathcal{F}^{(i+1)}(\mu)$ of linear systems of equations

$$\frac{d\varphi}{dt} = a(\varphi), \quad \frac{dy}{dt} A(\varphi)y + c(\varphi, u^{(i)}(\varphi,\mu),\mu), \quad t \neq t_j(\varphi),$$
$$\Delta y|_{t=t_j(\varphi)} = J_j(\varphi_t(\varphi), u^{(i)}(\varphi_t(\varphi),\mu),\mu)|_{t=t_j(\varphi)} \tag{20.19}$$
$$(j = \ldots, -1, 0, 1, \ldots).$$

The function $u^{(0)}(\varphi,\mu) \equiv 0 \in \mathfrak{M}$ is taken as the initial approximation.

Denote $Z(d,\mu_0) = N(d,\mu_0)d + \mathcal{L}\mu_0$. It now follows from (20.18) that $\|c(\varphi,y,\mu)\| \leq Z(d,\mu_0)$ and $\|J_j(\varphi,y,\mu)\| \leq Z(d,\mu_0)$, where $Z(d,\mu_0) \to 0$ as $d \to 0$ and $\mu_0 \to 0$. Here, $\|y\| \leq d$ and $\mu \leq \mu_0$.

Note that Theorem 20.2 can be applied to the linear system (20.19) and, thus, the invariant torus $\mathcal{F}^{(i+1)}: y = u^{(i+1)}(\varphi,\mu)$, $\mu \in \mathcal{F}_\infty$, exists, the function $u^{(i+1)}(\varphi,\mu)$ has the form

$$u^{(i+1)}(\varphi,\mu) = \int_{-\infty}^{\infty} G_0(\tau,\varphi)c(\varphi_\tau(\varphi), u^{(i)}(\varphi_\tau(\varphi),\mu),\mu)d\tau$$
$$+ \sum_{-\infty<t_j(\varphi)<\infty} G_0(t_j(\varphi),\varphi)[J_j(\varphi, u^{(i)}(\varphi,\mu),\mu)]_{\varphi_1=\varphi^{(j)}}, \tag{20.20}$$

and the function

$$u^{(i+1)}(\varphi_t,\mu)$$
$$= \int_{-\infty}^{\infty} G_t(\tau,\varphi)c(\varphi_\tau, u^{(i)}(\varphi_\tau,\mu),\mu)d\tau$$
$$+ \sum_{-\infty<t_j(\varphi)<\infty} G_t(t_j(\varphi),\varphi_t(\varphi))[J_j(\varphi_{t_j(\varphi)}, u^{(i)}(\varphi_{t_j(\varphi)},\mu),\mu)]$$

is a bounded solution of this system dependent on φ and μ (regarded as parameters).

Let us show that, for all $i = 0, 1, 2, \ldots$, we have

$$\|u^{(i+1)}(\varphi,\mu)\| \leq \frac{\beta}{1-\beta N(d,\mu)} Z(0,\mu), \tag{20.21}$$

where

$$\beta = \frac{2K}{\gamma}\Big[1 + \frac{\gamma}{1 - \exp\{-\gamma\theta\}}\Big].$$

Indeed, in view of representation (20.20), we obtain

$$\begin{aligned}\|u^{(1)}(\varphi,\mu)\| &\le \int_{-\infty}^{\infty} K\exp\{-\gamma\,|\,\tau\,|\,\}\|c(\varphi_\tau,0,\mu)\|d\tau \\ &\quad + \sum_{-\infty<t_j(\varphi)<\infty} K\exp\{-\gamma\,|\,\tau_j\,|\,\}\|J_j(\varphi,0,\mu)\| \\ &\le \beta Z(0,\mu) \le \frac{\beta}{1-\beta N(d,\mu)} Z(0,\mu),\end{aligned}$$

provided that $\beta N(d,\mu) < 1$.

Assume that inequality (20.11) is true for $i = 0, 1, 2, \ldots, k-1$. Let us show that it is also true for $i = k$. Indeed,

$$\begin{aligned}&\|u^{(k+1)}(\varphi,\mu)\| \\ &\quad \le \|u^{(k+1)}(\varphi,\mu) - u^{(1)}(\varphi,\mu)\|\|u^{(1)}(\varphi,\mu)\| \\ &\quad \le \int_{-\infty}^{\infty} K\exp\{-\gamma\,|\,\tau\,|\,\}N(d,\mu)\|u^{(k)}(\varphi_\tau,\mu)\|d\tau \\ &\quad\quad + N(d,\mu)\sum_{-\infty<t_j<\infty} K\exp\{-\gamma\,|\,t_j\,|\,\}\|u^{(k)}(\varphi,\mu)\| + \beta Z(0,\mu) \\ &\quad \le \frac{2K}{\gamma} N(d,\mu)\frac{\beta}{1-\beta N(d,\mu)} Z(0,\mu) \\ &\quad\quad + N(d,\mu)\frac{2K}{1-\exp\{-\gamma\theta\}}\frac{\beta}{1-\beta N(d,\mu)} Z(0,\mu) + \beta Z(0,\mu) \\ &\quad = \frac{\beta}{1-\beta N(d,\mu)} Z(0,\mu).\end{aligned}$$

We choose $\mu^0 \le \mu_0$ and $d_0 \le d$ so that

$$\beta N(d_0,\mu^0) < 1 \quad \text{and} \quad \frac{\beta}{1-\beta N(d_0,\mu^0)} Z(0,\mu^0) \le d_0. \tag{20.22}$$

This can always be done because $N(d,\mu) \to 0$ as $d \to 0$ and $\mu \to 0$, $Z(0,\mu_0) \to 0$ as $\mu_0 \to 0$, and the function $N(d_0,\mu)$ increases with μ.

Thus, by induction, we conclude that inequality (20.11) is true in the domain $D\colon \varphi \in \mathcal{F}_\infty,\ \|y\| \le d_0,\ \mu \in [0,\mu^0]$, i.e., the sequence of invariant tori $\mathcal{F}^{(i+1)}(\mu)$ of the linear systems of equations (20.19) is uniformly bounded by a constant $d_0 \le d$. Let us show that this sequence converges as $i \to \infty$ uniformly in φ.

Indeed, we have

$$\|u^{(1)}(\varphi,\mu) - u^{(0)}(\varphi,\mu)\| \le \frac{\beta}{1-\beta N(d,\mu)} Z(0,\mu),$$

$$\begin{aligned}
&\|u^{(k+1)}(\varphi,\mu) - u^{(k)}(\varphi,\mu)\| \\
&\quad \le \int_{-\infty}^{\infty} K\exp\{-\gamma\,|\,\tau\,|\,\}N(d,\mu)\|u^{(k)}(\varphi,\mu) - u^{(k-1)}(\varphi,\mu)\|_0 d\tau \\
&\qquad + KN(d,\mu) + \sum_{-\infty<t_j<\infty} \exp\{-\gamma\,|\,t_j\,|\,\}\|u^{(k)}(\varphi,\mu) - u^{(k-1)}(\varphi,\mu)\|_0 \\
&\quad \le \frac{2K}{\gamma}\Big(1 + \frac{\gamma}{1-\exp\{-\gamma\theta\}}\Big)N(d,\mu)\|u^{(k)}(\varphi,\mu) - u^{(k-1)}(\varphi,\mu)\|_0 \\
&\quad \le (\beta N(d,\mu))^{k-1}\frac{\beta}{1-\beta N(d,\mu)} Z(0,\mu),
\end{aligned}$$

where

$$\|u^{(k)}(\varphi,\mu) - u^{(k-1)}(\varphi,\mu)\|_0 = \sup_{\varphi\in\mathcal{F}_\infty} \|u^{(k)}(\varphi,\mu) - u^{(k-1)}(\varphi,\mu)\|.$$

By using these inequalities and inequalities (20.22), we conclude that the sequence $\mathcal{F}^{(i)}(\mu)$ converges as $i \to \infty$ uniformly in the variables φ and $\mu \in [0,\mu^0]$.

We now denote $\lim\limits_{i\to\infty} u^{(i)}(\varphi,\mu) = u(\varphi,\mu)$ and prove that $\mathcal{F}\colon y = u(\varphi,\mu)$, $\varphi \in \mathcal{F}_\infty$, is an invariant torus of the system of equations (20.17).

Since $\mathcal{F}^{(i+1)}(\mu)$ is an invariant torus of system (20.19), the following equalities are true:

$$\frac{du^{(i+1)}(\varphi_t,\mu)}{dt} = A(\varphi_t)u^{(i+1)}(\varphi_t,\mu) + c(\varphi_t, u^{(i)}(\varphi_t,\mu),\mu), \quad t \ne t_j(\varphi),$$

$$\Delta u^{(i+1)}(\varphi_t,\mu)|_{t=t_j(\varphi)} = J_j(\varphi_{t_j}, u^{(i)}(\varphi_{t_j},\mu),\mu) \quad (j = \ldots,-1,0,1,\ldots),$$

whence, for $t \neq t_j(\varphi)$,

$$u^{(i+1)}(\varphi_t, \mu)$$

$$= u^{(i+1)}(\varphi, \mu) + \int_0^t \Big[A(\varphi_s)u^{(i+1)}(\varphi_s, \mu) + c(\varphi_s, u^{(i)}(\varphi_s, \mu), \mu)\Big]\, ds. \quad (20.23)$$

By using the inequalities

$$\begin{aligned} &\|A(\varphi_s)u(\varphi_s, \mu) + c(\varphi_s, u(\varphi_s, \mu), \mu)\| \\ &\quad -[A(\varphi_s)u^{(i+1)}(\varphi_s, \mu) + c(\varphi_s, u^{(i)}(\varphi_s, \mu), \mu)] \\ &\qquad \leq a^0\|u(\varphi_s, \mu) - u^{(i+1)}(\varphi_s, \mu)\| + \mathcal{L}\|u(\varphi_s, \mu) - u^{(i)}(\varphi_s, \mu)\| \end{aligned} \quad (20.24)$$

and choosing a number i such that, for all $\varphi \in \mathcal{F}_\infty$,

$$\|u(\varphi, \mu) - u^{(i)}(\varphi, \mu)\| \leq \min\Big\{\frac{\varepsilon}{2a^0}; \frac{\varepsilon}{2\mathcal{L}}\Big\},$$

we conclude that, in this case, the expression on the right-hand side of inequality (20.24) becomes smaller than ε for all $s \in [0, t]$.

Thus, in relation (20.23), we can pass to the limit as $i \to \infty$ and obtain

$$u(\varphi_t, \mu) = u(\varphi, \mu) + \int_0^t [A(\varphi_s)u(\varphi_s, \mu) + c(\varphi_s, u(\varphi_s, \mu), \mu)]ds, \quad t \neq t_j(\varphi).$$

It follows from the continuity of J_j $(j = \ldots, -1, 0, 1, \ldots)$ as functions of y that

$$\begin{aligned} \Delta u(\varphi_t, \mu)|_{t=t_j(\varphi)} &= \lim_{i\to\infty} J_j(\varphi_t, u^{(i)}(\varphi_{t_j(\varphi)}, \mu), \mu) \\ &= J_j(\varphi_{t_j(\varphi)}(\varphi), u(\varphi_{t_j(\varphi)}, \mu), \mu) \quad (j = \ldots, -1, 0, 1, \ldots). \end{aligned}$$

The last two representations demonstrate that the function $y = u(\varphi, \mu)$ specifies an invariant torus of the system of equations (20.17).

Since $\lim\limits_{\mu\to 0} \|u^{(i)}(\varphi, \mu)\| = 0$, we have

$$\|u(\varphi, \mu)\| = \lim_{i\to\infty} \|u^{(i)}(\varphi, \mu)\| \to 0 \quad \text{as} \quad \mu \to 0,$$

i.e., for $\mu = 0$, the system of equations (20.17) possesses the trivial invariant torus $y = 0$ due to the conditions imposed on J_j and c in the description of this system.

Let us now consider a quasilinear system of equations

$$\frac{d\varphi}{dt} = \omega + \mu a(\varphi), \quad \frac{dy}{dt} = A(\varphi)y + \mu c(\varphi, y),$$

$$\Delta y|_{\langle r,\varphi\rangle=2\pi k} = \mu H_k(\varphi, y) \quad (k = \ldots, -1, 0, 1, \ldots). \tag{20.25}$$

Here, we use the same notation as in the case of system (20.13), (20.14) and assume that system (20.25) satisfies the same conditions as the system of equations (20.13),(20.14) for $\|y\| \leq d$, $\mu \in [0, \mu_0]$, and $\varphi \in \mathcal{F}_m$.

The function $y = u(\varphi, \mu)$ specifies an invariant torus of the system of equations (20.25) if it is 2π-periodic in φ_i $(i = \overline{1, m})$, bounded in the norm $\|\cdot\|$, and satisfies the relations

$$\frac{du(\varphi_t(\varphi), \mu)}{dt} = A(\varphi_t(\varphi))u(\varphi_t(\varphi), \mu) + \mu c(\varphi_t(\varphi), u(\varphi_t(\varphi), \mu)), \quad t \neq t_k(\varphi),$$

$$\Delta u(\varphi_t(\varphi), \mu)|_{t=t_k(\varphi)} = \mu H_k(\varphi_{t_k(\varphi)}(\varphi), u(\varphi_{t_k(\varphi)}, \mu)), \tag{20.26}$$

where $t = t_k(\varphi)$ is a solution of equation (20.15).

We now establish conditions for the existence of an invariant torus $\mathcal{F}_{(\mu)}: y = u(\varphi, \mu)$, $\varphi \in \mathcal{F}_m$, of system (20.26).

Theorem 20.5. *Assume that the system of equations (20.25) is such that, for all $\varphi \in \mathcal{F}_m$, $\|y\| \leq d$, $\mu \in [0, \mu_0]$, and $k = \ldots, -1, 0, 1, \ldots$, the conditions of Theorem 20.3 are satisfied with $c(\varphi, y)$ and $H_k(\varphi, y)$ replaced by $c(\varphi)$ and $H_k(\varphi)$, respectively, and the functions $c(\varphi, y)$ and $H_k(\varphi, y)$ satisfy the Lipschitz condition with respect to y uniformly in φ and k.*

Then there exists $0 < \mu^0 \leq \mu_0$ such that, for all $\mu \in [0, \mu^0]$, system (20.25) possesses an invariant torus $\mathcal{F}(\mu)$ such that

$$\lim_{\mu \to 0} \|u(\varphi, \mu)\| = 0. \tag{20.27}$$

Proof. We construct an iterative process similar to that studied in Theorem 20.4. In this case, the role of the torus $\mathcal{F}^{(i+1)}(\mu): y = u^{(i+1)}(\varphi, \mu)$, $\varphi \in \mathcal{F}_m$, is played by the invariant torus of the system of equations

$$\frac{d\varphi}{dt} = \omega + \mu a(\varphi), \quad \frac{dy}{dt} = A(\varphi)y + \mu c(\varphi, u^{(i)}(\varphi, \mu)), \quad t \neq t_k(\varphi),$$

$$\Delta y|_{\langle r,\varphi\rangle=2\pi k} = \mu H_k(\varphi, u^{(i)}(\varphi, \mu)) \tag{20.28}$$

obtained by setting $u^{(0)}(\varphi, \mu) = 0$. Denote $\mu^* = \dfrac{\langle \omega, r\rangle}{Mm\|r\|}$, where $M > 0$ is the same constant as in Theorem 20.4. Then, for $0 \le \mu < \min\{\mu_0, \mu^*\}$, we arrive at the following representation for the invariant torus $\mathcal{F}^{(i+1)}(\mu)$:

$$\begin{aligned} u^{(i+1)}(\varphi, \mu) &= \mu \int_{-\infty}^{\infty} G_0(s, \varphi) c(\varphi_s, u^{(i)}(\varphi_s, \mu))\, ds \\ &\quad + \mu \sum_{-\infty < t_k < \infty} G_0(t_k(\varphi), \varphi)[H_k(\varphi, u^{(i)}(\varphi, \mu))]_{\langle r, \varphi\rangle = 2\pi k}. \end{aligned} \tag{20.29}$$

It remains to show that the sequence $\{u^{(i)}(\varphi, \mu)\}$ is uniformly bounded in norm by a constant d and converges as $i \to \infty$ uniformly in φ.

Let us show that, for all μ satisfying the inequality

$$\mu < \min\Big\{\mu_0, \mu^*, \frac{1}{\mathcal{L}\beta}\Big\},$$

we have

$$\|u^{(i+1)}(\varphi, \mu)\| \le \frac{\mu M \beta}{1 - \mu \mathcal{L} \beta} \quad (i = 0, 1, 2, \ldots), \tag{20.30}$$

where β is the same positive constant as in (20.21) and $\mathcal{L}$ is the constant from the Lipschitz condition with respect to y for the function $c(\varphi, y)$.

We have

$$\begin{aligned} \|u^{(1)}(\varphi, \mu)\| &\le \mu K M \Big[\int_{-\infty}^{\infty} \exp\{-\gamma |s|\}\, ds + \sum_{-\infty < t_k < \infty} \exp\{-\gamma |t_k|\}\Big] \\ &\le \Big[\frac{2}{\gamma} + \frac{2}{1 - \exp\{-\gamma\theta\}}\Big] \mu K M = \mu M \beta \le \frac{\mu M \beta}{1 - \mu \mathcal{L} \beta}. \end{aligned}$$

Assume that inequality (20.30) is true for $i = 1, 2, \ldots, j$. Let us show that it is true for $i = j + 1$. Indeed,

$$\begin{aligned} &\|u^{(i+1)}(\varphi, \mu)\| \\ &\quad \le \|u^{(j+1)}(\varphi, \mu) - u^{(1)}(\varphi, \mu)\| + \|u^{(1)}(\varphi, \mu)\| \\ &\quad \le \mu \mathcal{L} \frac{\mu M \beta}{1 - \mu \mathcal{L} \beta} \Big[\int_{-\infty}^{\infty} \exp\{-\gamma |s|\} ds + \sum_{-\infty < t_k < \infty} \exp\{-\gamma |t_k|\}\Big] + \mu M \beta \\ &\quad \le \frac{\mu^2 \mathcal{L} M \beta^2}{1 - \mu \mathcal{L} \beta} + \mu M \beta = \frac{\mu M \beta}{1 - \mu \mathcal{L} \beta}. \end{aligned}$$

Thus, inequality (20.30) is proved by induction. It is clear that the inequality

$$\frac{\mu_* M\beta}{1-\mu_*\mathcal{L}\beta} < 1$$

holds for sufficiently small μ_*. This inequality is also true for all $\mu < \mu_*$. This means that, for $\mu < \min\Big\{\mu_0, \mu^*, \mu_*, \dfrac{1}{\mathcal{L}\beta}\Big\}$, the sequence $\{u^{(i)}(\varphi,\mu)\}$ is bounded in norm by a constant d.

Let us show that this sequence converges uniformly in $\varphi \in \mathcal{F}_m$. Indeed,

$$\begin{aligned}
&\|u^{(i+1)}(\varphi,\mu) - u^{(i)}(\varphi,\mu)\| \\
&\quad\le \mu\mathcal{L}K\int_{-\infty}^{\infty}\exp\{-\gamma\,|\,s\,|\,\}\|u^{(i)}(\varphi_s,\mu) - u^{(i)}(\varphi_s,\mu)\|ds \\
&\quad+\mu\mathcal{L}K\sum_{-\infty<t_k<\infty}\exp\{-\gamma\,|\,t_k\,|\,\}\|u^{(i)}(\varphi_{t_k},\mu) - u^{(i-1)}(\varphi_{t_k},\mu)\| \\
&\quad\le \|u^{(i)}(\varphi,\mu) - u^{(i-1)}(\varphi,\mu)\|_0\,\mu\mathcal{L}K\left[\frac{2}{\gamma} + \frac{2}{1-\exp\{-\gamma\theta\}}\right]
\end{aligned}$$

and, therefore,

$$\begin{aligned}
\|u^{(i+1)}(\varphi,\mu) - u^{(i)}(\varphi,\mu)\|_0 &\le [\mu\mathcal{L}\beta]^i\|u^{(1)}(\varphi,\mu)\|_0 \\
&\le [\mu\mathcal{L}\beta]^i\mu M\beta.
\end{aligned}$$

Since $\mu\mathcal{L}\beta < 1$, sequence (20.29) converges uniformly in φ and μ for

$$\mu \le \mu^0 < \min\Big\{\mu_0, \mu^*, \mu_*, \frac{1}{\mathcal{L}\beta}\Big\}. \tag{20.31}$$

The validity of the last condition is sufficient for passing from system (20.28) to system (20.26).

For systems of equations (20.17) and (20.25), we now formulate an assertion concerning the behavior of these systems in the neighborhood of their invariant tori similar to Theorem 1.13 for system (12.37).

Assume that the matrizant $\Omega_\tau^t(\varphi)$ of the equation $\dfrac{dy}{dt} = A(\varphi_t(\varphi))y$ satisfies the condition of exponential decay for systems (20.17) and (20.25). Then the following theorem is true:

Theorem 20.6. *Assume that the system of equations (20.17) [(20.25)] is such that the conditions of Theorem 20.4 (20.5) are satisfied. Then, for sufficiently small μ, the trajectory of the solution $y_t = y(y_0, t, \varphi, \mu)$, $y_0 = (y_0, 0, \varphi, \mu)$, of this system is attracted as $t \to +\infty$ to the trajectory located on its invariant torus according to the exponential law.*

We prove Theorem 20.6 for the system of equations (20.25). We set $0 < \mu < \mu^0$, where μ^0 is specified by relation (20.31), and perform the change of variables

$$y = y(y_0, t, \varphi, \mu) = u(\varphi_t(\varphi), \mu) + \mu x(t, \varphi),$$

where the function $u(\varphi, \mu)$ determines the invariant torus $\mathcal{F}: y = u(\varphi, \mu)$, $\varphi \in \mathcal{F}_m$, of the system of equations (20.25).

For $x_t = x(t, \varphi)$, we obtain the following system:

$$\frac{dx}{dt} = A(\varphi_t(\varphi))x + \Big\{c(\varphi_t(\varphi), u(\varphi_t(\varphi), \mu) + \mu x) - c(\varphi_t(\varphi), u(\varphi_t(\varphi), \mu))\Big\},$$

$$t \neq t_k(\varphi),$$

$$\Delta x|_{\langle r, \varphi_t(\varphi)\rangle = 2\pi k} = H_k(\varphi_t, u(\varphi_t(\varphi), \mu) + \mu x) - H_k(\varphi_t, u(\varphi_t(\varphi), \mu)). \quad (20.32)$$

The solution of the system of equations (20.32) is sought in the form of the limit as $n \to \infty$ of the sequence $\{\overset{(n)}{x}t, \varphi)\}$ of solutions of the system of equations

$$\frac{d\overset{(n)}{x}}{dt} = A(\varphi_t(\varphi))\overset{(n)}{x}$$

$$+\Big\{c(\varphi_t(\varphi), u(\varphi_t(\varphi), \mu) + \mu \overset{(n-1)}{x}) - c(\varphi_t(\varphi), u(\varphi_t(\varphi), \mu))\Big\},$$

$$t \neq t_k(\varphi),$$

$$\Delta \overset{(n)}{x}|_{\langle r, \varphi_t(\varphi)\rangle = 2\pi k}$$

$$= H_k(\varphi_t(\varphi), u(\varphi_t(\varphi), \mu) + \mu \overset{(n-1)}{x}) - H_k(\varphi_t(\varphi), u(\varphi_t(\varphi), \mu)) \quad (20.33)$$

obtained by taking $\overset{(0)}{x} = 0 \in \mathfrak{M}$ as the initial approximation.

The solution of the system of equations (20.33) has the form

$$\begin{aligned}\overset{(n)}{x}(t,\tau,\varphi) &= \Omega_0^t(\varphi)x_0\\ &+\int_0^t \Omega_s^t(\varphi)\Big\{c(\varphi_s(\varphi),u(\varphi_s(\varphi),\mu)+\mu\overset{(n-1)}{x}(s,\tau,\varphi))\\ &\qquad -c(\varphi_s(\varphi),u(\varphi_s(\varphi),\mu))\Big\}\,ds\\ &+\sum_{0<t_k<t}\Omega_{t_k}^t(\varphi)\Big\{H_k(\varphi_{t_k}(\varphi),u(\varphi_{t_k}(\varphi),\mu)\\ &+\mu\overset{(n-1)}{x}(t,\tau,\varphi))-H_k(\varphi_{t_k}(\varphi),u(\varphi_{t_k}(\varphi),\mu))\Big\}. \end{aligned} \tag{20.34}$$

We set

$$s(\mu)=\mu\mathcal{L}K\left(\frac{2}{\gamma}+\frac{1}{1-\exp\left\{-\frac{\gamma}{2}\theta\right\}}\right)$$

and denote by μ_1^* the value of μ such that $s(\mu_1^*)=1$. Let us show that, for $\mu\in[0,\mu_1^*)$, the estimate

$$\|\overset{(n)}{x}(t,\varphi)\|\le\frac{K}{1-s(\mu)}\|x_0\|\exp\left\{-\frac{\gamma}{2}t\right\},\quad t\ge 0, \tag{20.35}$$

is true uniformly in $n=0,1,2,\dots$. Indeed, for $t\ge 0$,

$$\|\overset{(1)}{x}(t,\varphi)\|\le K\|x_0\|\exp\{-\gamma t\}\le K\|x_0\|\exp\left\{-\frac{\gamma}{2}t\right\}$$

and, hence, inequality (20.35) is true for $n=1$. Further, we assume that this inequality is true for all $n=2,3,\dots,j$ and prove its validity for $n=j+1$. Since, for $t\ge 0$,

$$\begin{aligned}\sum_{0<t_k<t}\exp\left\{-\gamma(t-t_k)-\frac{\gamma}{2}t_k\right\} &\le \exp\{-\gamma t/2\}\sum_{s=0}^{\infty}\exp\left\{-\frac{\gamma}{2}s\theta\right\}\\ &=\frac{\exp\left\{-\frac{\gamma}{2}t\right\}}{1-\exp\left\{-\frac{\gamma}{2}\theta\right\}},\end{aligned}$$

we can write

$$
\begin{aligned}
&\|\overset{(j+1)}{x}(t,\varphi)\| \\
&\quad \le \|\Omega_0^t(\varphi)\|\,\|x_0\| + \int_0^t \|\Omega_s^t(\varphi)\|\mathcal{L}\mu\|\overset{(j)}{x}(s,\varphi)\|\,ds \\
&\qquad + \sum_{0<t_k<t} \|\Omega_{t_k}^t(\varphi)\|\mathcal{L}\mu\|\overset{(j)}{x}(t_k,\varphi)\| \\
&\quad \le K\|x_0\|\exp\left\{-\frac{\gamma}{2}t\right\} + \frac{K^2\mathcal{L}\mu}{1-s(\mu)}\|x_0\| \\
&\qquad \times \left\{\int_0^t \exp\Big\{-\gamma(t-s)-\frac{\gamma}{2}s\Big\}\,ds + \sum_{0<t_k<t}\exp\Big\{-\gamma(t-t_k)-\frac{\gamma}{2}t_k\Big\}\right\} \\
&\quad \le K\|x_0\|\exp\{-\frac{\gamma}{2}t\}\left[1+\frac{K\mathcal{L}\mu}{1-s(\mu)}\left(\frac{2}{\gamma}+\frac{1}{1-\exp\{-\frac{\gamma}{2}\theta\}}\right)\right] \\
&\quad = \frac{K}{1-s(\mu)}\|x_0\|\exp\left\{-\frac{\gamma}{2}t\right\}, \quad t\ge 0.
\end{aligned}
$$

By induction, we conclude that relation (20.35) holds for all $n = 0, 1, 2, \ldots$.

Let us show that the sequence $\{\overset{(n)}{x}(t,\varphi)\}$ is uniformly convergent as $n \to \infty$. We set

$$
s_1(\mu) = \mu K\mathcal{L}\Big(\frac{1}{\gamma} + \frac{1}{1-\exp\{-\gamma\theta\}}\Big).
$$

Then

$$
\begin{aligned}
&\|\overset{(n+1)}{x}(t,\varphi) - \overset{(n)}{x}(t,\varphi)\| \\
&\quad \le \int_0^t K\exp\{-\gamma(t-s)\}\mathcal{L}\mu\|\overset{(n)}{x}(s,\varphi) - \overset{(n-1)}{x}(s,\varphi)\|\,ds \\
&\qquad + \sum_{0<t_k<t} K\exp\{-\gamma(t-t_k)\}\mathcal{L}\mu\|\overset{(n)}{x}(t_k,\varphi) - \overset{(n-1)}{x}(t_k,\varphi)\| \\
&\quad \le \mu\mathcal{L}K\|\overset{(n)}{x}(t,\varphi) - \overset{(n-1)}{x}(t,\varphi)\|_0\left\{\frac{1}{\gamma} + \frac{1}{1-\exp\{-\gamma\theta\}}\right\} \\
&\quad = \|\overset{(n)}{x}(t,\varphi) - \overset{(n-1)}{x}(t,\varphi)\|_0 s_1(\mu),
\end{aligned}
$$

and, hence,

$$\|\overset{(n+1)}{x}(t,\varphi) - \overset{(n)}{x}(t,\varphi)\| \le [s_1(\mu)]^n \|\overset{(1)}{x}(t,\varphi)\|_0.$$

Since $s_1(\mu) < s(\mu) < 1$ for $\mu \in [0, \mu_1^*)$, this estimate means that the sequence $\{\overset{(n)}{x}(t,\varphi)\}$ converges uniformly in $t \in R^+$ in the strong sense to a function $x(t,\varphi)$.

The functions $c(\varphi_t(\varphi), u(\varphi_t(\varphi),\mu) + \mu x)$ and $H_k(\varphi_t(\varphi), u(\varphi_t(\varphi),\mu) + \mu x)$ are obviously continuous in x in the strong sense for all integer k, and the series $\sum\limits_{j=1}^{\infty} a_{ij}(\varphi_t(\varphi))\overset{(n)}{x}_j(t,\varphi)$ converge absolutely and uniformly in t for $t \ge 0$. This enables us to pass to the limit as $n \to \infty$ in equalities (20.33) and show that $x(t,\varphi)$ is a solution of the system of equations (20.32) for $t \ge 0$ and, in addition,

$$\|x(t,\varphi)\| \le \frac{K}{1 - s(\mu)} \|x_0\| \exp\left\{-\frac{\gamma}{2}t\right\}, \quad t \ge 0. \tag{20.36}$$

Inequality (20.30) implies that $\|u(\varphi,\mu)\|$ is a monotonically decreasing function of μ such that $\|u(\varphi,\mu)\| \to 0$ as $\mu \to 0$. Then, for any $0 < \rho < d$, one can indicate a value $\mu = \mu_{**}$ such that $\|u(\varphi,\mu)\| \le d - \rho$ for any $\mu \in [0, \mu_{**}]$. Thus, we denote

$$\bar{\mu} = \min\{\mu^0, \mu_1^*, \mu_{**}\}.$$

If we now return to the original variables and use inequality (20.36), then we get

$$\|y_t - u(\varphi_t(\varphi),\mu)\| \le \frac{K}{1 - s(\mu)} \|y_0 - u(\varphi,\mu)\| \exp\left\{-\frac{\gamma}{2}t\right\}$$

for all $t \in R^+$, $\mu \in [0, \bar{\mu})$, and $\|y_0 - u(\varphi,\mu)\| \le \rho$. This completes the proof of Theorem 20.6.

4.21. Periodic Solutions for Impulsive Systems with Small Parameter

We consider a system of equations with pulse action

$$\frac{dx}{dt} = \varepsilon f(t,x), \quad t \ne t_j,$$

$$\Delta x|_{t=t_j} = \varepsilon H_j(t_j, x) \quad (j = \ldots, -1, 0, 1, \ldots), \tag{21.1}$$

where $x \in \mathfrak{M}$, $f(t,x)$ and $H_j(t,x)$ are countably dimensional functions periodic in t with period T (T-periodic functions) and defined in a domain D^*:

$$(t,x) \in R^1 \times D = (-\infty, \infty) \times \{x \in \mathfrak{M} \mid \|x\| \le R = \text{const}\},$$

and ε is a small positive parameter.

We pose the problem of finding a control (u_1, u_2) guaranteeing the T-periodicity of the solution of the impulsive system of equations

$$\frac{dx}{dt} = \varepsilon f(t,x) - u_1, \quad t \ne t_j,$$
$$\Delta x|_{t=t_j} = \varepsilon H_j(t_j, x) - u_2 \quad (j = \ldots, -1, 0, 1, \ldots) \tag{21.2}$$

that takes a value $x = x_0 \in \mathfrak{M}$ for $t = \tau$.

For the function H_j and the times of pulses t_j, we assume that $H_{j+d} = H_j$ and $t_{j+d} - t_j = T$, where d is a positive integer equal to the number of pulses per period.

It is also assumed that, in the domain D^*, the function $f(t,x)$

(a) is continuous in t,

(b) satisfies the Lipschitz condition with respect to x,

(c) is such that $\|f(t,0)\| \le p = \text{const} < \infty$.

Then, the Cauchy theorem on the existence and uniqueness of the solution $x = x(t, \tau, x_0)$ of equation $\dfrac{dx}{dt} = \varepsilon f(t,x)$ taking initial values τ and x_0 is true in the domain D^*, and the indicated solution admits the following representation:

$$x = x(t,\tau,x_0) = x_0 + \varepsilon \int_\tau^t f(s, x(s,\tau,x_0))\, ds.$$

For jump functions H_j bounded in the norm $\|\cdot\|$, this enables us to write the solution of the system of equations (21.2) in the form

$$x(t,\tau,x_0) = x_0 + \int_\tau^t [\varepsilon f(s, x(s,\tau,x_0)) - u_1] ds$$
$$+ \sum_{\tau < t_j < t} [\varepsilon H_j(t_j, x(t_j,\tau,x_0)) - u_2], \quad t \ge \tau. \tag{21.3}$$

Note that the function $f(t,x)$ with properties (a)–(c) is continuous in both arguments and bounded in norm in the domain D^*.

If the solution $x = x(t,\tau,x_0)$ is T-periodic, then

$$x(\tau + T, \tau, x_0) = x(\tau,\tau,x_0) = x_0.$$

Substituting the quantity $\tau + T$ for t in relation (21.3), we obtain

$$\int_{\tau}^{\tau+T} [\varepsilon f(s, x(s,\tau,x_0)) - u_1]ds + \sum_{\tau<t_j<\tau+T} [\varepsilon H_j(t_j, x(t_j,\tau,x_0)) - u_2] = 0. \tag{21.4}$$

The quantity (u_1, u_2) guaranteeing the validity of this relation can be chosen in different ways. In what follows, we consider three procedures used to determine the required control (u_1, u_2). We set $u_i = \varepsilon\mu_i$ $(i = 1, 2)$. The vector (μ_1, μ_2) is also regarded as control.

1°. Equality (21.4) is true for

$$\mu_1 = \frac{1}{T}\int_{\tau}^{t+T} f(s, x(s,\tau,x_0))ds,$$
$$\mu_2 = \frac{1}{pT}\sum_{\tau<t_j<\tau+T} H_j(t_j, x(t_j,\tau,x_0)), \quad p = \frac{d}{T}. \tag{21.5}$$

Denote $\mu_1 = \overline{f(t, x(t,\tau,x_0))}$ and $\mu_2 = \overline{H_j(t_j, x(t_j,\tau,x_0))}$.

In this case, the T-periodic solution of the system of equations (21.2) for (μ_1, μ_2) given by relations (21.5) must satisfy, in the interval $[\tau, \tau + T]$, the equation

$$x(t,\tau,x_0) = x_0 + \int_{\tau}^{t} [\varepsilon f(s, x(s,\tau,x_0)) - \varepsilon\overline{f(t, x(t,\tau,x_0))}\,]ds + \sum_{\tau<t_j<t} [\varepsilon H_j(t_j, x(t_j,\tau,x_0)) - \varepsilon\overline{H_j(t_j, x(t_j,\tau,x_0))}\,]. \tag{21.6}$$

Thus, if there exists a T-periodic function $x(t,\tau,x_0)$ $(x(\tau,\tau,x_0) = x_0)$ satisfying equation (21.6) on the segment $[\tau, \tau + T]$ and this equation possesses a unique solution with initial conditions τ and x_0, then this function is the required T-periodic solution of the system of equations (21.2).

Note that it is impossible to find the control (μ_1, μ_2) by using relations (21.5) because we do not know the solution $x = x(t, \tau, x_0)$.

In what follows, we use the following auxiliary statement:

Lemma 21.1. *Assume that the integrable function $f(t)$ and the function $H_j(t)$ are bounded in norm on the segment $[\tau, \tau + T]$ by a constant $M > 0$. Then*

$$\Big\|\int_\tau^t \Big(f(t) - \frac{1}{T}\int_\tau^{\tau+t} f(s)ds\Big)dt$$

$$+ \sum_{\tau<t_j<t}\Big(H_j(t_j) - \frac{1}{pT}\sum_{\tau<t_j<\tau+T} H_j(t_j)\Big)\Big\| \le M\alpha(t)$$

for all $t \in [\tau, \tau + T]$, *where* $\alpha(t) = 2(t - \tau)\left(1 - \dfrac{t-\tau}{T}\right) + 2d.$

The proof of the lemma directly follows from the inequalities

$$\Big\|\int_\tau^t \Big(f(t) - \frac{1}{T}\int_\tau^{\tau+t} f(s)ds\Big)dt + \sum_{\tau<t_j<t}\Big(H_j(t_j) - \frac{1}{pT}\sum_{\tau<t_j<\tau+T} H_j(t_j)\Big)\Big\|$$

$$\le \Big(1 - \frac{t-\tau}{T}\Big)\int_\tau^t \|f(s)\|ds + \frac{t-\tau}{T}\int_t^{\tau+T} \|f(s)\|ds$$

$$+\Big\|\sum_{\tau<t_j<t}\Big(H_j(t_j) - \frac{1}{pT}\sum_{\tau<t_j<\tau+T} H_j(t_j)\Big)\Big\|$$

$$\le 2M\Big(1 - \frac{t-\tau}{T}\Big)(t-\tau) + 2dM.$$

We now determine the sequence of T-periodic functions $\{x_m(t, \tau, x_0)\}$ given on the segment $[\tau, \tau + T]$ by the recurrence relations

$$x_{(0)}(t, \tau, x_0) = x_0, \quad t \in [\tau, \tau + T],$$

$$
\begin{aligned}
&x_{m+1}(t,\tau,x_0)\\
&\quad = x_0 + \varepsilon \int_{\tau}^{t} \Big[f(t, x_m(t,\tau,x_0)) - \frac{1}{T}\int_{\tau}^{\tau+t} f(s, x_m(s,\tau,x_0))\, ds\Big]\, dt\\
&\qquad + \varepsilon \sum_{\tau<t_j<t} \Big[H_j(t_j, x_m(t_j,\tau,x_0))\\
&\qquad - \frac{1}{pT} \sum_{\tau<t_j<\tau+T} H_j(t_j, x_m(t_j,\tau,x_0))\Big]. \qquad (21.7)
\end{aligned}
$$

Theorem 21.1. *Assume that the functions $f(t,x)$ and $H_j(t,x)$ defined in the domain D^* are continuous in t and satisfy the following conditions:*

(i) $\max\limits_{t\in[\tau,\tau+T]} \{\|f(t,x)\|, \|H_j(t,x)\|\} \le M = \text{const} < \infty$, $(j = \ldots, -1, 0, 1, \ldots)$;

(ii) $\|f(t,\bar{x}) - f(t,\bar{\bar{x}})\| \le L\|\bar{x}-\bar{\bar{x}}\|$ *and* $\|H_j(t,\bar{x}) - H_j(t,\bar{\bar{x}})\| \le L\|\bar{x}-\bar{\bar{x}}\|$, *where* $0 < L = \text{const}$, $\bar{x}, \bar{\bar{x}} \in D$.

Then, for any $0 < \varepsilon < \min\Big\{\dfrac{R}{M(T/2+2d)}, \dfrac{1}{L(T/2+2d)}\Big\}$ *and all initial values* $\tau \in (-\infty,\infty)$ *and* $x_0 \in D_f = \Big\{x \in D \,\big|\, \|x\| < R - \varepsilon M\big(T/2+2d\big)\Big\} \subset D$, *there exists a unique control* (u_1, u_2) *for which the system of equations (21.2) possesses a T-periodic solution* $x = x^0(t,\tau,x_0)$, $x^0(\tau,\tau,x_0) = x_0$ *and, in addition,*

$$
\begin{aligned}
u_1 &= \varepsilon \overline{H_j(t, x^0(t,\tau,x_0))},\\
u_2 &= \varepsilon \overline{H_j(t, x^0(t_j,\tau,x_0))}. \qquad (21.8)
\end{aligned}
$$

Proof. It is clear that the conditions of Theorem 21.1 imply the validity of conditions (a)–(c) for the function $f(t,x)$ in the domain D^*. Let us show that the sequence $\{x_m\}$ defined by relation (21.7) is uniformly bounded, i.e., $x_m(t,\tau,x_0) \in D$ for any natural m, provided that $x_0 \in D_f$.

By using Lemma 21.1, we estimate the difference $x_1(t,\tau,x_0) - x_0$. As a result, we obtain

$$
\|x_1(t,\tau,x_0) - x_0\| \le \varepsilon M\Big(2(t-\tau)\Big(1 - \frac{t-\tau}{T}\Big) + 2d\Big) \le \varepsilon M\Big(\frac{T}{2} + 2d\Big).
$$

Assume that

$$\|x_k(t,\tau,x_0)-x_0\| \le \varepsilon M\Big(\frac{T}{2}+2d\Big)$$

for all $k = 1, 2, \ldots, n$.. Then

$$\begin{aligned}\|x_{n+1}(t,\tau,x_0)-x_0\| &\le \varepsilon\Big\|\int_\tau^t\Big[f(t,x_n)-\frac{1}{T}\int_\tau^{\tau+t} f(s,x_n)\,ds\Big]\,dt \\ &\quad+\varepsilon\sum_{\tau<t_j<t}\Big[H_j(t_j,x_n)-\frac{1}{pT}\sum_{\tau<t_j<\tau+T} H_j(t_j,x_n)\Big]\Big\| \\ &\le \varepsilon M\Big(\frac{T}{2}+2d\Big)\end{aligned}$$

because

$$\begin{aligned}\|x_n(t,\tau,x_0)\| &\le \|x_0\|+\varepsilon M\Big(\frac{T}{2}+2d\Big) \\ &\le R-\varepsilon M\Big(\frac{T}{2}+2d\Big)+\varepsilon M\Big(\frac{T}{2}+2d\Big)=R,\end{aligned}$$

i.e., $x_n(t,\tau,x_0)$ belongs to the domain D.

By induction, we conclude that, for all natural m,

$$\|x_m(t,\tau,x_0)-x_0\| \le \varepsilon M\Big(\frac{T}{2}+2d\Big)$$

whenever $x_0 \in D_f$. Let us show that the sequence $\{x_m\}$ converges in the norm $\|\cdot\|$ uniformly in $t \in [\tau, \tau+T]$.

To this end, we estimate the difference $x_{m+1}(t,\tau,x_0)-x_m(t,\tau,x_0)$. By using condition (ii) of Theorem 21.1 and relation (21.7), we obtain

$$\begin{aligned}&\|x_{m+1}(t,\tau,x_0)-x_m(t,\tau,x_0)\| \\ &\quad\le \varepsilon\int_\tau^t\Big(L\|x_m(t,\tau,x_0)-x_{m-1}(t,\tau,x_0)\| \\ &\qquad+\frac{1}{T}\int_\tau^{\tau+T} L\|x_m(s,\tau,x_0)-x_{m-1}(s,\tau,x_0)\|ds\Big)dt \\ &\qquad+\varepsilon\sum_{\tau<t_j<t}\Big(L\|x_m(t_j,\tau,x_0)-x_{m-1}(t_j,\tau,x_0)\| \\ &\qquad-\frac{1}{pT}\sum_{\tau<t_j<\tau+T} L\|x_m(t_j,\tau,x_0)-x_{m-1}(t_j,\tau,x_0)\|\Big)\end{aligned}$$

$$\leq \varepsilon L2(t-\tau)\Big(1-\frac{t-\tau}{T}\Big)\|x_m(t,\tau,x_0)-x_{m-1}(t,\tau,x_0)\|_0$$
$$+2\varepsilon Ld\,\|x_m(t,\tau,x_0)-x_{m-1}(t,\tau,x_0)\|_0$$
$$\leq \varepsilon L\Big(\frac{T}{2}+2d\Big)\|x_m-x_{m-1}\|_0. \tag{21.9}$$

This yields

$$\|x_{m+1}(t,\tau,x_0)-x_m(t,\tau,x_0)\|$$
$$\leq \Big[\varepsilon L\Big(\frac{T}{2}+2d\Big)\Big]^m\|x_1(t,\tau,x_0)-x_0\|_0$$
$$\leq \Big[\varepsilon L\Big(\frac{T}{2}+2d\Big)\Big]^m\varepsilon M\Big(\frac{T}{2}+2d\Big) \tag{21.10}$$

and, hence, for $\varepsilon < \Big(L\Big(\frac{T}{2}+2d\Big)\Big)^{-1}$, the sequence $\{x_m\}$ converges in the norm $\|\cdot\|$ uniformly in $t\in[\tau,\tau+T]$.

Denote
$$\lim_{m\to\infty} x_m(t,\tau,x_0)=x^0(t,\tau,x_0).$$

Since the functions $x_m(t,\tau,x_0)$ are T-periodic, the function $x^0(t,\tau,x_0)$ is also T-periodic.

In relation (21.7), we pass to the limit as $m\to\infty$ and conclude that the function $x=x^0(t,\tau,x_0)$ is a solution of equation (21.6) on the segment $[\tau,\tau+T]$.

Let us show that equation (21.6) cannot have two or more solutions with the same initial values. Indeed, assume the contrary, i.e., that $y=y(t,\tau,x_0)$, $y(t,\tau,x_0)=x_0$, is a solution of equation (21.6) other than $x=x^0(t,\tau,x_0)$. In this case, we readily obtain the following estimate similar to inequality (21.9):

$$\|x^0(t,\tau,x_0)-y(t,\tau,x_0)\|_0\leq \varepsilon L\Big(\frac{T}{2}+2d\Big)\|x^0(t,\tau,x_0)-y(t,\tau,x_0)\|_0.$$

Since $\varepsilon L\Big(\frac{T}{2}+2d\Big)<1$, we have $\|x^0(t,\tau,x_0)-y(t,\tau,x_0)\|=0$ and, hence, $x^0(t,\tau,x_0)=y(t,\tau,x_0)$, i.e., we arrive at a contradiction.

It remains to prove the uniqueness of the control (u_1,u_2) satisfying equation (21.8). Denote $x^0(t,\tau,x_0)=x(t,\tau,x_0,u)$. Assume that there exists a control $v=(v_1,v_2)$ corresponding to a T-periodic solution $x=x(t,\tau,x_0,v)$ of the system of equations (21.2) and $x(\tau,\tau,x_0,v)=x_0$. Further, we assume that the quantities v and $x(t,\tau,x_0,v)$ satisfy relations (21.8). In this case, $x(t,\tau,x_0,v)$

is a solution of equation (21.6), which is impossible according to the assertion proved above. This means that $x(t,\tau,x_0,v) = x(t,\tau,x_0,u)$ and, moreover, relations (21.8) imply that $v_1 = u_1$ and $v_2 = u_2$.

2°. Equality (21.4) is satisfied for $\mu = (0, \mu_2)$, where

$$\mu_2 = \frac{1}{d}\Big\{ \int_{\tau}^{\tau+T} f(s, x(s,\tau,x_0))ds + \sum_{\tau<t_j<\tau+T} H_j(t_j, x(t_j,\tau,x_0))\Big\}. \quad (21.11)$$

This quantity is regarded as a control over the pulse part. Substituting relation (21.11) in (21.3), we obtain

$$\begin{aligned} x(t,\tau,x_0) &= x_0 + \int_{\tau}^{t} \varepsilon f(s, x(s,\tau,x_0))ds \\ &+ \sum_{\tau<t_j<t} \Big\{\varepsilon H_j(t_j, x(t_j,\tau,x_0)) - \frac{\varepsilon}{d}\Big[\int_{\tau}^{\tau+T} f(s, x(s,\tau,x_0))ds \\ &+ \sum_{\tau<t_j<\tau+T} H_j(t_j, x(t_j,\tau,x_0))\Big]\Big\}, \quad t \in [\tau, \tau+T]. \quad (21.12) \end{aligned}$$

In Case 2°, this equation plays the role of equation (21.6).

We define the sequence of T-periodic equation functions $\{x_m(t,\tau,x_0)\}$ given on the segment $[\tau, \tau+T]$ by the recurrence relation

$$\begin{aligned} &x_{m+1}(t,\tau,x_0) \\ &= x_0 + \int_{\tau}^{t} \varepsilon f(s, x_m(s,\tau,x_0))ds \\ &- \sum_{\tau<t_j<t} \frac{\varepsilon}{d} \int_{\tau}^{\tau+T} f(s, x_m(s,\tau,x_0))ds + \sum_{0<t_j<t} \varepsilon H_j(t_j, x_m(t_j,\tau,x_0)) \\ &- \frac{\varepsilon}{d} \sum_{\tau<t_j<t}\ \sum_{\tau<t_j<\tau+T} H_j(t_j, x_m(t_j,\tau,x_0)). \quad (21.13) \end{aligned}$$

Theorem 21.2. *Assume that the conditions of Theorem 21.1 are satisfied. Then, for any* $0 < \varepsilon < \min\Big\{\frac{R}{2M(T+d)}; \frac{1}{2L(T+d)}\Big\}$ *and all initial values*

$\tau \in (-\infty, \infty)$ and $x_0 \in D'_f = \{x \in D | \|x\| < R - 2\varepsilon M(T+d)\} \subset D$, there exists a unique control $(0, u_2)$ for which the system of equations (21.2) possesses a T-periodic solution $x = x_\infty(t, \tau, x_0)$ such that $x_\infty(\tau, \tau, x_0) = x_0$.

Proof. Since $x_0 \in D$, relation (21.13) implies the inequality

$$\|x_1 - x_0\| \leq \varepsilon M(t-\tau) + \frac{\varepsilon}{d} MT \sum_{\tau < t_j < t} 1$$

$$+\varepsilon M \sum_{\tau < t_j < t} 1 + \frac{\varepsilon}{d} \sum_{\tau < t_j < t} \sum_{\tau < t_j < \tau + T} M \leq 2\varepsilon M(T+d).$$

For $\varepsilon \leq \dfrac{R}{2M(T+d)}$, this yields

$$\|x_1(t, \tau, x_0)\| \leq \|x_0\| + 2\varepsilon M(T+d),$$

provided that $x_0 \in D'_f$.

Hence, $x_1(t, \tau, x_0) \in D$ for $x_0 \in D'_f$. Let $x_2(t, \tau, x_0)$, ..., $x_m(t, \tau, x_0)$ belong to D for $x_0 \in D'_f$. Then

$$\|x_{m+1}(t, \tau, x_0) - x_0\| \leq 2\varepsilon M(T+d)$$

and, hence, $x_{m+1}(t, \tau, x_0) \in D$. By induction, we conclude that $x_m(t, \tau, x_0) \in D$ for any natural m.

Let us show that the sequence $\{x_m(t, \tau, x_0)\}$ converges in the norm $\|\cdot\|$ uniformly in t on the segment $[\tau, \tau + T]$.

Thus,

$$\|x_{m+1}(t, \tau, x_0) - x_m(t, \tau, x_0)\|$$

$$\leq \varepsilon(t-\tau)\|x_m - x_{m-1}\|_0 + \varepsilon LT\|x_m - x_{m-1}\|_0$$

$$+\varepsilon Ld\|x_m - x_{m-1}\|_0 + \varepsilon L\|x_m - x_{m-1}\|_0$$

$$\leq 2\varepsilon L(T+d)\|x_m - x_{m-1}\|_0$$

and, hence, we arrive at the following inequality similar to (21.10):

$$\|x_{m+1}(t, \tau, x_0) - x_m(t, \tau, x_0)\| \leq [2\varepsilon L(T+d)]^m 2\varepsilon M(T+d).$$

For $\varepsilon < \dfrac{1}{2L(T+d)}$, this inequality guarantees the uniform convergence of the sequence $\{x_m(t, \tau, x_0)\}$ on the segment $[\tau, \tau + T]$.

Denote

$$\lim_{m\to\infty} x_m(t,\tau,x_0) = x_\infty(t,\tau,x_0).$$

Since the functions $x_m(t,\tau,x_0)$ are T-periodic, the function $x_\infty(t,\tau,x_0)$ is also T-periodic.

Passing to the limit as $m \to \infty$ in (21.13), we conclude that $x_\infty(t,\tau,x_0)$ is a solution of equation (21.12) on the segment $[\tau, \tau+T]$.

Equation (21.12) possesses a unique solution with initial values τ and x_0. Assume the contrary, i.e., assume that $y(t,\tau,x_0)$ is a solution of this equation and $y(t,\tau,x_0) = x_0$. This yields

$$\|x_\infty(t,\tau,x_0) - y(t,\tau,x_0)\|_0 \le 2\varepsilon L(T+d)\|x_\infty(t,\tau,x_0) - y(t,\tau,x_0)\|_0.$$

In view of the inequality $2\varepsilon L(T+d) < 1$, we conclude that $y(t,\tau,x_0) = x_\infty(t,\tau,x_0)$. This means that $x = x_\infty(t,\tau,x_0)$ is a T-periodic solution of the system of equations (21.2).

We now assume that there exists one more control over the pulse part $(0, v)$ for which the solution $x = x(t,\tau,x_0,v)$, $x(\tau,\tau,x_0,v) = x_0$ of the system of equations (21.2) is T-periodic. In this case, v and $x(t,\tau,x_0,v)$ must satisfy relation (21.11). Hence, the function $x(t,\tau,x_0,v)$ must satisfy equation (21.12), which, as shown above, possesses a unique solution. Therefore, $x(t,\tau,x_0,v) = x_\infty(t,\tau,x_0)$. By virtue of (21.11), this means that $v = u_2$, which completes the proof of Theorem 21.2.

3°. Equality (21.4) is satisfied for $\mu = (\mu_1, 0)$, where

$$\mu_1 = \frac{1}{T}\Big\{ \int_\tau^{\tau+T} f(s, x(x,\tau,x_0))ds + \sum_{\tau<t_j<\tau+T} H_j(t_j, x(t_j,\tau,x_0))\Big\}. \quad (21.14)$$

We say that this quantity plays the role of control over the differential part. Substituting relation (21.14) in (21.3), we obtain

$$\begin{aligned} x(t,\tau,x_0) &\\ &= x_0 + \varepsilon \int_\tau^t \Big(f(s, x(s,\tau,x_0)) - \frac{1}{T}\int_\tau^{\tau+T} f(\xi, x(\xi,\tau,x_0))d\xi\Big)ds \\ &\quad + \varepsilon\Bigg(\sum_{\tau<t_j<t} H_j(t_j, x(t_j,\tau,x_0)) \\ &\quad - \frac{t-\tau}{T} \sum_{\tau<t_j<\tau+T} H_j(t_j, x(t_j,\tau,x_0))\Bigg). \end{aligned} \quad (21.15)$$

Let us now define a sequence of T-periodic functions $\{x_m(t,\tau,x_0)\}$ given on the segment $[\tau,\tau+T]$ by the recurrence relations

$$x_{(0)}(t,\tau,x_0)=x_0,\quad t\in[\tau,\tau+T],$$

$$\begin{aligned} &x_{m+1}(t,\tau,x_0)\\ &\quad= x_0+\varepsilon\int_{\tau}^{t}\Big(f(s,x_m(s,\tau,x_0))-\frac{1}{T}\int_{\tau}^{\tau+T}f(\xi,x_m(\xi,\tau,x_0))d\xi\Big)ds\\ &\qquad+\varepsilon\Bigg(\sum_{\tau<t_j<t}H_j(t_j,x_m(t_j,\tau,x_0))\\ &\qquad-\frac{t-\tau}{T}\sum_{\tau<t_j<\tau+T}H_j(t_j,x_m(t_j,\tau,x_0))\Bigg). \end{aligned} \tag{21.16}$$

Theorem 21.3. *Assume that the conditions of Theorem 21.1 are satisfied. Then, for any* $0<\varepsilon<\min\Big\{R\Big[M\Big(\frac{T}{2}+2d\Big)\Big]^{-1};\Big[L\Big(\frac{T}{2}+2d\Big)\Big]^{-1}\Big\}$ *and all initial values* $\tau\in(-\infty,\infty)$ *and* $x_0\in D_f=\Big\{x\in D\,|\,\|x\|\le R-\varepsilon M(\frac{T}{2}+2d)\Big\}\subset D$*, there exists a unique control* $(u_1,0)$ *for which the system of equations (21.2) possesses a* T*-periodic solution* $x=x^\infty(t,\tau,x_0)$ *such that* $x^\infty(\tau,\tau,x_0)=x_0$.

Proof. This theorem is proved by analogy with Theorems 21.1 and 21.2. Thus, we have

$$\|x_1(t,\tau,x_0)-x_0\|\le\varepsilon M\Big(\frac{T}{2}+2d\Big).$$

For $x_0\in D_f$ and $\varepsilon<R\Big[M\Big(\frac{T}{2}+2d\Big)\Big]^{-1}$, this inequality implies that $x_1(t,\tau,x_0)$ belongs to D.

By induction, we conclude that the inequality

$$\|x_m(t,\tau,x_0)-x_0\|\le\varepsilon M\Big(\frac{T}{2}+2d\Big)$$

holds for all natural m. Thus, for $x_0\in D_f$, we have $x_m\in D$ for all $m=1,2,3\ldots$.

Moreover,

$$\|x_{m+1}(t,\tau,x_0) - x_m(t,\tau,x_0)\|$$

$$\leq \varepsilon\Big\{\int_\tau^t (\|f(s,x_m(s,\tau,x_0)) - f(s,x_{m-1}(s,\tau,x_0))\|$$

$$+\frac{1}{T}\int_\tau^{\tau+T} \|f(\xi,x_m(\xi,\tau,x_0)) - f(\xi,x_{m-1}(\xi,\tau,x_0))\|d\xi)ds\Big\}$$

$$+\varepsilon\Big\{\sum_{\tau<t_j<t} \|H_j(t_j,x_m(t_j,\tau,x_0)) - H_j(t_j,x_{m-1}(t_j,\tau,x_0))\|$$

$$+\frac{t-\tau}{T}\sum_{\tau<t_j<\tau+T} \|H_j(t_j,x_m(t_j,\tau,x_0)) - H_j(t_j,x_{m-1}(T_j,\tau,x_0))\|\Big\}$$

$$\leq \varepsilon L\Big(\frac{T}{2}+2d\Big)\|x_m(t,\tau,x_0) - x_{m-1}(t,\tau,x_0)\|_0, \quad t\in[\tau,\tau+T],$$

which yields inequality (21.10). This means that the sequence $\{x_m(t,\tau,x_0)\}$ given by the recurrence relation (21.16) converges as $m \to \infty$ to a function $x^\infty(t,\tau,x_0)$ uniformly in $t \in [\tau,\tau+T]$. This function is a solution of equation (21.15). The uniqueness of this solution and the uniqueness of control over the differential part are proved in exactly the same way as in Theorem 21.2.

Denote

$$\min\left\{\frac{R}{M(T/2+2d)};\frac{1}{L(T/2+2d)}\right\} = \gamma_{12},$$

$$\min\left\{\frac{R}{2M(T+d)};\frac{1}{2L(T+d)}\right\} = \gamma_2.$$

It is obvious that $\gamma_2 \leq \gamma_{12}$ and, in addition, $D'_f \subset D_f$.

This means that the pulse part can be controlled in a narrower domain and for lower values of the parameter ε than in the case of control over the differential part. In the case of simultaneous control over the differential and pulse parts (Case 1°), we can use the same domain as in the case of control over the differential part.

In what follows, systems of equations of the form (21.1) satisfying the conditions of Theorem 21.1 are called impulsive T-systems and the control (u_1,u_2) specified by relations (21.8) is called mixed.

We now formulate several corollaries of Theorems 21.1–21.3.

Corollary 21.1. *Under the conditions of Theorem 21.1, the following inequalities are true in the case of mixed control:*

$$\|x^0(t,\tau,x_0) - x_m(t,\tau,x_0)\| \le \zeta(m) = \frac{\varepsilon^{m+1}L^m M(T/2+2d)^{m+1}}{1-\varepsilon L(T/2+2d)}, \quad (21.17)$$

$$\|\mu_i - \mu_i^{(m)}\| \le L\zeta(m) \ \ (i=1,2), \quad (21.18)$$

where $x_m(t,\tau,x_0)$ is given by the recurrence relation (21.7),

$$\mu_1^{(m)} = \overline{f(t,x_m(t,\tau,x_0))}, \quad \textit{and} \quad \mu_2^{(m)} = \overline{H_j(t_j,x_m(t_j,\tau,x_0))}.$$

Proof. By using inequality (21.10), we obtain

$$\|x^0(t,\tau,x_0) - x_m(t,\tau,x_0)\|$$
$$\le \|x^0(t,\tau,x_0) - x_{m+1}(t,\tau,x_0)\| + \varepsilon^{m+1}L^m M\Big(\frac{T}{2}+2d\Big)^{m+1}.$$

Relations (21.6) and (21.7) yield

$$\|x^0(t,\tau,x_0) - x_{m+1}(t,\tau,x_0)\|_0 \le \varepsilon L\Big(\frac{T}{2}+2d\Big)\|x^0(t,\tau,x_0) - x_m(t,\tau,x_0)\|_0.$$

Inequality (21.17) now follows from these two inequalities.

It is clear that the mixed control satisfies the relations

$$\lim_{m\to\infty} \mu_i^{(m)} = \mu_i \ \ (i=1,2).$$

Furthermore,

$$\|\mu_1 - \mu_1^{(m)}\| \le \frac{1}{T}\int_\tau^{\tau+T} \|f(s,x^0(s,\tau,x_0)) - f(s,x_m(s,\tau,x_0))\|ds$$

$$\le \frac{L}{T}\int_\tau^{\tau+T} \|x^0(s,\tau,x_0) - x_m(s,\tau,x_0)\|ds,$$

$$\|\mu_2 - \mu_2^{(m)}\| \le \frac{1}{pT}\sum_{\tau<t_j<\tau+T}\Big\|H_j(t_j,x^0(t_j,\tau,x_0)) - H_j(t_j,x_m(t_j,\tau,x_0))\Big\|$$

$$\le \frac{L}{pT}\sum_{\tau<t_j<\tau+T}\|x^0(t_j,\tau,x_0) - x_m(t_j,\tau,x_0)\|.$$

This yields inequalities (21.18). It is also clear that $\zeta(m) \to 0$ as $m \to \infty$.

By analogy, we can prove the following two assertions:

Corollary 21.2. *Under the conditions of Theorem 21.2, the following inequalities are true in the case of control over the pulse part:*

$$\|x_\infty(t,\tau,x_0) - x_m(t,\tau,x_0)\| \le \sigma(m) = \frac{\varepsilon^{m+1}2^{m+1}ML^m(T+d)^{m+1}}{1-2\varepsilon L(T+d)},$$

$$\|\mu_2 - \mu_2^{(m)}\| \le L\sigma(m)\Big(\frac{T}{d}+1\Big),$$

where $x_m(t,\tau,x_0)$ *are given by the recurrence relation (21.13) and*

$$\mu_2^{(m)} = \frac{1}{d}\Big\{\int\limits_\tau^{\tau+T} f(s,x_m(s,\tau,x_0))ds + \sum_{\tau<t_j<\tau+T} H_j(t_j,x_m(t_j,\tau,x_0))\Big\}.$$

Corollary 21.3. *Under the conditions of Theorem 21.3, the following inequalities are true in the case of control over the differential part:*

$$\|x^\infty(t,\tau,x_0) - x_m(t,\tau,x_0)\| \le \zeta(m),$$

$$\|\mu_1 - \mu_1^{(m)}\| \le L(1+\frac{d}{T})\zeta(m), \tag{21.19}$$

where $x_m(t,\tau,x_0)$ *is given by the recurrence relation (21.16) and*

$$\mu_1^{(m)} = \frac{1}{T}\left\{\int\limits_\tau^{\tau+T} f(s,x_m(s,\tau,x_0))ds + \sum_{\tau<t_j<\tau+T} H_j(t_j,x_m(t_j,\tau,x_0))\right\}.$$

The solution $x = x(t,\tau,x_0)$ of the system of equations (21.1) for which $x(\tau,\tau,x_0) = x_0 \in D_f$ is T-periodic if and only if the control over its differential part (mixed control or control over the pulse part for $x_0 \in D'_f$) is equal to zero. This statement is a simple corollary of Theorems 21.1–21.3. Therefore, in what follows, it suffices to study one of the analyzed types of control. We restrict ourselves to the case of control over the differential part of the system of equations (21.1).

Corollary 21.4. *If an impulsive T-system (21.1) such that*

$$0 < \varepsilon < \min\left\{\frac{R}{M(T/2+2d)}; \frac{1}{L(T/2+2d)}\right\} \tag{21.20}$$

possesses a T-periodic solution $x = \varphi(t)$, $\varphi(\tau) = x_0 \in D_f$, then this solution is the strong limit of a uniformly convergent sequence of T-periodic functions $\{x_m(t, \tau, x_0)\}$ defined on the period $[\tau, \tau, +T]$ by the recurrence relation (21.16) and, in addition, $\|\varphi(t) - x_m(t, \tau, x_0)\| \leq \zeta(m)$, where $\zeta(m)$ is given by relation (21.17).

Proof. Since the solution $\varphi(t)$ is T-periodic, the expression for μ_1 on the right-hand side of relation (21.14) is equal to $0 \in \mathfrak{M}$. Hence, the function $\varphi(t)$ admits the representation

$$\varphi(t) = x_0 + \int_\tau^t \varepsilon f(t, \varphi(t))dt + \sum_{\tau<t_j<t} \varepsilon H_j(t_j, \varphi(t_j)) - \varepsilon\mu_1(t-\tau). \tag{21.21}$$

Substituting the value of μ_1 given by relation (21.14) in relation (21.21), we conclude that $\varphi(t)$ is a solution of equation (21.15). By virtue of Theorem 3.21, this equation possesses a unique T-periodic solution with initial values τ and x_0, and, hence,

$$\varphi(t) = x^\infty(t, \tau, x_0) = \lim_{m\to\infty} x_m(t, x_0).$$

We now denote

$$\Delta(x_0) = \frac{\varepsilon}{T}\left\{\int_\tau^{\tau+T} f(s, x^\infty(s, \tau, x_0))\, ds \right.$$

$$\left. + \sum_{\tau<t_j<\tau+T} H_j(t_j, x^\infty(t_j, \tau, x_0))\right\} \tag{21.22}$$

and fix the initial value τ. It is clear that, for $\Delta(x_0) = 0$, the function $x^\infty(t, \tau, x_0)$ is a T-periodic solution of the system of equations (21.1).

Thus, the problem of the existence of T-periodic solutions of the impulsive T-system (21.1) is reduced to the problem of finding zeros of the function $\Delta(x_0)$. However, it is practically impossible to determine this function. At the same time, the problem of zeros of the function $\Delta(x_0)$ can be solved by using the following function:

$$\Delta_m(x_0) = \frac{\varepsilon}{T}\Bigg\{ \int\limits_{\tau}^{\tau+T} f(s, x_m(s,\tau,x_0))\, ds + \sum_{\tau<t_j<\tau+T} H_j(t_j, x_m(t_j,\tau,x_0))\Bigg\}, \qquad (21.23)$$

where $x_m(t,\tau,x_0)$ are given by the recurrence relation (21.16). The corresponding result is formulated in the form of the following theorem, which can be regarded as an analog of Theorem 6.2 for T-systems without pulse action:

Theorem 21.4. *Assume that, in the domain D^*, the impulsive T-system (21.1) satisfies inequality (21.20) and the following conditions:*

(i) for some fixed $\tau \in (-\infty, \infty)$ and integer n $(n = 0, 1, 2, \ldots)$, the mapping $\Delta_n\colon D_f \to \mathfrak{M}$, where $\Delta_n(x_0)$ is given by relation (21.23), has a singular point x^0, i.e., $\Delta_n(x^0) = 0 \in \mathfrak{M}$;

(ii) there exists a closed bounded domain $D_1 \subset D_f$ containing the point x^0 such that Δ_n topologically maps D_1 onto $\Delta_n D_1$;

(iii) the inequality

$$\inf_{x\in\Gamma_{D_1}} \|\Delta_n(x)\| \geq \varepsilon L \left(1 + \frac{d}{T}\right)\zeta(n), \qquad (21.24)$$

where $\zeta(n)$ is given by relation (21.17), holds on the boundary Γ_{D_1} of the domain D_1.

Then the system of equations (21.1) possesses a T-periodic solution $x = x(t)$ such that $x(\tau) \in D_1$.

Proof. First, we show that the mapping $\Delta(x)$ is continuous on D_f. By using relation (21.22), we obtain

$$\|\Delta(x_1) - \Delta(x_2)\| \leq \frac{\varepsilon L}{T}\Bigg\{ \int\limits_{\tau}^{\tau+T} \|x^\infty(t,x_1) - x^\infty(t,x_2)\| dt + \sum_{\tau<t_j<\tau+T} \|x^\infty(t_j,x_1) - x^\infty(t_j,x_2)\|\Bigg\}, \qquad (21.25)$$

where $x_1, x_2 \in D_f$.

Denote $x^\infty(t,x_1) - x^\infty(t,x_2) = r(t)$. Since the functions $x^\infty(t,x_1)$ and $x^\infty(t,x_2)$ satisfy equation (21.15), we get

$$\|r(t)\| \leq \|x_1 - x_2\| + \varepsilon L\Big(1 - \frac{t-\tau}{T}\Big)\int_\tau^t \|r(s)\|_0\, ds$$

$$+ \frac{t-\tau}{T}\int_\tau^{\tau+T} \varepsilon L\|r(s)\|_0\, ds + \varepsilon L\Big\{\sum_{\tau<t_j<t}\|r(t)\|_0 + \sum_{\tau<t_j<\tau+T}\|r(t)\|_0\Big\}$$

$$\leq \|x_1 - x_2\| + \varepsilon L\Big(2d + \frac{T}{2}\Big)\|r(t)\|_0$$

and, hence,

$$\|r(t)\| \leq \frac{1}{1-\varepsilon L(T/2+2d)}\|x_1 - x_2\| \qquad \text{for} \quad \varepsilon L\Big(\frac{T}{2} + 2d\Big) < 1.$$

Further, it follows from relation (21.25) that

$$\|\Delta(x_1) - \Delta(x_2)\| \leq \frac{\varepsilon L(t+d)}{T[1-\varepsilon L(2d+T/2)]}\|x_1 - x_2\|, \tag{21.26}$$

i.e., the mapping $\Delta(x)$ is continuous on D_f. At the same time, we have $D_1 \subset D_f$, which means that $\Delta(x)$ is continuous on D_1.

By virtue of inequality (21.19) and Lemma 4.6, the set

$$\Delta_n D_1 - \varepsilon L\Big(1 + \frac{d}{T}\Big)\zeta(n) = D_2$$

is contained in ΔD_1. If the set D_2 contains $0 \in \mathfrak{M}$, then $0 \in \Delta D_1$. This is sufficient for the impulsive system (21.1) to have a T-periodic solution.

Thus, it remains to show that $0 \in D_2$. This fact is proved by using inequality (21.24) by analogy with the corresponding part of the proof of Theorem 6.2.

However, the procedure presented above does not enable us to find the initial values τ and x_0 of the T-periodic solution of system (21.1). This problem is in no case simple and can be solved without serious difficulties only in some special cases. In what follows, we study one of these cases.

Consider a system of equations

$$\frac{dx}{dt} = \varepsilon f(t,x), \quad t \neq t_j,$$

$$\Delta x\Big|_{t=t_j} = \varepsilon H(t)\Big|_{t=t_j} \qquad (j = \ldots, -1, 0, 1, \ldots), \tag{21.27}$$

for which the conditions of Theorem 21.1 are satisfied, i.e., the system of equations (21.27) is a special case of the impulsive T-system (21.1).

Theorem 21.5. *Assume that the system of equations (21.27) is such that*

(i) $H(-t) = -H(t)$ *and* $f(t,x) = -f(-t,x)$,

(ii) on the interval $\left(-\frac{T}{2}; \frac{T}{2}\right)$, *the times of pulses* t_j $(j = \overline{1,d})$ *are symmetric about zero.*

Then every point $x_0 \in D_f$ *is the initial value of a* T*-periodic solution* $x(t)$, $x(0) = x_0$, *of this system of equations.*

Proof. For $t \geq 0$, the solution $x(t)$ of the system of equations (21.27) satisfies the equation

$$x(t) = x_0 + \varepsilon \int_0^t f(s, x(s))ds + \sum_{0<t_j<t} \varepsilon H(t_j).$$

At the same time, for $t \leq 0$, this solution satisfies the equation

$$x(t) = x_0 + \varepsilon \int_0^t f(s, x(s))ds - \sum_{0<t_j<0} \varepsilon H(t_j).$$

Consider the recurrence relations

$$x_{m+1}(t) = \begin{cases} x_0 + \varepsilon \int_0^t f(s, x_m(s))ds + \sum_{0<t_j<t} \varepsilon H(t_j), & t \geq 0, \\ x_0 + \varepsilon \int_0^t f(s, x_m(s))ds - \sum_{t<t_j<0} \varepsilon H(t_j), & t < 0, \\ x_{(0)} = x_0. \end{cases} \tag{21.28}$$

Using these relations, we obtain

$$x_1(t) = x_0 + \varepsilon \int_0^t f(s, x_0)ds + \sum_{0<t_j<t} \varepsilon H(t_j), \quad t \geq 0, \tag{21.29}$$

$$x_1(t) = x_0 + \varepsilon \int_0^t f(s, x_0)ds - \sum_{t<t_j<0} \varepsilon H(t_j), \quad t < 0. \tag{21.30}$$

Let $t \geq 0$. Then $x_1(t)$ is given by relation (21.29), where the times of pulses t_j are positive. Since the times of pulses are symmetric about the origin and the function $H(t)$ is odd, by virtue of (21.30) we obtain

$$\begin{aligned} x_1(-t) &= x_0 + \varepsilon \int_0^{-t} f(s, x_0)\, ds - \sum_{-t<-t_j<0} \varepsilon H(-t_j) \\ &= x_0 + \varepsilon \int_0^{t} f(s, x_0) ds + \sum_{0<t_j<t} \varepsilon H(t_j) \\ &= x_1(t) \end{aligned}$$

because the values of the function $H(t)$ at the points symmetric about zero have different signs and, for the odd function $f(t, x_0)$, we have

$$\int_0^{-t} f(s, x_0) ds = \int_0^{t} f(s, x_0)\, ds.$$

Thus, the function $x_1(t)$ is even. Since

$$\sum_{t<t_j<t+T} \varepsilon H(t_j) = 0,$$

we have

$$\begin{aligned} x_1(t+T) &= x_0 + \varepsilon \int_0^{t} f(s, x_0)\, ds + \varepsilon \int_t^{t+T} f(s, x_0)\, ds + \sum_{0<t_j<t+T} \varepsilon H(t_j) \\ &= x_1(t), \end{aligned}$$

for $t \geq 0$ and, hence, $x_1(t-T) = x_1(t)$ for $t \leq 0$.

This means that the function $x_1(t)$ is T-periodic on the positive and negative semiaxes.

Let $0 < t < T$. If, in this case,

$$x_1(-t+T) = x_1(-t), \tag{21.31}$$

then the function $x_1(t)$ is T-periodic on the entire real axis.

It is clear that equality (21.31) is equivalent to the equality

$$x_1(t) = x_1(T-t), \tag{21.32}$$

where $T-t>0$. Using relation (21.29), we can write

$$x_1(T-t)=x_0+\varepsilon\int_0^{T-t} f(s,x_0)\,ds+\sum_{0<t_j<T-t}\varepsilon H(t_j).$$

Substituting this relation in (21.32), we obtain

$$\varepsilon\int_0^{t} f(s,x_0)\,ds+\sum_{0<t_j<t}\varepsilon H(t_j)$$

$$=\int_0^{-t}\varepsilon f(s,x_0)\,ds+\varepsilon\int_{-t}^{-t+T} f(s,x_0)\,ds+\sum_{0<t_j<T-t}\varepsilon H(t_j).$$

This yields

$$\sum_{0<t_j<t} H(t_j)-\sum_{0<t_j<T-t} H(t_j)=0.$$

For $t<T-t$, this equality takes the form

$$\sum_{t<t_j<T-t} H(t_j)=0. \tag{21.33}$$

At the same time, for $t>T-t>0$, it can be represented in the form

$$\sum_{T-t<t_j<t} H(t_j)=0. \tag{21.34}$$

In both these cases, the segment with the endpoints t and $T-t$ is symmetric about the point $\frac{T}{2}$. Since the times of pulses are also symmetric about this point, relations (21.33) and (21.34) are indeed true and, hence, relation (21.31) is also true.

Thus, the function $x_1(t)$ is even and T-periodic on the entire real axis.

By induction, we readily conclude that, for all $m\geq 1$, the functions $x_m(t)$ are defined for all $t\in R^1$, periodic in t with period T, and even.

Under the conditions of the theorem, the recurrence relation (21.28) for $t\geq 0$ can be rewritten in the form

$$x_{m+1}(t) \;=\; x_0+\varepsilon\int_0^{t}\Big(f(s,x_m(s))-\frac{1}{T}\int_0^{T} f(\xi,x_m(\xi))d\xi\Big)ds$$

$$+\,\varepsilon\left(\sum_{0<t_j<t} H(t_j)-\frac{1}{T}\sum_{0<t_j<T} H(t_j)\right)$$

because

$$\int_0^T f(\xi, x_m(\xi))\, d\xi = 0 \qquad \text{and} \qquad \sum_{0<t_j<T} H(t_j) = 0.$$

Note that this relation has the same form as relation (21.16) for $\tau = 0$ and, hence, by virtue of Theorem 21.3, the sequence of functions $\{x_m(t)\}$ converges to an even T-periodic function $x(t), x(0) = x_0$, which is a solution of the system of equations (21.27) because, in this case, relation (21.14) for the control μ_1 has the form

$$\mu_1 = \frac{1}{T}\Big\{\int_0^T f(s, x(s))ds + \sum_{0<t_j<T} H(t_j)\Big\} = 0,$$

which completes the proof of Theorem 21.5.

Remark 21.1. In the system of equations (21.27), we can set $\varepsilon = 1$ and, hence, Theorem 21.5 can be reformulated for a system that is not necessarily an impulsive T-system. It suffices to preserve the conditions imposed on the pulses, their times, and the function $f(t, x)$ and guarantee the applicability of the method of successive approximations to the solution of this system of equations.

For a general impulsive T-system (21.1), the following assertion can be helpful in finding the initial values of its T-periodic solution:

Lemma 21.2. *In order that the domain $D_1 \subset D_f$ contain a point x^* at which the function $\Delta(\tau, x^*)$ of the impulsive T-system (21.1) is equal to zero, it is necessary that the following inequality be true for some τ, all integer m, and all $x_1 \in D_1$:*

$$\|\Delta_m(\tau, x_1)\| \leq \frac{\varepsilon L(T+d)}{T}\left\{\zeta(m) + \frac{\sup\limits_{x \in D_1} \|x - x_1\|}{1 - \varepsilon L(2d + T/2)}\right\}, \qquad (21.35)$$

where $\zeta(m)$ is given by relation (21.17).

Proof. Assume that $\Delta(\tau, x^*) = 0$ at the point $x^* \in D_1$. Then, by using inequality (21.26), we obtain

$$\|\Delta(\tau, x_1)\| \leq \frac{\varepsilon L(T+d)}{T[1 - \varepsilon L(T/2 + 2d)]}\|x^* - x_1\|.$$

At the same time, according to inequalities (21.19), we have

$$\|\Delta_m(\tau, x_1)\| \leq \|\Delta(\tau, x_1)\| + \varepsilon L(1 + \frac{d}{T})\zeta(m)$$

and, hence,

$$\|\Delta_m(\tau, x_1)\| \leq \varepsilon L(1 + \frac{d}{T})\zeta(m) + \frac{\varepsilon L(T + d)}{T[1 - \varepsilon L(2d + T/2)]}\|x^* - x_1\|,$$

which yields inequality (21.35).

By virtue of Lemma 21.2, to find the initial values of T-periodic solutions of the impulsive T-system (21.1), it is necessary to act as follows: The set D_f is split into finitely many subsets D_i. In each D_i, we take a point $x = x^{(i)}$ and compute $\Delta_m(\tau, x^{(i)})$ for some m. After this, we remove the sets D_i for which $\Delta_m(\tau, x^{(i)})$ does not satisfy inequality (21.35) because these sets do not contain the points crossed by T-periodic solutions for $t = \tau$. The union of all remaining sets D_i is denoted by $\mathfrak{M}_i^{(m)}$. In the case where $m \to \infty$ and the diameters of the sets D_i contained in $\mathfrak{M}_i^{(m)}$ tend to zero (in this case, $i \to \infty$), the set $\mathfrak{M}_i^{(m)}$ approaches the set of initial values of T-periodic solutions of the system of equations (21.1).

4.22. Approximate Solution of the Periodic Problem of Control

In the present section, we study the possibility of reduction of the problem of control considered in the previous section for the system with control (21.1) to the finite-dimensional case. Note that, for some linear impulsive systems of equations, the problem of control can be solved with any *a priori* given accuracy. Clearly, the results presented in what follows are also valid for countable systems with small parameter but without pulse action.

We now consider the truncated system of equations

$$\frac{d\overset{(n)}{x}}{dt} = \varepsilon \overset{(n)}{f}(t, \overset{(n)}{x}), \quad t \neq t_j,$$

$$\Delta \overset{(n)}{x}\Big|_{t=t_j} = \varepsilon \overset{(n)}{H}_j(t_j, \overset{(n)}{x}) \quad (j = \ldots, -1, 0, 1, \ldots) \tag{22.1}$$

corresponding to system (21.1). Here,

$$\overset{(n)}{x} = (x_1, \ldots, x_n), \quad \overset{(n)}{f} = (f_1, \ldots, f_n, 0, 0 \ldots),$$

$$H_j = (H_1^{(j)}, H_2^{(j)}, \ldots), \quad \text{and} \quad \overset{(n)}{H_j} = (H_1^{(j)}, \ldots, H_n^{(j)}).$$

Since all Theorems in Section 4.21 remain valid for finite-dimensional systems of equations, we conclude that, for any natural n, there exists a unique mixed control $(\overset{(n)}{u_1}, \overset{(n)}{u_2})$ such that the solution $\overset{(n)}{x} = \overset{(n)}{x}(t, \tau, \overset{(n)}{x_0})$ of the impulsive system of equations

$$\frac{d\overset{(n)}{x}}{dt} = \varepsilon \overset{(n)}{f}(t, \overset{(n)}{x}) - \overset{(n)}{u_1}, \quad t \neq t_j,$$

$$\Delta \overset{(n)}{x}\Big|_{t=t_j} = \varepsilon \overset{(n)}{H_j}(t_j, \overset{(n)}{x}) - \overset{(n)}{u_2} \quad (j = \ldots, -1, 0, 1, \ldots) \tag{22.2}$$

with initial values $\tau, \overset{(n)}{x_0} \in D_f$ is T-periodic.

We now present conditions under which $\lim\limits_{n\to\infty} \overset{(n)}{x}(t) = x^0(t, \tau, x_0)$ at least weakly. Note that this problem cannot be solved by applying the theorem on truncation because the system of equations (22.2) cannot be regarded as a truncated system corresponding to system (21.2). In what follows, we restrict ourselves to the segment $[\tau, \tau + T]$ because, for any natural n, the solution $\overset{(n)}{x} = \overset{(n)}{x}(t, \tau, \overset{(n)}{x})$ of the system of equations (22.2) is T-periodic.

Further, we set $\overset{(n)}{u_i} = \varepsilon \overset{(n)}{\mu_i}$ $(i = 1, 2)$ and write the inequalities

$$\| \overset{(n)}{\mu_1} \| \leq \lim_{m\to\infty} \frac{1}{T} \int\limits_{\tau}^{\tau+T} \| \overset{(n)}{f}(s, \overset{(n)}{x_m}(s, \tau, \overset{(n)}{x_0})) \| \, ds \leq M,$$

$$\begin{aligned} \| \overset{(n)}{\mu_2} \| \; &\leq \; \lim_{m\to\infty} \frac{1}{pT} \sum_{\tau < t_j < \tau+T} \| \overset{(n)}{H_j}(t_j, \overset{(n)}{x_m}(t_j, \tau, \overset{(n)}{x_0})) \| \\ &\leq \; \frac{1}{pT} Md = M, \end{aligned}$$

where $\overset{(n)}{x_m}(t, \tau, \overset{(n)}{x_0})$ are determined for fixed n by the recurrence relation (21.7).

Lemma 22.1. *The sequences* $\{\overset{(n)}{\mu_1}\}$ *and* $\{\overset{(n)}{\mu_2}\}$ *contain weakly convergent subsequences* $\{\overset{(k)}{\mu_1}\}$ *and* $\{\overset{(k)}{\mu_2}\}$ *with limits* μ_1^0 *and* μ_2^0*, respectively, from the space* $\mathfrak{M}$.

Proof. Since $\| \overset{(n)}{\mu_1} \| \leq M$ $(n = 1, 2, \ldots)$, we can apply the method of diagonalization and extract a subsequence $\{\overset{(s)}{\mu_1}\}$ of the sequence $\{\overset{(n)}{\mu_1}\}$ such that $\overset{(s)}{\mu_1} \to \overset{(0)}{\mu_1}$ as $s \to \infty$ in the weak sense. Then we consider a subsequence $\{\overset{(s)}{\mu_2}\}$ of the sequence $\{\overset{(n)}{\mu_2}\}$. It is clear that $\| \overset{(s)}{\mu_2} \| \leq M$ for all s. Thus, applying the principle of diagonalization once again, we extract a subsequence $\{\overset{(k)}{\mu_2}\}$ of the sequence $\{\overset{(s)}{\mu_2}\}$ such that $\overset{(k)}{\mu_2} \to \overset{(0)}{\mu_2}$ as $k \to \infty$. Since the sequence $\{\overset{(k)}{\mu_1}\}$ is a subsequence of the sequence $\{\overset{(s)}{\mu_1}\}$, we conclude that $\overset{(k)}{\mu_1} \to \overset{(0)}{\mu_1}$ as $k \to \infty$.

We say that a function $f(t, x)$ belongs to the space $\widehat{C}_{\text{Lip}}(x)$ in the domain D^* if, in this domain, it satisfies the strong Cauchy–Lipschitz condition with respect to x (see Section 1.2). The following assertion is true:

Theorem 22.1. *Assume that all conditions of Theorem 21.1 are satisfied and, in addition, the functions $f(t, x)$ and $H_j(t, x)$ belong to the space $\widehat{C}_{\text{Lip}}(x)$ in the domain D^* for all integer j.*

Then the T-periodic solution $\overset{(n)}{x} = \overset{(n)}{x}(t, \tau, \overset{(n)}{x_0})$ of the system of equations (22.2) with initial conditions τ, $(x_{1_0}, x_{2_0}, \ldots, x_{n_0}) = \overset{(n)}{x_0} \in D_f$ converges weakly as $n \to \infty$ to the T-periodic solution $x^0(t, \tau, x_0)$ of the system of equations (21.2) such that $x^0(\tau, \tau, x_0) = x_0 = (x_{1_0}, x_{2_0}, \ldots) \in D_f$ and, moreover,

$$\mu_k = \lim_{n \to \infty} \overset{(n)}{\mu_k} \quad (k = 1, 2) \tag{22.3}$$

also in the weak sense.

Proof. We consider a subsequence of systems of equations from the sequence of systems (22.2):

$$\frac{d\overset{(k)}{x}}{dt} = \varepsilon \overset{(k)}{f}(t, \overset{(k)}{x}) - \varepsilon \overset{(k)}{\mu_1}, \quad t \neq t_j,$$

$$\Delta \overset{(k)}{x}\Big|_{t=t_j} = \varepsilon \overset{(k)}{H_j}(t_j, \overset{(k)}{x}) - \varepsilon \overset{(k)}{\mu_2} \quad (j = \ldots, -1, 0, 1, \ldots), \tag{22.4}$$

where the control $(\overset{(k)}{\mu_1}, \overset{(k)}{\mu_2})$ is specified by Lemma 22.1.

We write a solution $\overset{(k)}{x}(t, \tau, \overset{(k)}{x_0})$ of the system of equations (22.4) in the form

$$\overset{(k)}{x}(t,\tau,\overset{(k)}{x_0}) = \overset{(k)}{x_0} + \int_\tau^t [\varepsilon \overset{(k)}{f}(\sigma, \overset{(k)}{x}(\sigma,\tau,\overset{(k)}{x_0})) - \overset{(k)}{\mu_1}\varepsilon] d\sigma$$

$$+ \sum_{\tau<t_j<t} \Big[\varepsilon \overset{(k)}{H_j}(t_j, \overset{(k)}{x}(t_j,\tau,\overset{(k)}{x_0})) - \varepsilon \overset{(k)}{\mu_2}\Big]. \tag{22.5}$$

The norm of this solution can be estimated as follows:

$$\|\overset{(k)}{x}\| \le \|x_0^{(k)}\| + \varepsilon \Big\| \int_\tau^t \Big[\overset{(k)}{f}(\sigma, \overset{(k)}{x}) - \frac{1}{T}\int_\tau^{\tau+T} \overset{(k)}{f}(s, \overset{(k)}{x}) ds\Big] d\sigma$$

$$+ \sum_{\tau<t_j<t} \Big[\overset{(k)}{H_j}(t_j, \overset{(k)}{x}) - \frac{1}{pT}\sum_{\tau<t_j<\tau+T} \overset{(k)}{H_j}(t_j, \overset{(k)}{x}\Big]\Big\|$$

because the mixed control $(\overset{(k)}{u_1}, \overset{(k)}{u_2})$ is connected with the T-periodic solution $\overset{(k)}{x}(t,\tau,\overset{(k)}{x_0})$ by relations of the form (21.3).

In this case, by virtue of Lemma 21.1, we obtain

$$\|\overset{(k)}{x}(t,\tau,\overset{(k)}{x_0}\| \le \|\overset{(k)}{x_0}\| + \varepsilon M\alpha(t) \le \|x_0\| + \varepsilon M\alpha(t) \tag{22.6}$$

for all $t \in [\tau, \tau+T]$ and $k = 1, 2, 3, \ldots$.

It follows from inequality (22.6) that the sequence $\{\overset{(k)}{x}\}$ is uniformly bounded in norm on the segment $[\tau, \tau+T]$.

Let $\bar{t}$ and $\bar{\bar{t}}$ belong to the segment $[\tau, t_1]$, where t_1 is the time of the first pulse on the segment $[\tau, \tau+T]$, and let $\bar{t} > \bar{\bar{t}}$. Then

$$\overset{(k)}{x}(\bar{t}) - \overset{(k)}{x}(\bar{\bar{t}}) = \int_\tau^{\bar{t}} [\varepsilon \overset{(k)}{f}(\sigma, \overset{(k)}{x}) - \varepsilon \overset{(k)}{\mu_1}] d\sigma - \int_\tau^{\bar{\bar{t}}} [\varepsilon \overset{(k)}{f}(\sigma, \overset{(k)}{x}) - \varepsilon \overset{(k)}{\mu_1}] d\sigma$$

$$= \int_{\bar{\bar{t}}}^{\bar{t}} [\varepsilon \overset{(k)}{f}(\sigma, \overset{(k)}{x}) - \overset{(k)}{\mu_1}\varepsilon] d\sigma.$$

This yields

$$\|\overset{(k)}{x}(\bar{t}) - \overset{(k)}{x}(\bar{\bar{t}})\| \le \int_{\bar{\bar{t}}}^{\bar{t}} \|\varepsilon \overset{(k)}{f}(\sigma, \overset{(k)}{x}) - \varepsilon \overset{(k)}{\mu_1}\| d\sigma$$

$$\le (\varepsilon M + \varepsilon M)|\bar{t} - \bar{\bar{t}}| = 2N|\bar{t} - \bar{\bar{t}}|$$

for all $\bar{\bar{t}}, \bar{t} \in [\tau, t_1]$, where $\varepsilon M = N$.

The equicontinuity of the sequence $\{\overset{(k)}{x}\}$ on $[\tau, t_1]$ follows from inequality (22.7). According to the Arzelà theorem, we can extract a subsequence $\overset{(\alpha_1)}{x_1}, \overset{(\alpha_2)}{x_1}, \ldots, \overset{(\alpha_m)}{x_1}, \ldots$ of the sequence $\{\overset{(k)}{x_1}\}$ that converges uniformly in t on the segment $[\tau, t_1]$. In the sequence $\overset{(\alpha_1)}{x_2}, \overset{(\alpha_2)}{x_2}, \ldots, \overset{(\alpha_m)}{x_2}, \ldots$, we select a convergent subsequence $\overset{(\beta_1)}{x_2}, \overset{(\beta_2)}{x_2}, \ldots, \overset{(\beta_m)}{x_2}, \ldots$, etc. This procedure is repeated infinitely many times. By using the method of diagonalization, we extract a subsequence $\overset{(\alpha_1)}{x}, \overset{(\beta_2)}{x}, \overset{(\gamma_3)}{x}, \ldots$ of the sequence $\{\overset{(k)}{x}\}$ that converges uniformly in the coordinates on the segment $[\tau, t_1]$. This subsequence is denoted by $\{\overset{(k_1)}{x}\}$.

Now consider the sequence $\{\overset{(k_1)}{x}\}$ on the segment $[t_1+0, t_2]$, where t_2 is the time of the second pulse on the segment $[\tau, \tau+T]$ and $t_2 > t_1$. For all $\bar{t}$ and $\bar{\bar{t}}$ from $[t_1+0, t_2]$, the sequence $\{\overset{(k_1)}{x}\}$ satisfies inequalities (22.6) and (22.7). Therefore, the sequence $\{\overset{(k_1)}{x}\}$ contains a subsequence $\{\overset{(k_2)}{x}\}$ that converges uniformly in the coordinates on the segments $[\tau, t_1]$ and $[t_1+0, t_2]$.

After finitely many steps, we arrive at a sequence $\{\overset{(s)}{x}\}$. This sequence is a subsequence of the sequence $\{\overset{(k)}{x}\}$. Furthermore, $\{\overset{(s)}{x}\}$ converges to a T-periodic function $x^0(t)$ uniformly in the coordinates on each segment that lies in $[\tau, \tau+T]$ and does not contain the times of pulses t_j.

Further, we consider a sequence of impulsive systems of equations

$$\frac{d\overset{(s)}{x}}{dt} = \varepsilon \overset{(s)}{f}(t, \overset{(s)}{x}) - \overset{(s)}{u_1}, \quad t \neq t_j,$$

$$\Delta \overset{(s)}{x}\Big|_{t=t_j} = \varepsilon \overset{(s)}{H_j}(t_j, \overset{(s)}{x}(t_j)) - \overset{(s)}{u_2} \quad (s = 1, 2, \ldots,\ j = \ldots, -1, 0, 1, \ldots) \tag{22.8}$$

and prove that (uniformly in the coordinates)

$$x^0(t) = \lim_{s\to\infty} \overset{(s)}{x}(t, \tau, \overset{(s)}{x_0}) \equiv x^0(t, \tau, x_0), \quad t \in R^1,$$

where $x^0(t, \tau, x_0)$ is a T-periodic solution of the system of equations (21.2). The sequence $\{\overset{(s)}{x}(t)\}$ can be represented in the form

$$\begin{pmatrix} \overset{(1)}{x_1}(t) \\ \overset{(1)}{x_2}(t) \\ \vdots \end{pmatrix}, \quad \begin{pmatrix} \overset{(2)}{x_1}(t) \\ \overset{(2)}{x_2}(t) \\ \vdots \end{pmatrix}, \quad \ldots, \quad \begin{pmatrix} \overset{(n)}{x_1}(t) \\ \overset{(n)}{x_2}(t) \\ \vdots \end{pmatrix}, \quad \ldots .$$

Note that, beginning with a certain number, all elements of each column are identically equal to zero for all t.

For a fixed number l, we write the inequality

$$| f_l(t, \overset{(n)}{x}_1(t), \overset{(n)}{x}_2(t), \ldots) - f_l(t, x_1^0(t), x_2^0(t), \ldots) |$$

$$\leq | f_l(t, \overset{(n)}{x}_1(t), \overset{(n)}{x}_2(t), \ldots) - f_l(t, x_1^0(t), \ldots, x_g^0(t), \overset{(n)}{x}_{g+1}(t), \overset{(n)}{x}_{g+2}(t), \ldots) |$$

$$+ | f_l(t, x_1^0(t), \ldots, x_g^0(t), \overset{(n)}{x}_{g+1}(t), \overset{(n)}{x}_{g+2}(t), \ldots) - f_l(t, x_1^0(t), x_2^0(t), \ldots) | .$$

The differences on the right-hand side are denoted by $A(l, g)$ and $B(l, g)$, respectively. Since $f \in \widehat{C}_{\mathrm{Lip}}(x)$, we have

$$B(l, g) \leq \delta(t) \sup\{|x_{g+1}^0(t) - \overset{(n)}{x}_{g+1}(t)|, \ldots\} \varepsilon_l(g),$$

where $\varepsilon_l(g) \to 0$ as $g \to \infty$ and, hence, for any arbitrarily small positive number ν, there exists a number g^0 such that $\varepsilon_l(g) < \nu$. By using the fact that all functions $\overset{(s)}{x}(t)$ and, hence, the function $x^0(t)$ are uniformly bounded in norm by a constant K^0, we arrive at the inequality

$$B(l, g^0) \leq 2\delta(t) K^0 \varepsilon_g(g^0) < 2\delta(t) K^0 \nu. \tag{22.9}$$

We fix $g = g^0$. Then the following estimate is true:

$$A(l, g^0) \leq \delta(t) \varepsilon_l(g^0) \sup\{|x_1^0(t) - \overset{(n)}{x}_1(t)|, \ldots, |x_{g^0}^0(t) - \overset{(n)}{x}_{g^0}(t)|\}. \tag{22.10}$$

At the same time, $\overset{(s)}{x}(t)$ converges weakly to $x^0(t)$ uniformly in t. This means that there exists a number $N(l, \nu)$ such that, for $n \geq N(l, \nu)$, we have

$$\sup\{|x_1^0(t) - \overset{(k)}{x}_1(t)|, \ldots, |x_{g^0}^0(t) - \overset{(n)}{x}_{g^0}(t)|\} < \nu.$$

Thus, by using inequality (22.10), we get

$$A(l, g^0) + B(l, g^0) < \delta(t)(2K^0 + \varepsilon_l(0))\nu.$$

Let t be an arbitrary value of the argument from the segment $[\tau, \tau + T]$ that does not coincide with the time of pulse action and let σ be a segment that contains t, does not contain the times of pulses, and is such that $\sigma \subset [\tau, \tau + T]$. Denote $\delta = \max_{\sigma} \delta(t)$. Then

$$| f_l(t, \overset{(n)}{x}_1(t), \overset{(n)}{x}_2(t), \ldots) f_l(t, x_1^0(t), x_2^0(t), \ldots) | < \delta(2K^0 + \varepsilon_l(0))\nu$$

uniformly in $t \in \sigma$ whenever $n \geq N(l,\nu)$. This means that $f_l(t, \overset{(s)}{x}(t)) \to f_l(t, x^0(t))$ uniformly in $t \in \sigma$ as $s \to \infty$. It is clear that the last assertion is true for all $l = 1, 2, 3, \ldots$.

Passing to the limit as $s \to \infty$ in the first equation of system (22.8), we obtain

$$\lim_{s\to\infty} \frac{d\overset{(s)}{x}(t)}{dt} = \frac{d}{dt}\lim_{s\to\infty} \overset{(s)}{x}(t) = \varepsilon f(t, x^0(t)) - \varepsilon\mu_1^0 \quad \text{for} \quad t \neq t_j.$$

Assume that $H_j(t,x) \in \widehat{C}_{\mathrm{Lip}}(x)$. Using the same reasoning as above, we conclude that

$$\lim_{s\to\infty} H_j(t_j, \overset{(s)}{x}(t_j)) = H_j(t_j, x^0(t_j)).$$

Thus, we have

$$\frac{dx^0(t)}{dt} = \varepsilon f(t, x^0(t)) - \mu_1^0\varepsilon, \quad t \neq t_j,$$

$$\Delta x^0\Big|_{t=t_j} = \varepsilon H_j(t_j, x^0(t_j)) - \varepsilon\mu_2^0 \quad (j = \ldots, -1, 0, 1, \ldots).$$

Since

$$\overset{(s)}{\mu}_1 = \overline{\overset{(s)}{f}(t, \overset{(s)}{x}(t, \tau, \overset{(s)}{x}_0))} \quad \text{and} \quad \overset{(s)}{\mu}_2 = \overline{\overset{(s)}{H}_f(t, \overset{(s)}{x}(t_j, \tau, \overset{(s)}{x}_0))},$$

we can pass to the limit as $s \to \infty$ in these equalities and show that the control (u_1^0, u_2^0) is connected with $x^0(t)$ by relations of the form (21.8). By virtue of Theorem 21.1, this gives

$$\mu_1^0 = \mu_1, \quad \mu_2^0 = \mu_2, \quad \text{and} \quad x^0(t) = x_0(t, \tau, x_0).$$

Let us now return to the sequence of systems (22.2). We consider an arbitrary subsequence of this sequence:

$$\frac{d\overset{(r)}{x}}{dt} = \varepsilon \overset{(r)}{f}(t, \overset{(r)}{x}) - \varepsilon \overset{(r)}{\mu}_1, \quad t \neq t_j,$$

$$\Delta \overset{(r)}{x}\Big|_{t=t_j} = \varepsilon \overset{(r)}{H}_j(t_j, \overset{(r)}{x}) - \varepsilon \overset{(r)}{\mu}_2 \quad (j = \ldots, -1, 0, 1, \ldots).$$

Note that the reasoning presented above remains valid for this sequence of systems. This means that there exists a subsequence $\{\overset{(l)}{x}(t, \tau, \overset{(l)}{x}_0)\}$ of the sequence $\{\overset{(r)}{x}(t, \tau, \overset{(r)}{x}_0)\}$ that also converges (in the weak sense) to the function $x^0(t, \tau, x^0)$. At the same time, we have $\{\overset{(l)}{\mu}_i\} \to \mu_i$ $(i = 1, 2)$ as $l \to \infty$ in the weak sense.

Let us now show that the entire sequence $\{\overset{(n)}{x}(t,\tau,\overset{(n)}{x_0})\}$ also converges to $x^0(t,\tau,x_0)$ in the weak sense, and $\{\overset{(n)}{\mu_k}\} \to \mu_k$ as $n \to \infty$ $(k = 1, 2,)$.

Assume the contrary, i.e., assume that the relation

$$\lim_{n\to\infty} \overset{(n)}{x}(t,\tau,\overset{(n)}{x_0}) = x^0(t,\tau,x_0) \tag{22.11}$$

(understood in the weak sense) is not true for $\overset{(n)}{x_\rho}(t,\tau,\overset{(n)}{x_0})$, where $\overset{(n)}{x_\rho}$ is the ρth coordinate of the vector $\{\overset{(n)}{x}\}$, at the point $\bar{t} \in [\tau, \tau + T]$. This means that there exists a number $\varepsilon_0 > 0$ such that, for any arbitrarily large number $P > 0$, one can find a number $m \geq P$ for which

$$|\overset{(0)}{x_\rho}(\bar{t},\tau,x_0) - \overset{(m)}{x_\rho}(\bar{t},\tau,\overset{(m)}{x_0})| \geq \varepsilon_0. \tag{22.12}$$

We select an infinitely increasing sequence of positive numbers $N_1, N_2, \ldots,$ $N_k, \ldots$ and, by using this sequence, determine the sequence of natural numbers $m_k \geq N_k$ $(k = 1, 2, \ldots)$ for which inequality (22.12) is true with m replaced by m_k. At the same time, the sequence $\{\overset{(m_k)}{x_\rho}\}$ is a subsequence of $\{\overset{(n)}{x_\rho}\}$ and, hence, it contains a subsequence $\{\overset{(m_l)}{x_\rho}\}$ that converges to x^0_ρ. This means that there exists a number N_0 such that, for any $m_l \geq N_0$, inequality (22.12) cannot be true for $\{\overset{(m_l)}{x_\rho}\}$. Thus, we arrive at a contradiction. This completes the proof of relation (22.11).

The equality $\mu_k = \lim\limits_{n\to\infty} \overset{(n)}{\mu_k}$ $(k = 1, 2)$ is proved by analogy.

Remark 22.1. Condition (ii) of Theorem 21.1 and the requirement that $f(t,x)$, $H_j(t,x) \in \widehat{C}_{\text{Lip}}(x)$ can be replaced by a single inequality, namely

$$\max_{t\in[\tau,\tau+T]} \{\|f(t,x) - f(t,\bar{x})\|,\ \|H_j(t,x) - H_j(t,\bar{x})\|\} \leq \varepsilon(n)\|x - \bar{x}\|,$$

where the points $x = (x_1, \ldots, x_n, x_{n+1}, \ldots)$ and $\bar{x} = (x_1, \ldots, x_n, \bar{x}_{n+1}, \ldots)$ belong to the domain D and $\varepsilon(n) \to 0$ as $n \to \infty$.

By analogy with Theorem 22.1, one can easily prove the following two theorems:

Theorem 22.2. *Assume that all conditions of Theorem 21.2 are satisfied and $f(t,x), H_j(t,x) \in \widehat{C}_{\text{Lip}}(x)$ in the domain D^* for all integer j. Then the T-periodic solution $x = x_\infty(t,\tau,x_0)$ of the system of equations (21.2) is the weak* *limit of the sequence* $\{\overset{(n)}{x}(t,\tau,\overset{(n)}{x}_0)\}$ *as* $n \to \infty$, *where* $\overset{(n)}{x}(t,\tau,\overset{(n)}{x}_0)$ *is the* T-*periodic solution of the system of equations*

$$\frac{d\overset{(n)}{x}}{dt} = \varepsilon \overset{(n)}{f}(t,\overset{(n)}{x}), \quad t \neq t_j,$$

$$\Delta \overset{(n)}{x}\Big|_{t=t_j} = \varepsilon \overset{(n)}{H}_j(t_j,\overset{(n)}{x}) - \overset{(n)}{u}_2 \quad (j = \ldots,-1,0,1,\ldots) \tag{22.13}$$

obtained from the truncated system of equations (22.1) by using a control over its pulse part $(0,\overset{(n)}{u}_2)$. *Furthermore,*

$$\overset{(n)}{x}(\tau,\tau,\overset{(n)}{x}_0) = \overset{(n)}{x}_0 = (x_{1_0},\ldots,x_{n_0}), \quad u_2 = \lim_{n\to\infty} \overset{(n)}{u}_2, \quad \textit{and} \quad u_1 = 0.$$

Theorem 22.3. *Assume that all conditions of Theorem 21.3 are satisfied and* $f(t,x)$, $H_j(t,x) \in \widehat{C}_{\text{Lip}}(x)$ *in the domain* D^* *for all integer* j. *Then the* T-*periodic solution* $x = x^\infty(t,\tau,x_0)$ *of the system of equations (21.2) is the weak limit of the sequence* $\{\overset{(n)}{x}(t,\tau,\overset{(n)}{x}_0)\}$ $(n \to \infty)$, *where* $\overset{(n)}{x}(t,\tau,\overset{(n)}{x}_0)$ *is the* T-*periodic solution of the system of equations*

$$\frac{d\overset{(n)}{x}}{dt} = \varepsilon \overset{(n)}{f}(t,\overset{(n)}{x}) - \overset{(n)}{u}_1, \; t \neq t_j,$$

$$\Delta \overset{(n)}{x}\Big|_{t=t_j} = \varepsilon \overset{(n)}{H}_j(t_j,\overset{(n)}{x}) \quad (j = \ldots,-1,0,1,\ldots) \tag{22.14}$$

obtained from the truncated system of equations (22.1) by using a control over its differential part $(\overset{(n)}{u}_1,0)$. *Furthermore,*

$$\overset{(n)}{x}(\tau,\tau,\overset{(n)}{x}_0) = \overset{(n)}{x}_0, \quad u_1 = \lim_{n\to\infty} \overset{(n)}{u}_1, \quad \textit{and} \quad u_2 = 0.$$

We now consider the case of a T-periodic linear system of equations of the form (19.1) satisfying conditions 1°–3° from Section 4.19. The corresponding truncated system of equations has the form (19.6). The following assertion is true for systems of this sort:

Lemma 22.2. *Assume that the system of equations (19.1) is such that the following inequalities hold on the segment* $[\tau,\tau+T]$:

$$\|A(t) - \overset{(n)}{A}(t)\|_0 \leq \xi(n), \ \|B_j - \overset{(n)}{B_j}\| \leq \beta(n) \quad (j = 1, 2, \ldots, d), \tag{22.15}$$

where d is the number of pulses on the segment indicated.

Then any solution of system (19.1) $x = x(t, \tau, x_0)$, $x(\tau, \tau, x_0) = x_0 = (x_{1_0}, x_{2_0}, \ldots)$, *such that*

$$\|(0, \ldots, 0, x_{n+1_0}, x_{n+2_0}, \ldots)\| \leq \delta(n), \tag{22.16}$$

satisfies the inequality

$$\|x - \overset{(n)}{x}\|$$
$$\leq \{\delta(n) + K[\xi(n)T + \beta(n)d]\}\Big(1 + \max_{j=\overline{1,d}} \|B_j\|\Big)^d \exp\{\Psi_{[\tau,\tau+T]}T\}, \tag{22.17}$$

where

$$K = \delta(0)\Big(1 + \max_{j=\overline{1,d}} \|B_j\|\Big)^d \exp\{\Psi_{[\tau,\tau+T]}T\}, \quad t \in [\tau, \tau + T],$$

and $\overset{(n)}{x}$ *is a solution of the truncated system of equations of the form (19.6) with initial values* τ *and* $(x_{1_0}, x_{2_0}, \ldots, x_{n_0})$.

Inequality (22.17) immediately implies that if $\xi(n)$, $\beta(n)$, and $\delta(n)$ approach zero as $n \to \infty$, then

$$\lim_{n\to\infty} \overset{(n)}{x}(t, \tau, \overset{(n)}{x_0}) = x(t, \tau, x_0), \quad t \in [\tau, \tau + t],$$

in the strong sense (with respect to the norm $\|\cdot\|$).

Proof. In view of relations (19.2) and (19.7) and Statement 19.V, we obtain

$$\max_{t\in[\tau,\tau+T]} \{\|x(t, \tau, x_0)\|, \|\overset{(n)}{x}(t, \tau, \overset{(n)}{x_0})\|\}$$
$$\leq \delta(0)\Big(1 + \max_{j=\overline{1,d}} \|B_j\|\Big)^d \exp\{\Psi_{[\tau,\tau+T]}T\} = K.$$

By using inequalities (22.15) and (22.16), we get

$$\begin{aligned}\|x - \overset{(n)}{x}\| &\leq \{\|x_0 - \overset{(n)}{x_0}\| + \xi(n)T\|\overset{(n)}{x}\| + \beta(n)\|\overset{(n)}{x}\|d\} \\ &\quad + \int_\tau^t \Psi_{[\tau,\tau+T]}\|x - \overset{(n)}{x}\|dt + \sum_{\tau<\tau_j<t}\|B_j\|\|x - \overset{(n)}{x}\| \\ &\leq \{\delta(n) + [\xi(n)T + \beta(n)d]K\} \\ &\quad + \int_\tau^t \Psi_{[\tau,\tau+T]}\|x - \overset{(n)}{x}\|dt + \sum_{\tau<\tau_j<t}\max_{j=\overline{1,d}}\|B_j\|\|x - \overset{(n)}{x}\|,\end{aligned}$$

whence, by virtue of Statement 19.V, we obtain inequality (22.17).

Denote

$$\eta(n) = \max\{\delta(n), \xi(n), \beta(n)\} \quad \text{and} \quad b = \max_{j=\overline{1,d}}\|B_j\|.$$

Then, to guarantee the required accuracy $\|x(t) - \overset{(n)}{x}(t)\| \leq p$, we choose the order of truncation n from the condition

$$\eta(n) \leq \frac{p}{[1 + K(d+T)](1+b)^d \exp\{T\Psi_{[\tau,\tau+T]}\}}. \tag{22.18}$$

As an illustration, we consider a simple example of the system of equations (19.1) with

$$A(t) = \begin{bmatrix} \sin t & \dfrac{\cos t}{2} & \dfrac{\sin t}{4} & \cdots \\ \dfrac{\sin t}{2} & \dfrac{\cos t}{4} & \dfrac{\sin t}{8} & \cdots \\ \dfrac{\sin t}{4} & \dfrac{\cos t}{8} & \dfrac{\sin t}{16} & \cdots \\ \cdots & \cdots & \cdots & \cdots \end{bmatrix} \quad \text{and} \quad B_1 = \begin{bmatrix} \dfrac{1}{4} & \dfrac{1}{8} & \dfrac{1}{16} & \cdots \\ \dfrac{1}{8} & \dfrac{1}{16} & \dfrac{1}{32} & \cdots \\ \dfrac{1}{16} & \dfrac{1}{32} & \dfrac{1}{64} & \cdots \\ \cdots & \cdots & \cdots & \cdots \end{bmatrix}$$

and set $j = 1$.

The matrices $A(t)$ and B_1 satisfy the conditions of Lemma 22.2. Indeed, $\|A(t)\|_0 \leq 2$ and $\|B_1\| = \frac{1}{2}$. Moreover, $[\tau, \tau + T]$ is an arbitrary segment from R^1 and the functions $a_{ij}(t)$ $(i, j = 1, 2, \ldots)$ are continuous in R^1. Since $\|B_1\| < 1$, the matrix $(T + B_1)$ is invertible and, finally,

$$\|A(t) - \overset{(n)}{A}(t)\| \leq \frac{1}{2^{n-1}}, \quad \|B - \overset{(n)}{B}\| = \frac{1}{2^{n+1}}.$$

In this case, $\frac{1}{2^{n-1}} \to 0$ and $\frac{1}{2^{n+1}} \to 0$ as $n \to \infty$.

Denote $\eta(n) = \frac{1}{2^{n-1}}$, $\tau = 0$, $\tau + T = 2$, and $\tau_1 = 1$. For $d = 1$ (the case of a single pulse), we obtain inequality (22.18) in the form

$$\frac{1}{2^{n-1}} \leq \frac{p}{(1 + \frac{9}{2}e^4\delta(0))\frac{3}{2}e^4},$$

or

$$n \geq 1 + \frac{\ln[3e^4(2 + 9e^4\delta(0))] - \ln 4p}{\ln 2}.$$

We set $\delta(0) = 1$. Then, to find any solution of the system of equations (19.1) whose norm is smaller than one and the initial values satisfy inequality (22.16), where $\delta(n) \leq \eta(n)$ [e.g., $x_0 = ((1, \frac{1}{2}, \frac{1}{4}, \ldots)]$, with an accuracy to within $p = 0.001$ on the segment $[0, 2]$, it suffices to consider the truncated system of equations (19.6) of order $n = 26$.

Consider a linear impulsive system of equations with small parameter

$$\frac{dx}{dt} = \varepsilon A(t)x, \quad t \neq t_j,$$

$$\Delta x\Big|_{t=t_j} = \varepsilon B_j x(t_j - 0) \quad (j = \ldots, -1, 0, 1, \ldots). \tag{22.19}$$

This system is a special case of the system of equations (21.1). Here, $A(t) = [a_{is}(t)]_{i,s=1}^{\infty}$ and $B_j = [b_{is}^{(j)}]_{i,s=1}^{\infty}$ are infinite matrices and, moreover, the functions $a_{is}(t)$ are T-periodic and continuous in t on the entire real axis R^1.

Assume that the following conditions are satisfied:

(a) $B_{j+d} = B_j$ and $t_{j+d} - t_j = T$, where d is the number of pulses per period;

(b) $\max\{\|A(t)\|_G, \|B_j\|\} \leq M = \text{const} < \infty$, where $G = [\tau, \tau + T]$.

In this case, all conditions of Theorem 21.1 are satisfied and, furthermore, $L = M$ and, hence, there exists a unique mixed control (u_1, u_2) for which the system of equations

$$\frac{dx}{dt} = \varepsilon A(t)x - u_1, \quad t \neq t_j,$$

$$\Delta x\Big|_{t=t_j} = \varepsilon B_j x(t_j - 0) - u_2 \quad (j = \ldots, -1, 0, 1, \ldots) \tag{22.20}$$

possesses a unique T-periodic solution $x^0(t,\tau,x_0) \in D^*$ such that $x^0(\tau,\tau,x_0) = x_0 \in D_f$. Moreover, the indicated control and T-periodic solution satisfy the relations

$$u_1 = \frac{\varepsilon}{T}\int_{\tau}^{\tau+T} A(\sigma)x^0 d\sigma \quad \text{and} \quad u_2 = \frac{\varepsilon}{d}\sum_{\tau<t_j<\tau+T} B_j x^0(t_j - 0).$$

The truncated system of equations corresponding to system (22.19) has the form

$$\frac{dx}{dt} = \varepsilon \overset{(n)}{A}(t)\overset{(n)}{x}, \quad t \neq t_j,$$

$$\Delta \overset{(n)}{x}\Big|_{t=t_j} = \varepsilon \overset{(n)}{B_j}\overset{(n)}{x}(t_j - 0) \quad (j = \ldots,-1,0,1,\ldots). \tag{22.21}$$

Under conditions (a) and (b), for the system of equations (22.21) there exists a unique mixed control $(\overset{(n)}{u_1}, \overset{(n)}{u_2})$ such that the solution $\overset{(n)}{x} = \overset{(n)}{x}(t,\tau,\overset{(n)}{x_0})$ of the system of equations

$$\frac{d\overset{(n)}{x}}{dt} = \varepsilon \overset{(n)}{A}(t)\overset{(n)}{x} - \overset{(n)}{u_1}, \quad t \neq t_j,$$

$$\Delta \overset{(n)}{x}\Big|_{t=t_j} = \varepsilon \overset{(n)}{B_j}\overset{(n)}{x}(t_j - 0) - \overset{(n)}{u_2} \quad (j = \ldots,-1,0,1,\ldots) \tag{22.22}$$

with initial values τ and $\overset{(n)}{x_0}$ is T-periodic.

Let us formulate conditions under which relation (22.11), where the limit transition is understood in the strong sense, is true and estimate the accuracy of the approximation of the solution $x^0(t,\tau,x_0)$ by the function $\overset{(n)}{x}$.

Theorem 22.4. *Assume that the system of equations (22.19) is such that conditions (a) and (b) and inequalities (22.15) are satisfied. Then the T-periodic solution $x = x^0(t,\tau,x_0), x_0 \in D_f$, of the system of equations (22.20) for which inequality (22.16) is true satisfies relation (22.11), where the limit transition is understood in the strong sense and $\overset{(n)}{x}(t,\tau,\overset{(n)}{x})$ is the T-periodic solution of the system of equations (22.22) with initial values τ and $\overset{(n)}{x_0}$, provided that*

$$\eta(n) = \max\{\xi(n), \beta(n), \delta(n)\} \to 0 \quad \textit{as} \quad n \to \infty.$$

Furthermore,

$$\|x^0(t,\tau,x_0) - \overset{(n)}{x}(t,\tau,\overset{(n)}{x})\| \le \eta(n)N, \quad \textit{where} \quad N = \text{const} > 0.$$

Proof. It is clear that, for $i = 1, 2, \ldots$ and $j = \ldots, -1, 0, 1 \ldots$, we have

$$\sum_{s=n+1}^{\infty} \max_t \|a_{is}(t)\| \le \xi(n), \qquad \sum_{s=n+1}^{\infty} \|b_{is}^{(j)}\| \le \beta(n).$$

These inequalities imply that the functions $A(t)x$ and $B_j x$ belong to the space $\widehat{C}_{\text{Lip}}(x)$. Thus, by using conditions (a) and (b), we conclude that relation (22.11) holds in the weak sense.

Recall that, for $x_0 \in D_f$, the functions $x^0(t,\tau,x_0)$ and $\overset{(n)}{x}(t,\tau,\overset{(n)}{x_0})$ are bounded in the norm $\|\cdot\|$ by the constant

$$R^0 = \|x_0\| + \varepsilon M\Big(\frac{T}{2} + 2d\Big) \le R \quad (n = 1, 2, \ldots).$$

We have

$$\begin{aligned}\|x^0 - \overset{(n)}{x}\| \le \|x_0 - \overset{(n)}{x_0}\| &+ \int_\tau^t [\varepsilon\|A(\sigma)x^0 - \overset{(n)}{A}(\sigma)\overset{(n)}{x}\| + \|\varepsilon\mu_1 - \varepsilon\overset{(n)}{\mu_1}\|]d\sigma \\ &+ \sum_{\tau<t_j<t} \Big[\varepsilon\|B_j x^0(t_j - 0) - \overset{(n)}{B_j}\overset{(n)}{x}(t_j - 0)\| \\ &+ \varepsilon\|\mu_2 - \overset{(n)}{\mu_2}\|\Big]. \end{aligned} \tag{22.23}$$

The inequalities

$$\|A(t)x^0 - \overset{(n)}{A}(t)\overset{(n)}{x}\| \le M\|x^0 - \overset{(n)}{x}\| + R^0\xi(n),$$

$$\|B_j x^0(t_j - 0) - \overset{(n)}{B_j}\overset{(n)}{x}(t_j - 0)\| \le M\|x^0 - \overset{(n)}{x}\| + R^0\beta(n)$$

imply that

$$\|u_1 - \overset{(n)}{u_1}\| \le \frac{\varepsilon}{T} \int_{\tau}^{\tau+T} \|A(\sigma)x^0 - \overset{(n)}{A}(\sigma)\overset{(n)}{x}\|d\sigma \le \varepsilon R^0 \xi(n) + \varepsilon M \|x^0 - \overset{(n)}{x}\|_G,$$

$$\begin{aligned} \|u_2 - \overset{(n)}{u_2}\| &\le \frac{\varepsilon}{d} \sum_{\tau<t_j<\tau+T} \|B_j x^0(t_j - 0) - \overset{(n)}{B_j}\overset{(n)}{x}(t_j - 0)\| \\ &\le \frac{\varepsilon}{d}(M\|x^0 - \overset{(n)}{x}\|_G + R^0\beta(n))d. \end{aligned} \tag{22.24}$$

Further, by virtue of inequality (22.23), we obtain

$$\begin{aligned} &\|x^0 - \overset{(n)}{x}\|_G \\ &\le \|x_0 - \overset{(n)}{x_0}\| + 2\int_{\tau}^{t} \Big[\varepsilon M\|x^0 - \overset{(n)}{x}\|_G + \varepsilon\xi(n)R^0\Big]d\sigma \\ &\quad + \sum_{\tau<t_j<t} \Big[\varepsilon M\|x^0 - \overset{(n)}{x}\|_G + \varepsilon R^0\beta(n) + \varepsilon M\|x^0 - \overset{(n)}{x}\|_G + \varepsilon R^0\beta(n)\Big] \\ &\le [1 + 2\varepsilon R^0 T + 2\varepsilon R^0 d]\eta(n) + \int_{\tau}^{t} 2\varepsilon M\|x^0 - \overset{(n)}{x}\|_G d\sigma \\ &\quad + \sum_{\tau<t_j<t} 2\varepsilon M\|x^0 - \overset{(n)}{x}\|_G. \end{aligned}$$

By using Statement 19.V, we arrive at the estimate

$$\|x^0 - x^{(n)}\| \le \eta(n)N,$$

where

$$N = (1 + 2\varepsilon R^0 T + 2\varepsilon R^0 d)(2\varepsilon M)^d \exp\{2\varepsilon MT\} = \text{const} > 0.$$

This completes the proof of Theorem 22.4.

Slightly worsening the accuracy of approximation, we can identify the constant R^0 with the constant R from the definition of the domain D. Thus, by using inequalities (22.24), we obtain

$$\|u_1 - \overset{(n)}{u_1}\| \le \varepsilon R^0\xi(n) + \varepsilon M\eta(n)N \le \varepsilon\eta(n)(R^0 + MN),$$

$$\|u_2 - \overset{(n)}{u_2}\| \le \varepsilon(MN\eta(n) + R^0\beta(n)) \le \varepsilon\eta(n)(R^0 + MN),$$

where $\eta(n) \to 0$ as $n \to \infty$. These estimates determine the accuracy of approximation of the control (u_1, u_2).

Thus, if it is necessary to find the solution $x^0(t,\tau,x_0)$ with an accuracy to within σ, then the order n of a truncated system of equations of the form (22.21) should be chosen from the condition

$$\eta(n) \le \frac{\sigma_1}{N}, \quad 0 < \sigma_1 \le \sigma,$$

and the approximate value of the T-periodic solution of the system of equations (22.22) $\overset{(n)}{x}_m(t,\tau,\overset{(n)}{x}_0)$ given by the recurrence relation (21.7) should be found by the method of mixed control. In this case, it is necessary to repeat the outlined procedure as many times as necessary to guarantee the validity of the inequality

$$\| \overset{(n)}{x} - \overset{(n)}{x}_m \| < \sigma_2, \quad \sigma_1 + \sigma_2 = \sigma.$$

The required number of steps can be found by using inequality (21.17). Since the choice of the parameters σ_1 and σ_2 is ambiguous, it is necessary to solve the problem of its optimization. The indicated problem of optimization can be solved numerically. This is actually a modification of the numerical-analytic method presented in Section 1.6.

In conclusion, we note that Theorem 4.22 remains valid for the cases of control over the pulse $[(0,u_2)]$ and differential $[(u_1,0)]$ parts of the system of equations.

Let us now estimate the accuracy of approximation of the T-periodic solution and control for the order of truncation n in these cases.

1. Control over Pulse Part. In this case, $x_0 \in D'_f$ and $t \in [\tau, \tau+T]$. The inequalities

$$\begin{aligned}
\|u_2 - \overset{(n)}{u}_2\| &\le \frac{\varepsilon}{d}\Big\{ \int_\tau^{\tau+T} \|A(t)x_\infty(t) - \overset{(n)}{A}(t)\,\overset{(n)}{x}_\infty(t)\|_G\,dt \\
&\quad + \sum_{\tau<t_j<\tau+T} \|B_j x_\infty(t) - \overset{(n)}{B}_j\,\overset{(n)}{x}_\infty(t)\|_G \Big\} \\
&\le \frac{\varepsilon}{d}\Big\{ \int_\tau^{\tau+T} (M\|x_\infty(t) - \overset{(n)}{x}_\infty(t)\|_G + R_0\xi(n))dt \\
&\quad + \sum_{\tau<t_j<\tau+T} (M\|x_\infty(t) - \overset{(n)}{x}_\infty(t)\|_G + R_0\beta(n)) \Big\} \\
&\le \varepsilon M\|x_\infty(t) - \overset{(n)}{x}_\infty(t)\|_G\Big(\frac{T}{d}+1\Big) + \varepsilon R_0\Big(\frac{T}{d}\xi(n) + \beta(n)\Big),
\end{aligned}$$

where

$$R_0 = \|x_0\| + 2\varepsilon M(T+d),$$

imply that

$$\begin{aligned}
&\|x_\infty(t) - \overset{(n)}{x}_\infty(t)\|_G \\
&\quad \le \|x_0 - \overset{(n)}{x_0}\| + \varepsilon \int_\tau^t \|A(t)x_\infty(t) - \overset{(n)}{A}(t)\overset{(n)}{x}_\infty(t)\|_G\, dt \\
&\qquad + \sum_{\tau<t_j<t} \{\varepsilon \|B_j x_\infty(t) - \overset{(n)}{B_j}\overset{(n)}{x}_\infty(t)\|_G + \|u_2 - \overset{(n)}{u_2}\|\} \\
&\quad \le (1 + 2\varepsilon R_0 T + 2\varepsilon R_0 d)\eta(n) + \int_\tau^t \varepsilon M \|x_\infty(t) - \overset{(n)}{x}_\infty(t)\|_G\, dt \\
&\qquad + \sum_{\tau<t_j<t} \Big\{\Big(\frac{T}{d}+2\Big)\varepsilon M \|x_\infty(t) - \overset{(n)}{x}_\infty(t)\|_G\Big\}.
\end{aligned}$$

By using Statement 19.V, we conclude that

$$\begin{aligned}
\|x_\infty(t) - \overset{(n)}{x}_\infty(t)\| &\le \eta(n)[1 + 2\varepsilon R_0(T+d)]\Big[\varepsilon M\Big(\frac{T}{d}+2\Big)\Big]^d \exp\{\varepsilon MT\} \\
&= \eta(n)P
\end{aligned}$$

and, hence,

$$\|u_2 - \overset{(n)}{u_2}\| \le \eta(n)\varepsilon\Big(\frac{T}{d}+1\Big)(MP + R_0).$$

2. *Control over Differential Part.* In this case, $x_0 \in D_f$ and $t \in [\tau, \tau+T]$. The inequalities

$$\begin{aligned}
\|u_1 - \overset{(n)}{u_1}\| &\le \frac{\varepsilon}{T}\Big\{ \int_\tau^{\tau+T} \|A(t)x^\infty(t) - \overset{(n)}{A}(t)\overset{(n)}{x}{}^\infty(t)\|_G dt \\
&\qquad + \sum_{\tau<t_j<\tau+T} \|B_j x^\infty(t) - \overset{(n)}{B_j}\overset{(n)}{x}{}^\infty(t)\|_G\Big\} \\
&\le \varepsilon M\Big(1 + \frac{d}{T}\Big)\|x^\infty(t) - \overset{(n)}{x}{}^\infty(t)\|_G + \varepsilon R^0\left(\xi(n) + \beta(n)\frac{d}{T}\right)
\end{aligned}$$

imply that

$$\|x^{\infty}(t) - \overset{(n)}{x^{\infty}}(t)\|_G$$

$$\leq \|x_0 - \overset{(n)}{x_0}\| + \varepsilon \int_{\tau}^{t} \|A(t)x^{\infty}(t) - \overset{(n)}{A}(t)\,\overset{(n)}{x^{\infty}}(t)\|_G$$

$$+ \int_{\tau}^{t} \|u_1 - \overset{(n)}{u_1}\|dt + \sum_{\tau<t_j<t} \varepsilon\|B_j x^{\infty}(t) - \overset{(n)}{B_j}\,\overset{(n)}{x^{\infty}}(t)\|_G$$

$$\leq [1 + 2\varepsilon R^0(T + d)]\eta(n) + \int_{\tau}^{t} \varepsilon M\Big(2 + \frac{d}{T}\Big)\|x^{\infty}(t) - \overset{(n)}{x^{\infty}}(t)\|_G\,dt$$

$$+ \sum_{\tau<t_j<t} \varepsilon M\|x^{\infty}(t) - \overset{(n)}{x^{\infty}}(t)\|_G.$$

By using Statement 19.V, we obtain

$$\|x^{\infty}(t) - \overset{(n)}{x^{\infty}}(t)\|_G$$

$$\leq \eta(n)[1 + 2\varepsilon R^0(T + d)](\varepsilon M)^d \exp\Big\{\varepsilon M\Big(2 + \frac{d}{T}\Big)\Big\} = P_1\eta(n)$$

and, hence,

$$\|u_1 - \overset{(n)}{u_1}\| \leq \eta(n)\varepsilon\Big(\frac{d}{T} + 1\Big)(MP_1 + R^0).$$

REFERENCES

Andrianova, L. Ya.

[And] "On the reducibility of linear differential equations with quasiperiodic coefficients," *Vestn. Leningrad. Univ., Ser. Mat., Mekh., Astr.,* No. 7, 14–24 (1962).

Andronov, A. A., Vitt, A. A., and Khaikin, S. É.

[AVK] *Theory of Oscillations,* Moscow University, Moscow (1971) [in Russian].

Arnol'd, V. I.

[Arn1] "Small denominators. I. On the mapping of a circle onto itself," *Izv. Akad. Nauk SSSR, Ser. Mat.,* **25**, No. 1, 21–86 (1961).

[Arn2] "Small denominators. II. Proof of the Kolmogorov theorem on conservation of conditionally periodic motions in the case of a small perturbation of the Hamilton function," *Usp. Mat. Nauk,* **18**, Issue 5, 13–40 (1963).

[Arn3] "Small denominators and problems of stability of motion in classical and celestial mechanics," *Usp. Mat. Nauk,* **18**, Issue 6, 91–192 (1963).

[Arn4] *Additional Chapters of the Theory of Ordinary Differential Equations,* Nauka, Moscow (1979) [in Russian].

Artem'ev, N. A.

[Art] "Method for determination of characteristic indices and its application to two problems of celestial mechanics," *Izv. Akad. Nauk SSSR,* **8**, No. 2, 61–100 (1944).

Barbashin, E. A.

[Bar1] *Introduction to the Theory of Stability,* Nauka, Moscow (1968) [in Russian].

[Bar2] *Lyapunov Functions,* Nauka, Moscow (1970) [in Russian].

Baris, Ya. S.

[Bar] "On the integral manifold of an irregularly perturbed differential system," *Ukr. Mat. Zh.,* **20**, No. 4, 439–448 (1968).

Baris, Ya. S. and Forchuk, V. I.

[BaF] "Investigation of bounded solutions of nonlinear irregularly perturbed systems by the method of integral manifolds," *Ukr. Mat. Zh.,* **22**, No. 1, 3–11 (1970).

Belaga, É. G.

[Bel] "On the reducibility of a system of ordinary differential equations in the neighborhood of conditionally periodic motion," *Dokl. Akad. Nauk SSSR,* **143**, No. 2, 255–258 (1962).

Bibikov, Yu. N. and Pliss V. A.

[BiP] "On the existence of invariant tori in the neighborhood of the zero solution of a system of ordinary differential equations," *Differents. Uravn.,* **3**, No. 11, 1864–1881 (1967).

Birkhoff, G. D.

[Bir] *Dynamical Systems,* American Mathematical Society, New York (1927).

Blinov, I. N.

[Bli1] "Analytic representation for solutions of a system of linear differential equations with almost periodic coefficients depending on parameters," *Differents. Uravn.,* **1**, No. 8, 1041–1053 (1965).

[Bli2] "On the problem of reducibility of systems of linear differential equations with quasiperiodic coefficients," *Izv. Akad. Nauk SSSR, Ser. Mat.,* **31**, No. 2, 349–354 (1967).

Bogolyubov, N. N.

[Bog1] *On Statistical Methods in Mathematical Physics,* Ukrainian Academy of Sciences, Kiev (1945) [in Russian].

[Bog2] "On quasiperiodic motions in problems of nonlinear mechanics," in: *Proceedings of the First Mathematical School,* Vol. 1, Institute of Mathematics, Ukrainian Academy of Sciences, Kiev (1964), pp. 11–101 [in Russian].

Bogolyubov, N. N. and Mitropolsky, Yu. A.

[BoM1] "Methods of integral manifolds in nonlinear mechanics," in: *Proceedings of the International Symposium on Nonlinear Oscillations,* Institute of Mathematics, Ukrainian Academy of Sciences, Kiev (1963), pp. 93–154 [in Russian].

[BoM2] *Asymptotic Methods in the Theory of Nonlinear Oscillations,* Nauka, Moscow (1974) [in Russian].

Bogolyubov, N. N., Mitropolsky, Yu. A., and Samoilenko, A. M.

[BMS] *Method of Accelerated Convergence in Nonlinear Mechanics,* Naukova Dumka, Kiev (1969) [in Russian].

Bohr, H. A.

[Boh] *Almost Periodic Functions,* Chelsea, New York (1947).

Brodovskii, V. G.

[Bro] "On the construction of solutions of an infinite system of differential equations analytic with respect to parameters," *Izv. Akad. Nauk Kaz. SSR, Ser. Mat. Fiz.,* **1**, 84–94 (1966).

Bronshtein, I. U.

[Bro1] "Weak regularity and the Green function of linear expansions of dynamical systems," *Differents. Uravn.,* **19**, No. 12, 2031–2039 (1983).

[Bro2] *Nonautonomous Dynamical Systems,* Shtiintsa, Kishinev (1984) [in Russian].

Bylov, B. F.

[Byl1] "On the structure of solutions of systems of linear differential equations with almost periodic coefficients," *Mat. Sb.,* **66**, No. 2, 215–229 (1965).

[Byl2] "On the reduction of a system of linear equations to the diagonal form," *Mat. Sb.,* **66**, No. 3, 338–344 (1965).

Coddington, E. A. and Levinson, N.

[CoL] *Theory of Ordinary Differential Equations,* McGraw-Hill, New York (1955).

Daletskii, Yu. L. and Krein, M. G.

[DaK] *Stability of Solutions of Differential Equations in a Banach Space,* Nauka, Moscow (1970) [in Russian].

Daletskii, Yu. L. and Krein, S. G.

[DaKr] "On differential equations in a Hilbert space," *Ukr. Mat. Zh.,* **4**, No. 4, 71–74 (1950).

Danjoy, A.

[Dan] "Sur les courbes définies par les équations différentielles à la surface du tore," *J. Math. Pure Appl.,* **II**, No. 4, 333–375 (1932).

Ellis, R. and Johnson, R.

[ElJ] "Topological dynamics and linear differential systems," *J. Different. Equat.,* **44**, No. 1, 23–39 (1982).

Ermolaev, L. A.

[Erm] "On the uniform stability of countable almost linear and nonlinear systems of differential equations in the first approximation," *Izv. Akad. Nauk Kaz. SSR, Ser. Astr. Fiz., Mat., Mekh.,* **1**, No. 16, 88–114 (1952).

Erugin, N. P.

[Eru] *Linear Systems of Ordinary Differential Equations,* Belorussian Academy of Sciences, Minsk (1963) [in Russian].

Feshchenko, S. F.

[Fes] "Asymptotic solution of an infinite system of differential equations with slowly varying coefficients," *Dopov. Akad. Nauk Ukr. RSR,* No. 2, 82–86 (1954).

Filippov, M. G.

[Fil] *Reducibility of Almost Periodic Systems,* Author's Abstract of the Candidate-Degree Thesis (Physics and Mathematics), Kiev (1990) [in Russian].

Gantmakher, F. R.

[Gan] *Theory of Matrices,* Nauka, Moscow (1967) [in Russian].

Gel'man, A. E.

[Gel] "On the reducibility of systems with quasiperiodic matrix," *Differents. Uravn.,* **1**, No. 3, 283–294 (1965).

Golets, V. L.

[Gol] "On perturbation of stability of an invariant torus of a dynamical system," *Ukr. Mat. Zh.,* **23**, No. 1, 130–137 (1971).

Gorshin, S. I.

[Gor] "On stability of solutions of a countable system of differential equations with permanently acting perturbations," *Izv. Akad. Nauk Kaz. SSR, Ser. Mat. Mekh.,* Issue 3, 32–38 (1949).

Grebenikov, E. A. and Ryabov, Ya. A.

[GrR1] *New Qualitative Methods in Celestial Mechanics,* Nauka, Moscow (1971) [in Russian].

[GrR2] *Constructive Methods for Analysis of Nonlinear Systems,* Nauka, Moscow (1979) [in Russian].

Halanay, A. and Wexler, D.

[HaW] *Teoria Calitativă a Sistemelor cu Impulsuri,* Academiei Republicii Socialiste România, Bucharest (1968).

Kantorovich, L. V. and Krylov, V. I.

[KaK] *Approximate Methods of Higher Analysis,* Fizmatgiz, Moscow (1962) [in Russian].

Khalilov, Z. I.

[Kha1] "Cauchy problem for an infinite system of partial differential equations," *Dokl. Akad. Nauk SSSR,* **84**, 229–232 (1952).

[Kha2] "On stability of solutions of equations in a Banach space," *Dokl. Akad. Nauk SSSR,* **137**, No. 4, 797–799 (1961).

Kharasakhar, V. Kh.

[Kha] "On fundamental solutions of countable systems of differential equations," *Izv. Akad. Nauk Kaz. SSR, Ser. Mat. Mekh.,* Issue 4, 98–108 (1950).

Kolmogorov, A. N.

[Kol1] "On preservation of conditionally periodic motions under small variation of the Hamilton function," *Dokl. Akad. Nauk SSSR,* **98**, No. 4, 527–530 (1954).

[Kol2] "General theory of dynamical systems and classical mechanics," in: *Proceedings of the International Congress in Amsterdam,* Fizmatgiz, Moscow (1961), pp. 187–208 [in Russian].

Krein, M. G.

[Kre] *Lectures on the Theory of Stability of Solutions of Differential Equations in a Banach Space,* Institute of Mathematics, Ukrainian Academy of Sciences, Kiev (1964) [in Russian].

Krylov, N. M. and Bogolyubov, N. N.

[KrB] *New Methods in Nonlinear Mechanics,* ONTI, Moscow–Leningrad (1934) [in Russian].

Kulik, V. L.

[Kul1] "On the dependence of the Green function of the problem of invariant torus on a parameter," *Ukr. Mat. Zh.,* **30**, No. 4, 544–551 (1978).

[Kul2] "On the relation between quadratic forms and the Green function of a linear expansion of dynamical systems on a torus," *Ukr. Mat. Zh.,* **36**, No. 2, 258–262 (1984).

Kurtsvail', Ya.

[Kur] "Invariant manifolds of differential systems," *Differents. Uravn.,* **4**, No. 5, 785–7971 (1968).

Levitan, B. M.

[Lev] *Almost Periodic Functions,* Gostekhizdat, Moscow (1953) [in Russian].

Lyapunov, A. M.

[Lyap] *General Problem of Stability of Motion,* GITTL, Moscow (1950) [in Russian].

Lyashchenko, N. Ya.

[Lya] "An analog of the Floquet theorem for one particular case of linear homogeneous systems of differential equations with quasiperiodic coefficients," *Dokl. Akad. Nauk SSSR,* **III**, No. 2, 295–298 (1956).

Lykova, O. B. and Bogatyrev, V. M.

[LyB] "On reducibility of some differential equations in a Banach space," *Ukr. Mat. Zh.,* **20**, No. 5, 628–641 (1968).

Malkin, I. G.

[Mal] *Theory of Stability of Motion,* Nauka, Moscow (1966) [in Russian].

Massera, J. L. and Schaffer, J. J.

[MaS] *Linear Differential Equations and Function Spaces,* Academic Press, New York (1966).

Millionshchikov, V. M.

[Mil] "Structure of fundamental matrices with almost periodic coefficients," *Dokl. Akad. Nauk SSSR,* **171**, No. 2, 238–291 (1966).

Mitropolsky, Yu. A.

[Mit] *Averaging Method in Nonlinear Mechanics,* Naukova Dumka, Kiev (1971) [in Russian].

Mitropolsky, Yu. A. and Kulik, V. L.

[MiK1] "Bounded solutions of nonlinear systems of differential equations," *Ukr. Mat. Zh.,* **36**, No. 6, 720–729 (1984).

[MiK2] "On application of quadratic forms in the theory of invariant manifolds," *Ukr. Mat. Zh.,* **37**, No. 3, 306–316 (1985).

[MiK3] "Lyapunov function and bounded solutions of linear systems of differential equations," *Ukr. Mat. Zh.,* **38**, No. 1, 39–49 (1986).

Mitropolsky, Yu. A. and Lykova, O. B.

[MiL] *Integral Manifolds in Nonlinear Mechanics,* Nauka, Moscow (1973) [in Russian].

Mitropolsky, Yu. A. and Samoilenko, A. M.

[MiS1] "On the structure of trajectories on toroidal manifolds," *Dokl. Akad. Nauk Ukr. SSR,* **8**, 984–985 (1964).

[MiS2] "On the construction of solutions of linear differential equations with quasiperiodic coefficients," *Mat. Fiz.,* 185–198 (1967).

Mitropolsky, Yu. A., Samoilenko, A. M., and Kulik, V. L.

[MSK1] *Investigation of Linear Systems of Differential Equations Using Quadratic Forms,* Preprint No. 82.10, Institute of Mathematics, Ukrainian Academy of Sciences, Kiev (1982) [in Russian].

[MSK2] *Investigation of Dichotomy of Linear Systems of Differential Equations Using Lyapunov Functions,* Naukova Dumka, Kiev (1990) [in Russian].

Mitropolsky, Yu. A., Samoilenko, A. M., and Perestyuk, N. A.

[MSP] *Canonical Form of One Class of Differential Equations with Pulse Action,* Preprint No. 85.36, Institute of Mathematics, Ukrainian Academy of Sciences, Kiev (1985) [in Russian].

Moser, J.

[Mos1] "On invariant curves of area-preserving mappings of an annulus," *Nachr. Akad. Wiss. Göttingen Math. Phys.* KL II, No. 1, 1–20 (1962).

[Mos2] "On invariant surfaces and almost periodic solutions for ordinary differential equations," *Notices Amer. Math. Soc.,* **12**, No. 1 (1965).

[Mos3] "A rapidly convergent iteration method and nonlinear differential equations," *Usp. Mat. Nauk.,* **23**, No. 4, 179–238 (1968).

Myshkis, A. D. and Samoilenko, A. M.

[MyS] "A system with pushes at given times," *Mat. Sb.,* **74**, Issue 2, 202–208 (1967).

Palmer, K.

[Pal] "On the reducibility of almost periodic systems of linear differential equations," *J. Different. Equat.,* **36**, No. 2, 374–390 (1980).

Parasyuk, I. O.

[Par] *Construction and Investigation of Quasiperiodic Equations,* Author's Abstract of the Candidate-Degree Thesis (Physics and Mathematics), Kiev (1978) [in Russian].

Perestyuk, N. A.

[Per] "Invariant sets of one class of discontinuous dynamical systems," *Ukr. Mat. Zh.,* **36**, No. 1, 63–69 (1984).

Persidskii, K. P.

[Per1] "On stability of solutions of a countable system of differential equations," *Izv. Akad. Nauk Kaz. SSR, Ser. Mat. Mekh.,* Issue 2, 2–35 (1948).

[Per2] "On characteristic numbers of solutions of a countable system of linear differential equations," *Dokl. Akad. Nauk SSSR,* **3**, 229–232 (1948).

[Per3] "On the spectrum of characteristic numbers," *Prikl. Mat. Mekh.,* **14**, Issue 6, 635–650 (1950).

[Per4] *Infinite Systems of Differential Equations. Differential Equations in Nonlinear Spaces,* Nauka, Alma-Ata (1976) [in Russian].

Pliss, V. A.

[Pli1] "Principle of reduction in the theory of stability of motion," *Izv. Akad. Nauk SSSR, Ser. Mat.,* **28**, No. 6, 1297–1324 (1964).

[Pli2] *Nonlocal Problems in the Theory of Oscillations,* Nauka, Moscow–Leningrad (1964) [in Russian].

Poincaré, A.

[Poi] *On Curves Defined by Differential Equations,* OGIZ GITTL, Moscow–Leningrad (1947) [Russian translation].

Pontryagin, L. S.

[Pon] *Ordinary Differential Equations,* Nauka, Moscow (1970) [in Russian].

Reshetov, M. T.

[Res] "On the boundedness of solutions and characteristic numbers of a countable system of linear differential equations of triangular form," *Izv. Akad. Nauk Kaz. SSR, Ser. Mat. Mekh.,* Issue 4, 109–114 (1950).

Sacker, R.

[Sac] "A new approach to the perturbation theory of invariant surfaces," *Comm. Pure. Appl. Math.,* **18**, No. 4, 717–732 (1965).

Samoilenko, A. M.

[Sam1] "On the structure of trajectories on a torus," *Ukr. Mat. Zh.,* **16**, No. 6, 769–782 (1964).

[Sam2] "On reducibility of a system of ordinary differential equations in the neighborhood of a smooth integral manifold," *Ukr. Mat. Zh.*, **18**, No. 6, 41–64 (1966).

[Sam3] "A numerical-analytic method for the investigation of countable systems of periodic differential equations," *Mat. Fiz.*, Issue 2, 115–132 (1966).

[Sam4] "On reducibility of systems of linear differential equations with quasiperiodic coefficients," *Ukr. Mat. Zh.*, **20**, No. 2, 279–281 (1968).

[Sam5] "On the preservation of an invariant torus under perturbations," *Izv. Akad. Nauk SSSR, Ser. Mat.*, **34**, No. 6, 1219–1240 (1970).

[Sam6] "On the reduction of a dynamical system in the neighborhood of a smooth invariant torus to the canonical form," *Izv. Akad. Nauk SSSR, Ser. Mat.*, **35**, No. 1, 209–233 (1972).

[Sam7] "On the exponential stability of an invariant torus of a dynamical system," *Differents. Uravn.*, No. 5, 820–834 (1975).

[Sam8] "Green function of a linear expansion of a dynamical system on a torus, conditions of its uniqueness, and properties following from these conditions," *Ukr. Mat. Zh.*, **32**, No. 6, 791–797 (1980).

[Sam9] "Separatrix manifolds and decomposition of a linear expansion of dynamical systems on a torus," *Ukr. Mat. Zh.*, **33**, No. 1, 31–38 (1981).

[Sam10] *Elements of the Mathematical Theory of Multifrequency Oscillations. Invariant Tori,* Nauka, Moscow (1987) [in Russian]; ***English translation:*** Kluwer, Dordrecht (1991).

[Sam11] "Reducibility of a system of linear differential equations with quasiperiodic coefficients," *Ukr. Mat. Zh.*, **41**, No. 12, 1669–1680 (1989).

[Sam12] *Investigation of Dynamical Systems in the Neighborhood of a Quasiperiodic Trajectory,* Preprint No. 90.35, Institute of Mathematics, Ukrainian Academy of Sciences, Kiev (1990) [in Russian].

[Sam13] "Investigation of a dynamical system in the neighborhood of an invariant toroidal manifold," *Ukr. Mat. Zh.,* **42**, No. 4, 530–537 (1991).

Samoilenko, A. M. and Kulik, V. L.

[SaK1] "On the continuity of the Green function of the problem of invariant torus," *Ukr. Mat. Zh.,* **30**, No. 6, 779–783 (1978).

[SaK2] "Sets of invariant tori of linear expansions of dynamical systems," *Dokl. Akad. Nauk Ukr. SSR, Ser. A,* No. 8, 15–19 (1984).

[SaK3] "Decomposition of linear expansions of dynamical systems on a torus," *Dokl. Akad. Nauk Ukr. SSR, Ser. A,* No. 12, 23–27 (1984).

Samoilenko, A. M. and Perestyuk, N. A.

[SaP1] "On the existence of invariant toroidal sets of impulsive systems ," in: *Method of Integral Manifolds in Nonlinear Differential Equations,* Institute of Mathematics, Ukrainian Academy of Sciences, Kiev (1973), pp. 262–273 [in Russian].

[SaP2] "Invariant sets of impulsive systems in the standard form," *Ukr. Mat. Zh.,* **25**, No. 1, 129–134 (1973).

[SaP3] *Impulsive Differential Equations,* Vyshcha Shkola, Kiev (1987) [in Russian]; ***English translation:*** World Scientific, Singapore (1995).

Samoilenko, A. M. and Ronto, N. I.

[SaR] *Numerical-Analytic Methods for the Investigation of Periodic Solutions,* Vyshcha Shkola, Kiev (1976) [in Russian].

Samoilenko, A. M. and Teplinskii, Yu. V.

[SaT1] "Green function of the problem of invariant tori of countable systems of differential equations," *Dokl. Akad. Nauk Ukr. SSR, Ser. A,* No. 1, 26–30 (1981).

[SaT2] "On invariant tori of impulsive differential systems in the space of bounded numerical sequences," *Differents. Uravn.,* **21**, No. 8, 1353–1361 (1985).

[SaT3] *On Reducibility of Differential Systems in the Space of Bounded Numerical Sequences,* Preprint No. 89.44, Institute of Mathematics, Ukrainian Academy of Sciences, Kiev (1989) [in Russian].

[SaT4] "Reducibility of differential systems in the space of bounded numerical sequences," *Dokl. Akad. Nauk Ukr. SSR, Ser. A,* No. 10, 38–41 (1989).

[SaT5] "On reducibility of differential equations in the space of bounded numerical sequences," *Ukr. Mat. Zh.,* **41**, No. 2, 194–201 (1989).

[SaT6] "On reducibility of linear differential systems with almost periodic coefficients," in: *Methods for Investigation of Differential and Functional-Differential Equations,* Institute of Mathematics, Ukrainian Academy of Sciences, Kiev (1990), pp. 69–80 [in Russian].

[SaT7] "On stability of an invariant torus of a nonlinear countable system of equations," *Dokl. Akad. Nauk Ukr. SSR, Ser. A,* No. 7, 31–33 (1991).

[SaT8] "On periodic problem of control for impulsive countable systems with small parameter," *Dokl. Akad. Nauk Ukr. SSR, Ser. A,* No. 3, 7–11 (1992).

Samoilenko, A. M., Teplinskii, Yu. V., and Avdeyuk, P. I.

[STA] "Existence of exponentially stable invariant tori of impulsive differential systems in the space of bounded numerical sequences," *Dokl. Akad. Nauk Ukr. SSR, Ser. A,* No. 11, 21–25 (1985).

Samoilenko, A. M., Teplinskii, Yu. V., and Luchik, V. E.

[STL] "Reducibility of impulsive differential equations in the space of bounded numerical sequences," *Dokl. Akad. Nauk Ukr. SSR, Ser. A,* No. 5, 27–30 (1990).

Samoilenko, A. M., Teplinskii, Yu. V., and Tsyganovskii, N. S.

[STT1] "On the behavior of solutions of a quasilinear system of differential equations with pulses in the neighborhood of an invariant torus," *Differents. Uravn.,* **28**, No. 5, 833–839 (1982).

[STT2] *On Invariant Tori of Countable Systems of Differential Equations,* Preprint No. 83.30, Institute of Mathematics, Ukrainian Academy of Sciences, Kiev (1983) [in Russian].

Schwartz, L.

[Sch] *Analyse Mathématique,* II, Hermann, Paris (1967).

Shkil', M. I.

[Shk] *Asymptotic Methods in Differential Equations,* Vyshcha Shkola, Kiev (1971) [in Ukrainian].

Shtokalo, I. Z.

[Sht] *Linear Differential Equations with Periodic Coefficients,* Ukrainian Academy of Sciences, Kiev (1960) [in Russian].

Siegel, C. L.

[Sie1] "Note on differential equations on the torus," *Ann. Math.,* **46**, No. 2, 423–428 (1945).

[Sie2] "Über die Normalform analytischer Differentialgleichungen in der Nahe einer Cleichgewichtslosung," *Nachr. Akad. Wiss. Göttingen Math. Phys.* KL, 21–30 (1952).

Teplinskii, Yu. V.

[Tep1] "On the existence and smoothness of an invariant toroidal manifold of a nonlinear countable system of differential equations," *Ukr. Mat. Zh.,* **27**, No. 6, 847–851 (1975).

[Tep2] "On the existence of invariant toroidal manifolds of countable systems of differential equations," in: *Analytic Methods for the Investigation of Solutions of Nonlinear Differential Equations,* Institute of Mathematics, Ukrainian Academy of Sciences, Kiev (1975), pp. 164–170 [in Russian].

[Tep3] "On reducibility of a linear countable system of equations with quasiperiodic coefficients," in: *Approximate Methods for the Investigation of Nonlinear Systems,* Institute of Mathematics, Ukrainian Academy of Sciences, Kiev (1976), pp. 172–181 [in Russian].

[Tep4] "On reducibility of countable systems of differential equations," *Mat. Fiz.,* Issue 22, 47–54 (1977).

[Tep5] "On the existence of an invariant toroidal manifold of a countable system of differential equations with pulses," *Ukr. Mat. Zh.,* **29**, No. 6, 835–841 (1977).

[Tep6] "On invariant tori of countable systems of differential equations," *Dokl. Akad. Nauk Ukr. SSR, Ser. A,* No. 9, 796–800 (1978).

[Tep7] "On the smoothness of an invariant toroidal manifold of a linear countable system of differential equation," *Mat. Fiz.,* Issue 23, 32–39 (1978).

[Tep8] "On invariant tori and reducibility of countable systems," *Differents. Uravn.,* **15**, No. 4, 759–761 (1979).

[Tep9] "On reducibility of countable systems of differential equations with quasiperiodic coefficients," *Ukr. Mat. Zh.,* **31**, No. 4, 463–465 (1979).

[Tep10] "On invariant tori of linear systems of differential equations in the space $\mathfrak{M}$," *Ukr. Mat. Zh.,* **35**, No. 2, 194–199 (1983).

Teplinskii, Yu. V. and Avdeyuk, P. I.

[TeA1] "On the dependence of the Green function of the problem of invariant torus of a linear differential system of equations in the space $\mathfrak{M}$ on a parameter," in: *Approximate Methods for the Investigation of Nonlinear Oscillations,* Institute of Mathematics, Ukrainian Academy of Sciences, Kiev (1984), pp. 120–128 [in Russian].

[TeA2] "On exponential stability of an invariant torus of a nonlinear countable system of differential equations," *Ukr. Mat. Zh.,* **42**, No. 3, 401–405 (1990).

[TeA3] "Reduction of the problem of the existence of an invariant torus of an infinite differential system to the finite case," *Ukr. Mat. Zh.,* **43**, No. 9, 1251–1255 (1991).

Teplinskii, Yu. V. and Luchik, V. E.

[TeL] "On reducibility of impulsive differential equations in the space of bounded numerical sequences," *Ukr. Mat. Zh.,* **42**, No. 10, 1376–1382 (1990).

Teplinskii, Yu. V. and Tsyganovskii, N. S.

[TeT] "On one periodic control problem for impulsive differential equations in the space of bounded numerical sequences," *Ukr. Mat. Zh.,* **42**, No. 2, 271–275 (1990).

Tikhonov, A. N.

[Tik] "On infinite systems of differential equations," *Mat. Sb.,* **41**, Issue 4, 551–560 (1934).

Trofimchuk, S. I.

[Tro] "Necessary condition for the existence of an invariant manifold of a linear expansion of a dynamical system on a compact manifold," *Ukr. Mat. Zh.,* **36**, No. 3, 390–393 (1984).

Umbetzhanov, D. U.

[Umb] *Almost Multiperiodic Solutions of Partial Differential Equations,* Nauka, Alma-Ata (1979) [in Russian].

Urabe, M.

[Ura1] "Existence theorems of quasiperiodic solutions to nonlinear differential systems," *Funkcial. Ekvac,* **15**, 75–100 (1972).

[Ura2] "On modified Galerkin's procedure for nonlinear quasiperiodic differential systems," in: *Equations Differentielles et Functionelees non Lineiares,* Paris (1973), pp. 223–258.

Valeev, K.G. and Zhautykov, O. A.

[VaZ] *Infinite Systems of Differential Equations,* Nauka, Alma-Ata (1974) [in Russian].

Volosov, V. M.

[Vol] "Averaging in systems of ordinary differential equations," *Usp. Mat. Nauk,* **13**, Issue 6, 3–126 (1962).

Wasow, W.

[Was] *Asymptotic Expansions for Ordinary Differential Equations,* Interscience, New York (1965).

Zadiraka, K. V.

[Zad] "On the integral manifold of a system of differential equations with small parameter," *Dokl. Akad. Nauk SSSR,* **115**, No. 4, 646–649 (1957).

Zhautykov, O. A.

[Zha1] "On the construction of an integral of a linear first-order partial differential equation with a countable set of independent variables," *Uchen. Zap. Kaz. Ped. Inst., Ser. Fiz.-Mat.,* Issue 8, 32–41 (1957).

[Zha2] "On a countable system of differential equations with variable parameters," *Mat. Sb.,* **49**, Issue 3, 317–330 (1957).

[Zha3] "Application of the averaging principle in nonlinear mechanics to countable systems of equations," *Ukr. Mat. Zh.,* **17**, No. 1, 39–46 (1965).

Zubov, V. I.

[Zub] *Stability of Motion,* Vysshaya Shkola, Moscow (1973) [in Russian].